The Bigouden region in south-west Brittany may appear to be extremely 'traditional', but over the centuries it has witnessed much social and economic change, with the introduction of commercial fishing and a canning industry in the 1880s and, more recently, the development of tourism and restructuring of agriculture. Following a community of Breton peasants over fifteen generations, Martine Segalen traces the effects of these economic changes on family life and analyses the strategies of marriage alliance and inheritance which were used to shore up social hierarchies. She thus reveals the importance of kinship networks in social intercourse, both today and in the past.

The value of Dr Segalen's study lies both in the case material, which is of interest for what it reveals about the social history of the French peasantry and peasants in general, and, more particularly, in the methodology she applies which combines anthropological, historical, and demographic approaches.

Dr Segalen is the author of *Historical Anthropology of the Family* and several ethnographic studies of French rural society. She has been the Director of the Centre d'ethnologie française since 1986.

Cambridge Studies in Social and Cultural Anthropology

74

Fifteen generations of Bretons

For other titles in this series turn to page 320

Cet ouvrage est publié dans le cadre de l'accord de co-édition passé en 1977 entre la Fondation de la Maison des sciences de l'homme et le Press Syndicate of the University of Cambridge. Toutes les langues européennes sont admises pour les titres couverts par cet accord, et les ouvrages collectifs peuvent paraître en plusieurs langues.

Les ouvrages paraissent soit isolément, soit dans l'une des séries que la Maison des sciences de l'homme et Cambridge University Press ont convenu de publier ensemble. La distribution dans le monde entier des titres ainsi publiés conjointement par les deux établissements est assurée par Cambridge University Press.

This book is published as part of the joint publishing agreement established in 1977 between the Fondation de la Maison des sciences de l'homme and the Press Syndicate of the University of Cambridge. Titles published under this arrangement may appear in any European language or, in the case of volumes of collected essays, in several languages.

New books will appear either as individual titles or in one of the series which the Maison des sciences de l'homme and the Cambridge University Press have jointly agreed to publish. All books published jointly by the Maison des sciences de l'homme and the Cambridge University Press will be distributed by the Press throughout the world.

FIFTEEN GENERATIONS OF BRETONS

Kinship and society in Lower Brittany, 1720–1980

MARTINE SEGALEN
translated from the French by J.A. Underwood

CAMBRIDGE UNIVERSITY PRESS
Cambridge
New York Port Chester Melbourne Sydney
EDITIONS DE
LA MAISON DES SCIENCES DE L'HOMME
Paris

CAMBRIDGE UNIVERSITY PRESS
Cambridge, New York, Melbourne, Madrid, Cape Town, Singapore, São Paulo

Cambridge University Press
The Edinburgh Building, Cambridge CB2 8RU, UK

With Editions de la Maison des Sciences de l'Homme
54 Boulevard Raspail, 75270 Paris Cedex 06, France

Published in the United States of America by Cambridge University Press, New York

www.cambridge.org
Information on this title: www.cambridge.org/9780521333696

Originally published in French as *Quinze générations de Bas-Bretons*
by Presses Universitaires de France, 1985

First published in English by Editions de la Maison des Sciences de l'Homme and
Cambridge University Press 1991 as
Fifteen Generations of Bretons

This digitally printed version 2007

A catalogue record for this publication is available from the British Library

Library of Congress Cataloguing in Publication data
Segalen, Martine
[Quinze générations de Bas-Bretons. English]
Fifteen generations of Bretons: kinship and society in lower
Brittany, 1720–1980/Martine Segalen; translated from the French
by J.A. Underwood.
p. cm. – (Cambridge studies in social and cultural
anthropology; 74)
Includes bibliographical references.
1. Bigouden (France)–Social life and customs. 2. Family–France–
Bigouden–History. 3. Peasantry–France–Bigouden–Social
conditions. 4. Bigouden (France)–Social conditions. I. Title.
II. Series.
DC611.B5783S4413 1991
944′.1–dc20

ISBN 978-0-521-33369-6 hardback
ISBN 978-0-521-04055-6 paperback

Contents

Plates

Figures

Tables

Introduction

The Bigouden countryside today is liberally dotted with 'traditional Breton homes' – as hymned by architects and estate agents alike. With their white-washed walls and granite door and window surrounds, they positively flaunt themselves across open fields and hillsides. Yet for that very reason they can qualify as no more than distant cousins of the old farmhouses nestling in valley bottoms, shielded from prying eyes by a curtain of trees. The new homes are inhabited by the families of the builder, the plumber, the man who goes to work in Quimper. The old dwellings – often simply the latest in a series of houses erected on the same site by generation after generation – are farms. The new houses going up in such numbers in the towns and villages of the region are already eating away at all the agricultural land; the farms lie quietly enclosed amid their wooded fields or snuggle up against their neighbours when they lie at the centre of a *méjou*, the open fields that other parts of France call *campagnes* or *champagnes*. A fictitious architectural continuity masks a real social continuity.

This book deals with the passage from the one type of house to the other, the transition from a society based exclusively on agriculture to a society that is diversifying, playing on the new relationships between town and country, becoming 're-urbanised', to use the hideous term sociologists have coined to account for the fact that the countryside, having emptied, is now filling up again. Descendants of former farmers or outsiders now living in the village fill the seats on the local council but for the most part work elsewhere. How has the change come about? What upheavals aided it – but what continuities, too? How has social reproduction occurred in a socio-economic environment that evolved only slowly up until the beginning of the twentieth century but that, since that time, has been developing swiftly and unceasingly?[1]

The following study concerns the Bigouden region in the far west of

France, a region bounded on three sides by the ocean and often represented as the archetype of the smallest unit that can be said to constitute a cultural area. Characterised – if not caricatured – by the tall *coiffes* worn by its womenfolk, described by Pierre Hélias, the self-appointed poet of a sublimated culture, dissected by sociologists in the 1960s, analysed by many of its own residents,[2] Bigouden is not in fact the homogenous unit that many people like to claim it is. The people of the southern part of the region, around the town of Pont-L'Abbé, regard themselves as the only 'true Bigoudens', and it is a fact that, because of their geographical situation, they have always been less subject to the influence of Quimper, the region's capital, and hung on to their distinctive identity longer than their cousins to the north. The division of the region into two parts – North Bigouden and South Bigouden – is no mere product of mental perceptions; it is firmly anchored in social usage. For an inhabitant of Penmarc'h (pronounced 'Pen-mar'), Loctudy, Plobannalec, Plomeur, Le Guilvinec, Saint-Jean-Trolimon, Tréguennec, or Plonéour-Lanvern there is a geographical space outside which one does not look for a tenancy or for a spouse. Its frontier happens to coincide with the northern boundary of the large commune* of Plonéour.

Contained within this precisely defined area, reconstructed family genealogies tell the story of social reproduction, chiefly through the medium of forms of matrimonial alliance. Analysing them in terms of how they relate to social structures and economic change involves drawing on a number of disciplines, notably history, demography, and ethnology. In this way past and present cast light upon each other and facilitate cross-objectification of one's various sources. Oral genealogies very quickly dry up beyond the subject's grandparents. After that the researcher has to fall back on vital records and census returns to trace the genealogical links back generation by generation; here most of them led as far as the early years of the eighteenth century. It then becomes necessary to adopt the opposite approach as we travel back towards the present by way of the social, economic, and cultural characteristics of the group under investigation. To put social flesh on the genealogical bones it is necessary to understand the hierarchies of the group, the organisation of production, modes of agricultural operation, the nature of production, and the organisation of local markets. In addition to the usual wide range of documentary sources, including post-mortem inventories, the accounts of folklorists and travellers, and archive

* In France, the smallest territorial division on the administrative map. Above it come the *canton*, the *arrondissement*, and the *département* [Translator].

sources, there is verbal testimony gathered at first hand – usually referred to as 'oral history' – which sheds light on the changes the region has undergone during the most recent period. (The memories of today's old people do not go back much before the 1900s, and let us not make the mistake of supposing that their evidence relates to an immemorial, frozen past.)

Historical detachment might be prescribed by the subject under investigation, but the ethnological approach took precedence. That approach is now in fashion – a welcome sign of the interest currently being taken in the socio-cultural systems of the recent past and the present day. As applied to the study of the behaviour of living inhabitants of the same country as the observer, it is not without its problems.

To begin with, one of the premisses of the ethnological approach is the negation of the principle of 'participant observation'. Achieving an effective degree of integration within a village means establishing private ties with certain members of it, and ties of familiarity very quickly become pseudo-family ties. Catherine Daniel (not her real name), the wife of the former blacksmith (hence her Breton 'surname' *ar Marichal*), and her daughter were my hosts during the various visits I made over a ten-year period. I stayed in their house. The sincere friendship that grew up between us, the closeness we were able to share, and my having been associated with the family through periods of mourning as well as through marriages and births prohibit me from talking about them as if I had been observing them from outside. How do you give an academic account of people who take you into their home, feed you, surround you with their love, and become your dear friends? Your integration is so effective that it eventually destroys not only your faculty of detachment but all desire to exercise it.

In any case, working on the family and kinship and researching systems of inheritance and transmission of property, one is dealing with less neutral subjects than, say, studying a particular technique or a particular collective manifestation. On the contrary, 'talking family' is an exercise loaded with emotion and consequently with latent violence, even where the ethnologist claims to be keeping to the objective ground of the vocabulary of kinship, relations between affines and consanguines, and marriage rites. All things considered, it is not wholly unlike discussing witchcraft, which, as Jeanne Favret-Saada says in her study of the subject in Normandy, 'is power rather than knowledge or information'.[3] The reactions I received to my questions or simply to my presence and the comments made about me in my absence and painstakingly repeated to me later opened my eyes to the special status of this subject. Occasionally I had the feeling of having suddenly entered a forbidden zone, invading an intimacy in which I had no share. It was as if the more knowledge I gained by my intrusion, the less I left my interlocutor with. Reactions ranged from a mildly demurring 'don't put that down' (with a glance at my notebook) or, in another instance, 'you'll make my mother look ridiculous' to violent letters of protest and unceremonious ejections from people's houses. From 'information' I had slipped imperceptibly into the realm of 'power', understood as knowledge about others.

So it is fortunate that putting things into historical perspective gives the ethnologist back his faculty of detachment, as the study of documents enables him to objectify what people tell him, re-situate the individual in the general, and compare the way things are done with the way the rules say they should be done. Various historical approaches are available. First and foremost there is historical demography, which with its rigorous data provides a framework for the ways in which a population behaves. This may be considered either in terms of geographical mobility or demographic growth – phenomena that are traced at the level of the community – or in terms of marriage, birth, and death rates, which in a different fashion give a reading of the domestic group and its biological and social reproduction. What chance have we of understanding marriage strategies unless we know the ages at which men and women marry and the numbers of bachelors and spinsters? How shall we piece together the practices of the system of property transmission without knowing how many children survive in each generation? Brittany is noted for its egalitarian use of the partible inheritance system, unlike central or southern parts of France where the single-heir system operates. The region was able to retain the partible system for so long because, in contrast to other peasant societies, its farmers did not often own their own land. They had only movables to hand down from generation to generation, and movables are easier to share out than landed property. This is because a landowner will always seek by one means or another to keep up the size of his holding in order to protect its viability. Bigouden happens to have a complex system of land ownership known as the *domaine congéable*. Under this system, which is specific to Brittany, a farm will have two owners: the actual landowner (generally a member of the nobility, the bourgeoisie, or the clergy) and the holder of the 'reparative rights' (sometimes referred to as the *convenant*) covering the 'edifices and surfaces' – that is to say, the farm buildings and the arable stratum of the land, regarded as movables and hence as the property of the domanial tenant.

No peasant ownership of land, then – and very little permanence, either. There are parts of France where the transfer of tenancies down the same line of descendants means that a family will have occupied a particular farm virtually since time immemorial. In nineteenth-century Bigouden this was not often the case. In fact it was common for a small tenant farmer to quit his farm for another, either because his family had increased in size or because he could no longer afford the rent. There was considerable mobility up until the 1930s, but only within South Bigouden; people did not 'emigrate'. That mobility increased as a result of the upsurge in demographic growth. This was in stark contrast to the movements experienced in the rest of France. In general an exodus from rural areas slowly emptied villages of their inhabitants during the second half of the nineteenth century. Here it was the other way around. As pressure on the land mounted, the expanding population was enabled to stay within South Bigouden when the region rediscovered the sea and there was a revival of the fishing industry – and above all when, from the 1880s onward, ways were found of turning the products of that industry to commercial account on a large scale with the development of canning factories.

One of the minor paradoxes of the region is the fact that for two centuries the people of South Bigouden turned their backs on the sea.

Fig. 1. Commune boundaries in South Bigouden today.

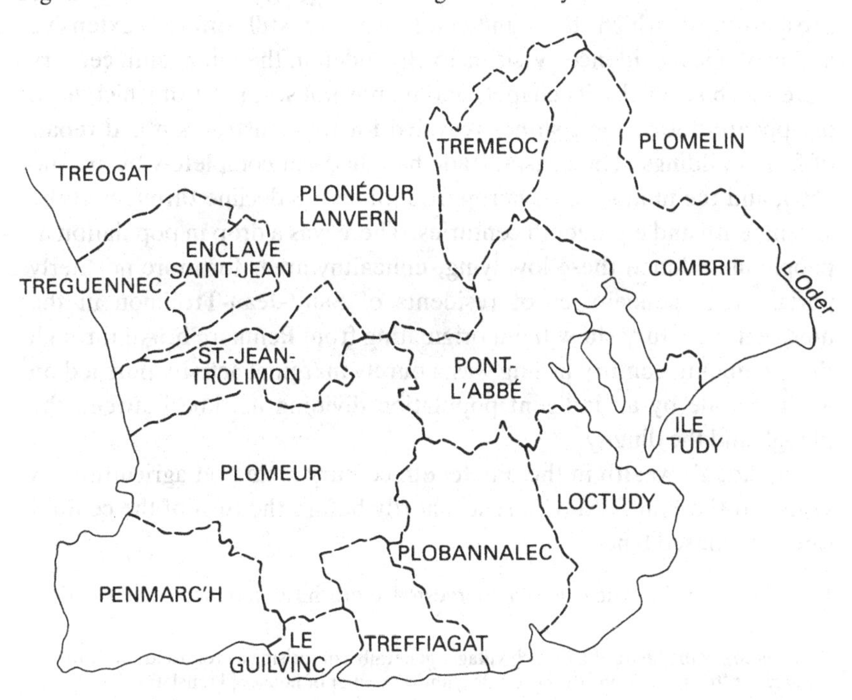

Between the high ground on which the *bourg** of Plonéour stands and the low-lying rocky and formerly marshy coast of Kérity-Penmarc'h the landscape consists of a series of alternate hills and valleys watered by little streams that flow into *étangs* opening into Audierne Bay. The bay is lined with a broad strip of sandy heath extending through the communes of Plomeur, Saint-Jean-Trolimon, and Tréguennec. In the case of Saint-Jean it takes up a third of the area of the commune. Despite low fertility and high winds, these areas (known as *palues*) had an important role in the agro-pastoral system, helping to contain the population explosion of the nineteenth century. This magnificent but (with its storms and fierce currents) hostile coast was largely unexploited by the local inhabitants. They gathered seaweed, which they mixed with dung and used to manure their land, and they picked up whatever was washed ashore from shipwrecks; but that was all. Yet in the fifteenth and sixteenth centuries the sea had been a source of wealth. Tréoultré-Penmarc'h was a prosperous port right up until the mid sixteenth century, based on the fishing and drying of conger eel, ling, and hake for Lenten fare. Seamen also maintained a coastal trade down to Nantes and carried the products of that city to Quimper and Saint-Malo. There were even coasters that called at Bordeaux.[4] Competition from the Newfoundland fishing grounds and France's maritime war with Spain combined to ruin a flourishing business, bringing to an end a level of prosperity of which the nineteenth century still offered extensive archaeological evidence. Visitors to Bigouden in the nineteenth century were much struck by its chapels and manor houses, most of which have disappeared now, their stones recycled for the construction and repair of farm buildings. The coastal trade had died out completely by around 1660, and the number of fishermen continued to decline throughout the seventeenth and eighteenth centuries. There was a drop in population as people abandoned these low-lying, unhealthy areas for more northerly parts; many genealogies of residents of Saint-Jean-Trolimon in the nineteenth century show them originating from Penmarc'h. All through the nineteenth century fishing was a purely marginal activity pursued on a small scale by an indigent population dividing its time between the plough and the dinghy.

Bigouden's wealth in the nineteenth century rested on agriculture. A visitor to the Pont-l'Abbé district shortly before the turn of the century saw 'a promised land':

In addition to the wheat that is harvested there in abundance, much barley,

* The *bourg* is the heart of a French village, generally comprising a crossroads, a church, a *mairie* or 'town hall', one or more cafés, and a number of houses [Translator].

buck-wheat, and oats is also found. The butters of this region are celebrated; fruit of all kinds is very common: delicious cherries, peaches, apricots, figs . . . There are a great many gardens filled with cabbages, onions, beans, asparagus, melons, artichokes, and parsnips. To obtain this wealth of produce, only the lightest ploughing is necessary.[5]

It was cereals that brought prosperity to Bigouden. Surplus production in the late eighteenth century enabled the region to export its 'noble' cereals and keep the poorer grades for its own consumption. Potato-growing was introduced quite late and did not radically alter the productive equilibrium of the farms of the region, which continued to practise a blend of polyculture and animal husbandry until the 1960s. The latter was never developed as extensively as in neighbouring Normandy, for example; it was restricted to domestic needs, and numbers of animals were always small. It was thus the fertility of the soil that accounted for the relative affluence of Bigouden throughout most of the nineteenth century. The present diversity of economic activities has only a hundred-year history behind it. Fishing came back into its own with the advent of new conservation techniques and the arrival of the railway at Quimper. The more deprived section of the population, most of whom came from the sandy coastal strip lining Audierne Bay, flocked to the coast as factories began to spring up and a flurry of economic activity developed around the boat-building and fish-haulage industries. In other words, the supposedly typical image of Bigouden cuisine with its proud emphasis on *fruits de mer* (sea food) is no more traditional than the architecture of recent years. There are still people alive today who remember a time when the *langoustine** was spurned and left to rot on the dunghill.

Now the situation is reversed. Maritime activities account for most of the region's wealth, while agriculture is in the middle of a process of reorganisation and the secondary and tertiary employment sectors are starting to expand. Yet despite these economic and social upheavals there has been no mixing of the population. Granted, there was emigration from the 1930s onwards, but the same names go on – Coïc, Le Berre, Daniel, Tanneau, Lelgoualch, Stephan – their history engraved in the Bigouden landscape as the names themselves are inscribed in the census records and notarised documents.

At the centre of both the Bigouden region and this study is the commune of Saint-Jean-Trolimon – the last part of the name comes from *Treff-Rumon* or 'sub-parish' (*treve*) of St. Rumon, a daughter church of the parish of Beuzec. Its territory is divided in two in a way

* A delicacy of French provincial cuisine resembling the Dublin Bay prawn [Translator].

Fig. 2. The commune of Saint-Jean-Trolimon.

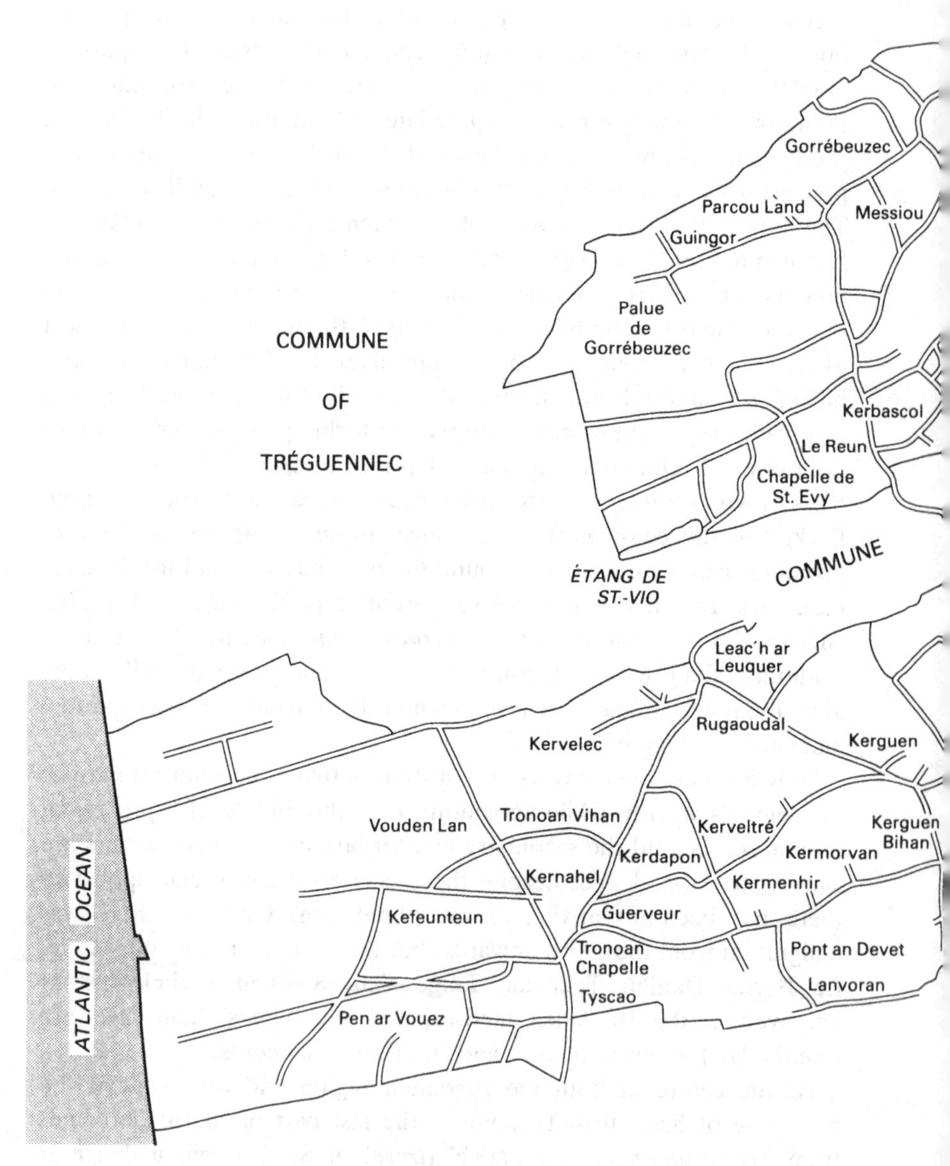

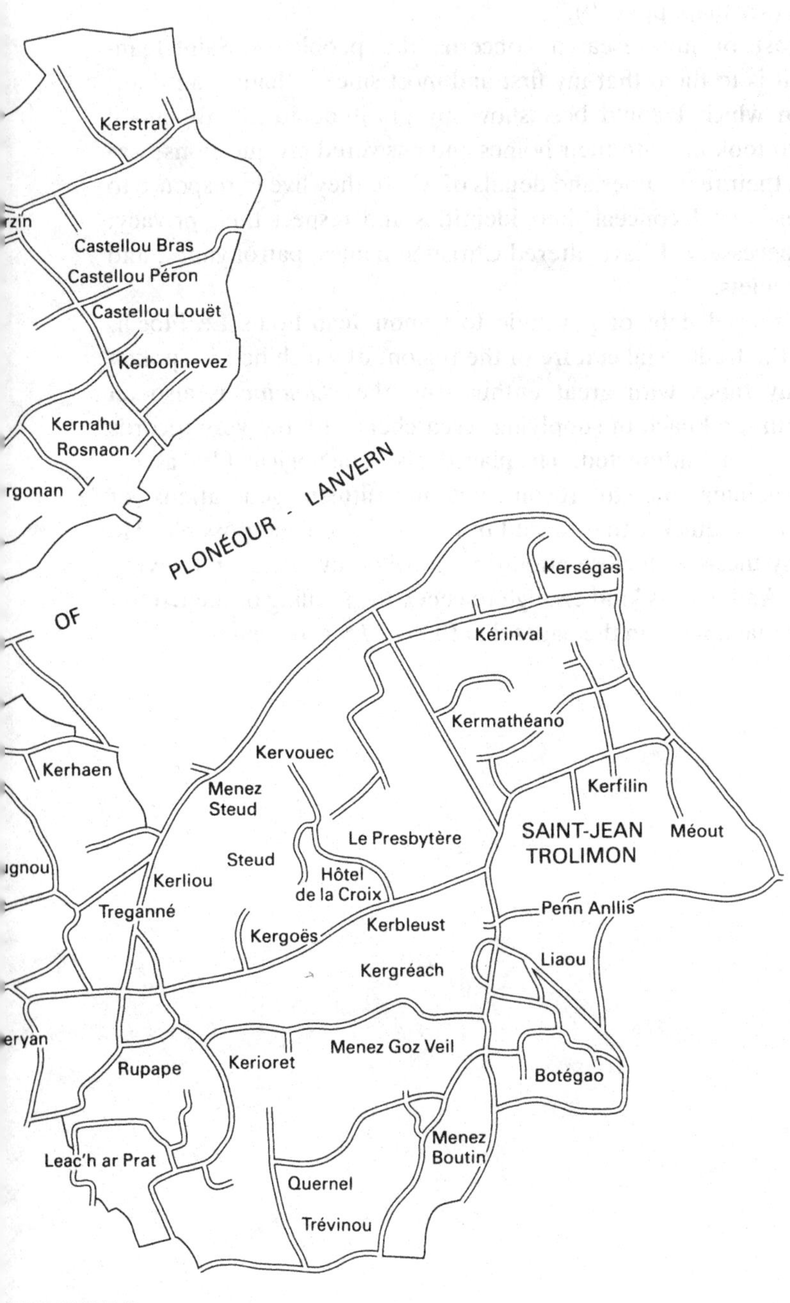
Kerstrat
Castellou Bras
Castellou Péron
Castellou Louët
Kerbonnevez
Kernahu
Rosnaon
PLONÉOUR - LANVERN
OF
Kerségas
Kérinval
Kermathéano
Kervouec
Kerhaen
Menez
Steud
Kerfilin
Le Presbytère
SAINT-JEAN
TROLIMON
Méout
Steud
Hôtel
de la Croix
Kerliou
Treganné
Penn Anllis
Kerbleust
Kergoës
Liaou
Kergréach
Menez Goz Veil
Kerioret
Rupape
Botégao
Menez
Boutin
Leac'h ar Prat
Quernel
Trévinou
PLOMEUR

that bears witness to historical changes to the parish and communal boundaries (see map. pp. 8–9).[6]

Since most of my research concerns the people of Saint-Jean-Trolimon, it is to them that my first and most sincere thanks are due.[7] The way in which I could best show my gratitude to my Bigouden friends, who took me into their homes and answered my questions, was to suppress their real names and details of where they live in response to their request that I conceal their identities and respect their privacy. Wherever necessary[8] I have altered Christian names, patronymics, and names of hamlets.

I owe a special debt of gratitude to Canon Jean-Louis Le Floc'h. Steeped in the traditional culture of the region, of which he has spoken to me many times with great enthusiasm, the *chanoine* is also an archivist with the knack of supplying researchers with the very records in which they are interested. He placed his genealogical files at my disposal, enabling me to reconstruct my fifteen generations of Bigoudens more quickly than would otherwise have been possible. He also read my thesis with close attention, drawing my attention to every inaccuracy. And he was kind enough to check the spelling of the Breton expressions that appear in the pages that follow. *Trugarez mad*!

1

A mobile population

Unlike certain research subjects, which carry their own localisation within them – analysing dance in Lower Brittany, for instance, or following the transformations of penitent brotherhoods in south-eastern France – the family is a universal phenomenon and as such may be observed anywhere. However, when one is studying the family in relation to socio-economic change, spatial limits need to be imposed. At the beginning of this research project, the commune of Saint-Jean-Trolimon, a large community with around 900 inhabitants in the nineteenth century, looked like offering a suitably defined area. As the research progressed, however, something that had appeared to be beyond dispute came to seem increasingly questionable. What emerged with great force from the very earliest interviews and dips into the archives was the frequency with which people moved house. This constituted a major social phenomenon, and it placed a big question mark against the limits of the spatial unit originally selected. Describing networks of marriage and kinship and analysing inheritance practices involved going beyond the administrative boundaries drawn on the local-government map. But how far should the limits of the field of observation be extended? An initial group of questions therefore had to do with the area within which such moves took place, based on an analysis of census returns and family geographies. Dealing with residential mobility also means dealing with social and professional mobility, since peasant society knows no social development outside the family. While the documentary record makes it possible to draw up an account of moves and family genealogies show significant swarming at the heart of South Bigouden, the oral record tells of the experience of mobility, representing on an imaginary, symbolic stage the contrast between the two Bigoudens – North and South.

Saint-Jean-Trolimon: a commune where few stay put

One tends to imagine the peasant as having had roots in his village, in his hamlet, and on his farm since time immemorial. It is as if one were deliberately seeking to reinforce the association between agriculture and stability. In Saint-Jean-Trolimon, as in Bigouden generally, the opposite was usually the case; most peasant families were obliged in the course of their lives to occupy a number of farms in succession.

This can be demonstrated by examining the nominal census lists drawn up by the authorities every five years from 1836 onwards. Record cards were completed – something like 1,000 of them for the whole commune over the period 1836–1975 – following each household (or domestic group) through those censuses. They show that, between 1836 and 1886, 13 per cent of households resided only for the period of a single census; between 1891 and 1946 the figure was 10 per cent. Various counts made census by census made it possible to establish that in the nineteenth century more than one-third of households changed every five years – a figure that includes both domestic groups leaving the commune and those moving into it. Among the latter, of course, two

Fig. 3. Example of a household file or record card.

①	Pierre le Berre* farmer, age 23 b. Saint-Jean	Kermatheano**	Anna Tanneau farmer, age 26 b. Plomeur	date of census: (1891)
	Members of domestic group present at this census: Pierre, 2, Marie Cossec, farmer and mother of head, 62, her son Henri, 19, and 1 servant			
②	in addition to the head of household and his wife: their children Pierre Jean, 7, Anne, 5, Marie-Jeanne, 3, Noël, 1; also Marie Cossec, 67, and 2 servants			(1896)
③	in addition to the head of household and his wife: Anna, 10, Marie-Jeanne, 8, Noël, 6, Henri, 4, Marie-Louise, 1; also Henri, 28, brother of Anna Tanneau, and 2 servants			(1901)
④	Anna Tanneau is a widow and declared as head farmer (*cultivateur-patron*). With her: Anna, 16, Noël, 9, Marie-Louise, 7, Marie-Anne, 4, and 2 servants			(1906)
⑤	Anna Tanneau, head farmer Noël, 16, Henri, 14, Marie-Louise, 11, Marie, 9, and 1 servant			(1911)
⑥	Anna Tanneau, head farmer Pierre Jean, 33, Noël, 26, Henri, 24, Marie-Louise, 21, Maria, 19			(1921)

* The couple wed at Plomeur on 19 October 1887.

** Same place of residence throughout the family life cycle.

groups need to be distinguished – children of couples present at earlier censuses naturally succeeding their deceased parents, and those having no close ties of kinship with the former residents. It is the latter that represent the true mobile households. Depending on the period, these made up between 15 and 20 per cent (sometimes more) of new households. Tendentially, the number of new households kept increasing until 1861. That, as we shall see, was the period of the first upsurge in population; after 1861 the rate of change slowed down until 1881. Subsequently the percentage of new households rose again until 1901 before dropping sharply in 1906 and 1911. After that period the figures are difficult to interpret, owing to the irregularity of census dates. What can be stated with confidence is that structural mobility of households was greater in the first half of the nineteenth century than in the second. This constant renewal of a portion of the population was both cause and effect of the system of land tenure, the mode of property transmission, and the structure of the domestic life cycle.

Households moved about within the commune, too. A good many farms were handed down from father to son or son-in-law in accordance with the laws of inheritance, depending on the stage the family had reached in its life cycle or as a result of demographic fluctuations. But an even larger number of farms saw a succession of tenants who stayed for the term of a lease (nine years) or for even shorter periods if it was a question of a verbal sub-lease that the tenant terminated for one reason or another. The instability of a few leases seems in fact to have guaranteed the stability of the rest, as if certain farms or hamlets acted as systematic turntables, providing temporary homes for domestic groups in search of better accommodation. Stability and instability in this case complemented each other. Of 179 houses recorded in 1841, spread among sixty-six hamlets, only fifty display the characteristics of stability as defined by the presence of three successive generations, from father to son or son-in-law. Those farms provided the base for the genealogical investigation underlying the present study. The remaining 120 or so show a high degree of mobility, with households staying for only a short time – one or at most two generations.

Here is one example of that mobility, observed at the former manorial domain of Kerbleust. Situated near the *bourg*, this consisted of a single tract of land extending over approximately fifteen hectares.* For no apparent reason (the whims of its owner, perhaps?) this farm was always changing hands, whereas most holders of such large tenancies managed to hang on to them and pass them down to their children.

* 1 hectare = 2.471 acres [Translator].

Resident at Kerbleust from 1836 to 1846 were Guillaume Calvez and Jeanne Peron, with one or two servants; they subsequently moved to the *bourg*. In 1851 the place was occupied by Louis Trebern and Marie-Louise Larnicol, with two servants. From 1856 to 1866 it was home to a widow Goulletquer from Plomeur, née Marie-Louise Gloaguen, with three children, a son-in-law, and several servants (two in some years, four in others). Between 1872 and 1876 the occupiers were Jean Lucas and Marie-Jeanne Quenet from Plonéour with one servant. Between 1881 and 1891 Michel Quiniou and Louisa Carval had the farm. In 1901 Jean-Louis Garrec and Marie-Anne Riou moved to Kerbleust after ten years at Kerstrad, staying long enough to be recorded there by the census before taking over a *bureau de tabac** in Plonéour. The tenancy was then shared by Jean and his brother Laurent Garrec and Laurent's wife Marie-Louise Derrien from Kersine; they were working Kerbleust jointly in 1901 and 1906. In 1911 there were different tenants again; Marie L., widow of Pierre-Jean C., had purchased the farm, which then stayed in the family until around 1960.

Apart from this wealthy farm, the poorest lands occupied by day labourers often exhibited the greatest mobility. At Botégao, also near the *bourg*, there was a hollow of very small, marshy farms occupied for short periods by small tenant farmers as well as by day labourers, weavers, and tailors who were conveniently placed for their customers, not too far from the *bourg*, from Pont-l'Abbé, and from the Plonéour-Plomeur road. Kerguen and Kerioret were other hamlets marked by a high degree of mobility, where only one farm was of any use for the purposes of coherent genealogical observation. Lastly, in the northern enclave of Saint-Jean-Trolimon, there were Kersine and Kerbascol. Distributed evenly throughout the commune, these hamlets offered small farmers short-term tenancies and the sons of wealthier farmers a chance of finding a temporary home. Safety valves for a system under pressure, perhaps? Mobility inside and outside the commune was not confined to the most deprived. While the turnaround of day labourers, amounting almost to vagrancy, was a product of poverty, the mobility of the children of well-to-do farmers went with the domestic life cycle and the mode of property transmission, the effects of which were amplified in the nineteenth century by demographic pressure. People felt they must 'leave the farm' and look for a tenancy elsewhere in Saint-Jean-Trolimon and its neighbour communes.

* Premises (often incorporating a café) licenced to sell tobacco products [Translator].

South Bigouden: a closed region

The standard method of studying the geographical area of individual changes of residence is to start with the residential data given on marriage certificates or in nominal census lists. The common phenomenon of local endogamy (in this case, within South Bigouden) seems to have been as manifest in Saint-Jean-Trolimon as in other parts of France, with 90 per cent of marriages, even up until recently, taking place between spouses born or residing in Saint-Jean or adjacent communes (see table 1). However, the middle part of the nineteenth century appears to have been more exogamous than the latter part, because the percentage of spouses born in Saint-Jean was lower between 1831 and 1850 than that observed between 1890 and 1910. The distance between place of birth and place of residence was also greater during the former period, indicating a higher degree of mobility. Incidentally, this tended to be a characteristic of male behaviour; brides were more often born in Saint-Jean-Trolimon and very much more often resident there than grooms at the end of the nineteenth century and in the mid twentieth century. Incoming brides were most often born in Plonéour and Plomeur and to a lesser extent in Penmarc'h, Pont-l'Abbé, and Plobannalec; non-resident grooms were most often born in Plomeur, Plonéour, and Penmarc'h and to a very much lesser extent in Tréguennec and Tréméoc.

We see that the geographical area of choice of spouse was confined to South Bigouden, with very few spouses coming from Tréogat and Plovan, which lie to the north of Tréguennec, and very few from east of Plonéour, Combrit, or Tréméoc. It should be emphasised that scarcely a single spouse came from the North Bigouden communes of Pouldreuzic, Plogastel, Plozévet, or Landudec. Spouses from coastal communes (Loctudy, Treffiagat) were also few in number and all came from farming rather than fishing families.

This analysis is confirmed by examination of the figures drawn from the 1872 and 1876 censuses. At that time although less than 40 per cent of men and no more than 40 per cent of women residing in Saint-Jean-Trolimon had been born there, they originated in the main from no farther afield than the two adjacent communes of Plomeur and Plonéour-Lanvern.

The change in the list of places of origin is very recent. The first thing to note is that, although women increasingly marry within their commune, that does not determine whether or not they will reside there. Most of them in fact emigrate, which robs the information on marriage certificates of much of its significance. Secondly, the distances from which husbands come are increasing, as regards both place of residence

Table 1. *Geographical origins of spouses marrying in Saint-Jean-Trolimon (in percentages)*

	1831–1850				1891–1910				1941–1960			
	Place of birth		Place of residence		Place of birth		Place of residence		Place of birth		Place of residence	
	M	W	M	W	M	W	M	W	M	W	M	W
Saint-Jean	28.7	56.7	59.3	75.0	36.3	54.4	42,3	86.1	18.2	75.9	20.9	97.8
Plomeur	24.2	14.0	14.8	3.5	17.4	6.5	16.7	4.4	11.6	3.5	10.3	
Plonéour	14.5	13.4	11.0	17.0	14.8	19.6	12.6	5.5	22.5	6.8	22.4	1.1
Zone A	23.4	12.2	11.8	3.5	18.3	8.5	19.8	1.0	26.0	8.6	31.6	
Zone B	8.2	3.2	2.6	1.0	10.2	6.0	5.1	1.0	9.7	2.8	6.7	
Zone C	0.5				0.5	1.0		1.0	2.8		1.8	
Other	0.5	0.5	0.5		2.5	4.0	3.5	1.0	9.2	2.4	6.3	1.1
Total no. of marriages	194				196				170			
No. of endogamous marriages (with both spouses born and resident in Saint-Jean)	18 (9.2%)				21 (10.7%				21 (12%)			

Source: 1831–1960: marriage certificates

Zone A: communes of Plobannalec, Penmarc'h, Pont-l'Abbé, Treffiagat, Combrit, Loctudy, Lesconil, Le Guilvinec (coastal South Bigouden)

Zone B: communes of Tréguennec, Plovan, Tréogat (South Bigouden, north of Saint-Jean-Trolimon)

Zone C: communes of Plogastel, Pouldreuzic, Plozévet (North Bigouden) and birthplace. Neighbouring villages have ceased to supply half the partners since 1960, and those born in South Bigouden are on the decrease.

However, few spouses come from North Bigouden. As has already been noted in numerous studies, once the traditional boundary of choice has been breached, spouses come from great distances. An 'aversion' to North Bigouden continues to show itself, spouses tending to come from Vannes, Nantes, or Rennes rather than from Plozévet.

Finally, there is no commune-based endogamy in Saint-Jean-Trolimon and even less within the *quartier*.* Analyses of the few marriages in which both spouses are born and reside in Saint-Jean show that they occur between people from all the hamlets in this area of scattered population; there is no self-contained marriage zone, either in the enclave or elsewhere, leaving aside the sandy coastal strip. Exchanges occur between all hamlets.

Commune-based endogamy appears even less marked, however, if we examine the combined birthplaces of spouses, starting from the 1872 census. As a source this gives a better indication of the population's residential attachment to the commune than marriage certificates, which often concern couples who, having celebrated their union in Saint-Jean-Trolimon, subsequently leave the place (see table 2). Sometimes fewer then 20 per cent of both spouses who were resident in Saint-Jean had been born there, as our reconstituted genealogies will confirm. The low incidence of commune-based endogamy does seem to be a feature of smaller communes. Samples taken from the registers of large and more populous communes such as Plomeur, Loctudy, and Plonéour for the period 1850–1859 reveal higher endogamy rates. On the other hand, in Saint-Jean's neighbour commune of Tréguennec, (with an even smaller area) we find the same phenomenon, namely a relatively low endogamy rate.

If we compare percentages of spouses marrying in the commune where they reside, we find very much higher figures in populous communes such as Loctudy, Plonéour, and Plomeur than in Saint-Jean-Trolimon. This trend is further reinforced when we specify whether we are interested in men or women, birthplace or place of residence. Between 1850 and 1859, for example, only 25 per cent of the men who married in Saint-Jean had been born there and only 57.5 per cent of the women, whereas in Plonéour the figures were 51.8 and 64.1 per cent respectively, in Plomeur 53.6 and 70 per cent, and in Loctudy 42.9 and 66.6 per cent.

* A sub-division of a commune, generally comprising a cluster of three or four hamlets [Translator].

Table 2. *Combined birthplaces of heads of households and their spouses in Saint-Jean-Trolimon (in percentages)*

	1872	1876	1906	1911	1921	1926	1931	1936	1975
Both spouses born in Saint-Jean	20	12.9	15.9	13.3	10.9	7.8	11.1	10.6	9.1
Husband born elsewhere, wife in Saint-Jean	17.5	19.8	17.4	12.3	10.9	12.8	13.4	15.0	19.5
Husband born in Saint-Jean, wife elsewhere	15	15.8	21.8	22.2	22.2	24.0	21.6	18.7	10.8
Both spouses born elsewhere	22	27.7	23.7	27.4	29.2	32.3	27.6	29.7	28.6
Female head of household (widow, spinster) born in Saint-Jean	4.5	5.7	5.4	6.2	6.9	7.5	9.3	8.6	9.5
Female head of household (widow, spinster) born elsewhere	9.7	5.7	10.1	8.4	13.0	10.3	12.3	14.4	13.0
Male head of household (widower, bachelor) born in Saint-Jean	4	3.5	3.8	3.7	1.4	0.8	1.1	1.4	6.5
Male head of household (widower, bachelor) born elsewhere	2.2	2.9	1.9	3.7	4.7	3.3	2.2	1.4	3.0
Unknown		6.2		2.8	0.8	1.2	1.4		

Source: Nominal census list (residents' birthplaces are indicated only from 1872 onwards and do not appear on all returns).

Not only is the territory of the commune too small within the limits imposed by administrative divisions to meet the requirements of social usage; because of the social hierarchies dividing the population into marriageable and non-marriageable partners, the population of the commune is also too small to provide an adequate pool of spouses. Hence the need to take them from other communes that, allowing for the scattered mode of habitation, are contiguous with Saint-Jean-Trolimon and where consanguines or affines probably already reside. Finding a partner in Plomeur or Loctudy does not mean marrying an outsider but marrying into a village where one already has relations either by blood or by marriage.

Plonéour-Lanvern often figures as one of the fringe communes in the mobility range of the inhabitants of South Bigouden. So it is appropriate to ask whether the people of Plonéour have their own matrimonial territory and whether their outside spouses come as much from North as from South Bigouden. Do we find something like an oil spill spreading from one group of communes to the other, or is there, so far as matrimonial territory is concerned, a seal along the northern boundary of Plonéour? In fact the latter is shown to be the case when we list the geographical origin of spouses marrying in Plonéour by whether they come from the north or south of the region (see table 3). To the north of Plonéour-Lanvern the seal is relatively tight. As a result we find a large and relatively enclosed zone around Saint-Jean-Trolimon with, at its heart, spouses and farm tenancies. We see a mirror-image of that enclosedness when we examine the corresponding data for North Bigouden, the matrimonial territory represented by the large communes of Pouldreuzic and Plozévet and the smaller communes that surround them. In Plozévet the area of matrimonial choice extends beyond the boundaries of the commune, but there are few marriages outside Bigouden; most are made with partners from Pouldreuzic and Landudec. André Burguière notes:

> When people from Plozévet marry outside the commune, they show a certain preference for the Bigouden region, but within a more closely defining frame – namely the immediate neighbourhood. If we look at all the marriages contracted between inhabitants of Plozévet and 'allogens' in the period 1850–1960, 82.3 per cent of those 'allogens' born before 1850 and 71.1 per cent of those born afterwards came from adjacent communes.
>
> . . . Out of 3,925 marriages celebrated in Plozévet between 1850 and 1960 we find percentages for endogamous marriages (both partners born in Plozévet) of 61.5 among persons born prior to 1850 and 54.1 among persons born after that date.[1]

Farther to the north, the population of the Cap Sizun region likewise

Table 3. *Geographical origins of spouses married in Plonéour between 1850 and 1859*

	Men		Women	
	Place of birth	Place of residence	Place of birth	Place of residence
Plomelin	1.4	0.3	1.1	
Tréogat	2.6	1.8	0.7	
Peumerit	4.3	1.4	2.6	1.1
Mahalon	0.3			
Plovan	0.7	0.7	0.7	
Plozévet	0.3			
Pont-Croix	0.7			
Pouldreuzic	1.4	0.3	1.1	
Plogastel	1.8	0.3	1.1	
Northern sector total	13.5	4.8	4.0	1.1
Tréguennec	5.5	3.3	4.4	1.6
Saint-Jean-Trolimon	6.7	5.9	7.4	0.7
Plobannalec	2.6	1.4	2.6	0.7
Pont-l'Abbé	6.3	4.4	2.2	
Tréméoc	2.9	1.8	4.3	0.3
Plomeur	3.3	0.7	2.2	1.1
Combrit	2.6	0.7	3.3	0.3
Penmarc'h	2.2	0.3	0.7	
Loctudy	0.7	0.7	0.7	
Treffiagat	0.3	0.3		
Southern sector total	33.1	19.5	27.8	4.7
Born in Plonéour	51.8	70.8	64.1	92.6
Total	98.4	95.1	95.9	98.4

remains enclosed within its geographical limits, with areas of interchange comparable to those observed in North and South Bigouden. The population of the cape, North Bigouden with Plozévet and Pouldreuzic, and South Bigouden from Plonéour to the southern coast – all these regions of Lower Brittany constitute 'isolates' in the standard sense of 'territories within which spouses are selected'. None of those isolates is hermetically sealed; a number of buffer zones such as the communes of Plovan, Tréogat, and Peumerit serve as 'air-locks' in a few rare cases. South Bigouden for its part forms an isolate within which the communes with the smallest populations (or the smallest surface areas) perform much the same role as the above buffer communes. These are Tréguennec and Saint-Jean-Trolimon. In the larger communes of Plonéour, Loctudy, Plomeur, and also Penmarc'h there is a significant level of geographical endogamy. That endogamy, however, is coupled with a high degree of mobility.

To label a region an 'isolate' may give the impression that all the inhabitants of that territory intermarry. Quite the reverse: our study of genealogies and matrimonial regularities will reveal the existence of segmentation within the population. It is not a question of all marrying all; the isolate is divided into sub-sections determined by socio-professional status.

So far our analysis has touched only on the nineteenth and twentieth centuries, but its findings are undoubtedly relevant to earlier centuries as well. Geographical isolation must have been greater still in the seventeenth and eighteenth centuries. It has been established that 92 per cent of those who got married in Penmarc'h between 1720 and 1790 were from that parish or from parishes lying within a five-kilometre radius of the *bourg*.[2] Our genealogies will further confirm that this pattern goes back a long way.

Exploded family geographies

It is possible, by tracing moving households to their destination and finding them again in census returns for other communes within the Bigouden region, to define the area within which people from Saint-Jean-Trolimon preferred to settle. The first thing this confirms is the gulf between South and North Bigouden, because no one settled to the north of Plonéour or east of Tréméoc. It also confirms the zone of endogamy, since the most frequent destinations were Plomeur, Plonéour, and Pont-l'Abbé (see table 4).

The third feature to emerge from these findings is the gradual shift that set in at the beginning of the twentieth century. In 1906 and 1911, as population pressure increased, geographical distribution underwent a change. Fewer couples settled in Plomeur, Plonéour, and Pont-l'Abbé, but the lower figures there are balanced by higher figures in Penmarc'h and in the new commune of Le Guilvinec, a coastal appendage of Plomeur that acquired a separate adminstrative identity in 1881.[3] This was the period of the rise of the fishing industry and associated activities, when many people left the land to become fishermen or take jobs in the canning factories that were springing up along the coast. This southward migration was confined to the most deprived section of the population, as is abundantly clear from a letter written by the parish priest of Penmarc'h, commenting on poverty in his parish in 1903. Serious enquiry had shown, he wrote,

> that the poverty and destitution of Penmarc'h stem from the immigration into Penmarc'h of the ruined, the wretched, and the idle of the parishes of Saint-Jean-Trolimon, Plonéour, Tréguennec, Plovan, Peumerit, etc. In fifteen years, 2,500 beggars have come to Penmarc'h to share the meagre pittance of

Table 4. *Households where either husband or wife is born in Saint-Jean-Trolimon and resides in the commune of:*

	1872	1876	1906	1911	1921	1926	1931	1936
Plomeur	67 24%	68 21.5%	72 16.3%	93 18.3%	90	80	64	44
Plonéour	73 26%	66 21%	65 14.7%	62 12.2%	62	71		
Penmarc'h	27 9.4%	29 9.2%	63 14.2%	69 13.6%	44	33		
Tréguennec	27 9.4%	17 5.3%	17 3.8%	22 4.3%	23	23		
Pont-l'Abbé	59 20.5%	76 24.1%	89 20%	95 18.7%	109	111		
Plobannalec	6 2.0%	7 2.2%	13 2.9%	17 3.3%	16	18		
Loctudy	8 2.7%	11 3.4%	17 3.8%	24 4.7%		15		
Le Guilvinec			46 10.4%	50 9.8%	43	43		
Ile Tudy	0	0	1 0.2%	0				
Combrit	9 3.1%	10 3.1%	11 2.4%	13 2.5%				
Treffiagat	0	6 1.9%	12 2.7%	20 3.9%	20	14		
Tréméoc	0	12 3.8%	19 4.3%	21 4.1%				
Plomelin	2	3 0.9%	1 0.2%	3 0.5%				

Tréogat	3	3	9	6
	1.0%	0.9%	2%	1.1%
Plovan	2	2	0	1
	0.6%	0.6%		0.1%
Peumerit	2	3	5	9
	0.6%	0.9%	1.1%	1.7%
Pouldreuzic	1	1	1	1
	0.3%	0.3%	0.2%	0.1%
Plozévet	1	1	0	0
	0.3%	0.3%		
Total	287	315	441	506
	63%	64%	68%	70%
Number of households resident in Saint-Jean	175	177	207	212
	37%	36%	32%	30%
	462	492	648	718

people who are already poor. At this year's First Communion, 39 out of 154 children had been baptised elsewhere than at Penmarc'h. There is nothing wrong with being poor, but not to that extent.[4]

Having left their village, did people from Saint-Jean-Trolimon all regroup in certain villages, the better to support one another – like immigrants to the cities who will crowd into houses together in order to preserve their cultural identity, help one another, and adapt to urban life? Not at all. Households that left the commune were scattered among all the hamlets of South Bigouden. The sole exception was the relatively large concentration of households occupying the sandy coastal strip in Plomeur and Tréguennec. These extremely poor farmers wandered from one sub-tenancy to another over an area that was wide open.

Communal boundaries, drawn for administrative purposes, foster no feeling of belonging or territorial identification. The hamlet within which the farm is situated and the *quartier* of several hamlets have always meant far more in this respect.

Let us follow one or two households on their wanderings. Jean Buannic, for example, was the son of a Kerbascol farmer. Born in 1862, he got married in 1895 to Marie Jacq (b.1866), the daughter of a day labourer from Gorré-Beuzec. At first, until around 1898, they lived at Plonéour; then they spent a year at Plomeur. Jean Buannic subsequently worked as a 'navvy' at Le Gouesnarc'h in Penmarc'h, but by 1911 he was farming again at the Kermathéano in Plomeur.

Sébastien Le Coz was the son of a Saint-Jean-Trolimon miller who worked at Rupape. Married to Perrine Corre of Tréguennec, he worked as a miller at Kerouille in Penmarc'h from 1872 until 1880, when he returned to Rupape to take his deceased father's place. Sébastien himself died in 1886, and his widow Perrine had to move to Kerguen with her two children, since a woman could not be a miller. Later, in 1911, she left Saint-Jean-Trolimon for Menmeur in Penmarc'h to live with a son-in-law who, born in Saint-Jean, had settled in Penmarc'h as a fisherman.

These examples of individual life cycles repeatedly subjected to the upheaval of moving show that the family territory remained circumscribed. One could reckon up in terms of hours on foot, taking all the short cuts, the distances between people's places of origin, where possibly their aged parents or their brothers and sisters still lived, and the places where they had settled as day labourers, tenant farmers, or builders.

Our examples also show that mobility was greater, the more modest the economic level of the household concerned. If a man inherited a large enough farm or tenancy, he could count on the assistance of servants in the early years of his marriage, replacing them as soon as his children were old enough to work. A farm of thirteen hectares, for example, required a permanent work-force of eleven persons, for whom conversely it provided a living. The tenant who found himself working such a farm was able to balance the number of servants he employed against the number of children of working age. Only a family emergency could compel him to quit his tenancy. The small tenant farmer, on the other hand, was obliged to move on as and when his family grew in size. Every nine years, come Michaelmas, he would comb the district – via notaries, kin, and neighbours – for the tenancy he needed in order to provide for his ever increasing progeny.

The fact that the head of a household might be recorded as a day labourer (*journalier*) in one census and a farmer (*cultivateur*) in the next is a sign that his status was unstable. Studying life cycles and mobility also shows how the fisherman class of Penmarc'h or Le Guilvinec grew out of that of the sons of day labourers. During that formative generation professional endogamy did not yet operate. On the contrary, it was through marriage that passage from one social group to another was effected. Endogamy was very slow to establish itself among the seagoing fishermen.

Tracing the wanderings of the poorest people, we find that geographical mobility and occupational mobility very often went together. The shift of occupation was still going both ways in the early years of the twentieth century, and if we find some fishermen going back to the land, it was because they were going home to Saint-Jean-Trolimon. After 1920 this was no longer possible.

Let us take the case of four siblings from Saint-Jean-Trolimon who moved to Le Guilvinec, undoubtedly in response to the call of the family network.

François Durand was a day labourer in Saint-Jean from 1872 to 1886. His eldest daughter, Marie, who was born around 1870, married Jean-Marie Le Bleis, a fisherman from Plobannalec. His second daughter, Marie-Jeanne, born in 1874, married another fisherman, Corentin Tanniou, who came from Plomeur. Louise, the third daughter, born in 1875, also married a fisherman from Plomeur, Joseph Le Corre. The 1906 census found all three households living in Le Guilvinec *bourg*, as did subsequent censuses up until 1921 or 1926. Family relationships will likewise have had something to do with the fact that their brother François moved (temporarily) to Penmarc'h. François Jr. married the daughter of a Saint-Jean farmer, and the couple worked on a farm in Saint-Jean from 1896 to 1898. In 1906

François was farming in Le Guilvinec, in the hamlet of Prat-en-Ilis, not far from the *bourg* where the household was living at the time of the 1906 census. From 1911 to 1931 he took a farm in Plomeur, in the hamlet of Kelarun.

Several similar cases suggest that people felt more confident about venturing into a maritime trade when a number of brothers went together. For instance, back in the days when Le Guilvinec was still part of Plomeur the two sons of Jacques Coïc, Jean and René, day labourers of Saint-Jean-Trolimon, both moved there as fishermen in 1872. Another case was that of the three Le Pape brothers, sons of Yves Le Pape and Marie Le Roux, farmers in the Saint-Jean *palue*. Michel, the eldest, born in 1877, settled in Saint-Jean in 1901 and was still there at the time of the 1906 census; he too was a day labourer. He subsequently moved to Penmarc'h, where we find him and his wife and four children in 1916, living in the hamlet of Kerameil. His brother Yves, immediately upon marrying a Saint-Jean girl, decided to try his luck at sea. In 1911 he was living in the hamlet of Kergadien, where we also find his younger brother Vincent, who had married his elder brother's wife's sister and was also, in 1911, a fisherman. Vincent was less successful and returned to settle at Kerbascol in 1921, resuming his father's trade of farmer. Here again family ties must have guided these moves, giving the newcomers a certain security and a start in a new walk of life.

These moves within a circumscribed territory established a family geography for each line. The destinations were like so many anchor points defining the new geography of the next generation. There was no household where a varying number of children did not leave the commune either of their parents' birth or of their parents' residence. This was possibly less true of large communes such as Plonéour. In all those of the size of Saint-Jean-Trolimon or smaller, however, children were subject to a law of mobility. Never did an entire sibling group find farms and jobs at home. No line was exempt from this law of dispersal.

Family geographies offer little opportunity of distinguishing between social classes. The son of a wealthy tenant farmer and the son of a poor day labourer would both seek a job and/or a wife within the same territory. Possibly all that can be said – and this is a more recent phenomenon, dating from the early twentieth century – is that the children of farmers who became fishermen or married fishermen were concentrated (necessarily, to some extent) in the coastal *bourgs*. The rest were scattered over a larger or smaller area that was the more unstable, the greater the degree of poverty involved.

Social usages and territorial perceptions
Family moves within the territory – be they economic or matrimonial or both – are something people speak about very readily. The first thing they do in conversation is to situate themselves in relation to where they live now; they were born in Saint-Jean-Trolimon, in such and such a hamlet, but their roots lay elsewhere in Saint-Jean or in Plobannalec. Their brothers and sisters live in Pont-l'Abbé or Plomeur (or nowadays Quimper, Rennes, or Paris), Their children have moved to Pont-l'Abbé or Le Guilvinec.

Implicit in these conversations is a feeling of having been uprooted or an awareness of a contrast between here and elsewhere. Beyond the statistics, which make it possible to measure the phenomenon, ethnological inquiry furnishes an understanding of how these multiple moves were experienced by the very people who underwent them and how the respective identities of North and South Bigouden came to be established.

Common sense might suggest that, as we move about the countryside today, using the local roads maintained by the commune and the farm access roads, we are tracing the journeys of the countrymen of a century ago. If we think that, however, we are deluding ourselves. Regrouping has altered the orientation of parcels of land and redistributed the roads running between them. The new roads correspond only very imperfectly to the old. Knowing the old roads would in fact help us to understand what it was like to experience this kind of mobility before the First World War, for example, when the bicycle was still a comparative rarity. Walking was the only means of transport that could be relied on in all weathers. The waterlogged roads might be hard work, but at least there was no risk of a cartwheel sinking in the mud, and one could always take a short cut across the fields.

Marie T. claimed that travelling from Kerbascol to Kervegit in Plobannalec, by way of Le Frout, took her no more than an hour on foot. (Her memory was clearly playing tricks with her; the two farms are at least eight kilometres apart!) She recalled her aunt coming to see her mother almost every evening, walking the five kilometres from Kergonda, just north of Plonéour *bourg*, to Kervouec, where her mother lived. Her nocturnal visits earned her the nickname *maoues an abardae* (woman of the evening), which was probably bestowed with a certain dread on account of the *maoues noz* or night women, evil spirits on the prowl.

For a longer journey, she might harness up a horse. It was accepted that women could drive horses, and a quiet animal would be chosen. This was what her cousin from Keryan, near Tréogat, did or the other cousin who lived at Trojoa in Plomelin.

So distance, whether covered on foot, behind a horse, or later on a bicycle, was no obstacle to family get-togethers. After all, the distances involved were not enormous. Even when people lived at some distance from their relatives, a very solid contact was preserved. On the other hand, the upheaval involved in changing farms necessitated a readjustment each time for the farmer and his wife and children. Every Michaelmas, the traditional expiry date for leases, the countryside would be dotted with carts bearing tenant farmers' families to their new homes. The expression was *ober gouel Mikel*, 'doing a Michaelmas', which meant moving, even when it occurred at a different time of year.

Neighbourhood ties were disturbed at each move and had to be renewed. People got to know their new neighbours when they faced a potentially difficult calving or foaling and needed to 'fetch reinforcements'. Women made neighbourly contact during one another's confinements and while working together at the hamlet's wash-place or communal oven. People stood by one another whenever there was a death in the hamlet. But social readjustment was not always easy. Marie-Thérèse L. tells how her mother was very sorry to leave Kersalous for Kerbleust. Her husband was over-fond of the bottle, and she dreaded the proximity of the *bourg* with its many bars. Moreover, she had made friends at Kersalous and missed seeing them every day.

A good way of assessing the area of territory familiar to a particular person is by looking at the places of residence of those who come to pay their last respects when he or she dies. The size of that area is often the same as the reach of a person's social relations; it is a well-known fact that honour in death is in proportion to the social situation of the deceased.

Before it became customary to put an announcement in the paper, the usual practice was to *faire prévenir* (to 'let people know'), and each village had one person whose job this was. Outside the commune, you 'dispatched' bicycles. Marie-Louise C. tells how, when Marie-Anne's father died in 1935,

> we dispatched four of them to notify neighbours, aunts, uncles, and first cousins, one towards Plovan, Plonéour, another to Loctudy, Plobannalec, another to Le Guilvinec, one to Pont-l'Abbé . . . We dispatched one to notify some cousins of my mother, an uncle on a farm at Kervenec, from Le Guilvinec to Plobannalec, at Kerandraon, a cousin and an aunt who kept [a hotel] in Loctudy. We dispatched a bicycle to go all round Pont-l'Abbé, another to go all round Saint-Jean to Kerbascol, Gorré-Beuzec, Kersine, one to Plonéour where there was family on the R. side, and to Plovan and Tréogat where there were cousins on the Le B. side . . . [Among those taking the invitations round] there was a man from the *bourg* who lived on the Plomeur road, Jean R.'s servant, the servant from Le Liaou farm. They were paid for the trip, and they were given

food and drink at the farms . . . And the people came, the mass was held in the morning, and afterwards we gave them soup and cold meat in a restaurant. All the neighbours helped: there were four of them to carry the candles, four to carry the coffin, four to carry the crosses.

In 1872, when the farmer Louis Loch died with no direct descendants but plenty of nephews and great-nephews scattered throughout South Bigouden, one Hervé Morvan, a farmer of Saint-Jean-Trolimon *bourg*, was paid three francs 'to announce the death of the said Le Loch [as he was sometimes called] to the whole family'.[5]

Rich and poor share the same space in South Bigouden. That does not mean each commune has forfeited its special identity to a sort of corporate feeling covering the whole region. It is possible, as in many parts of France, to collect proverbs, sayings, and popular nicknames here that, in referring to others in generally mocking or disparaging terms, underline differences between communes.

Saint-Jean-Trolimon is no exception to the rule. *Paotred San Yan fec'h ugent d'ober kant* is what they used to say about Saint-Jean fellows: 'it takes 120 to make a hundred'. The commune was nicknamed *San Yan Baoul*, 'Saint-Jean the poor', and even today people smilingly repeat a version of the name in French: 'Saint-Jean Monlitron'.*

Towards the end of the last century L. -F. Sauvé collected these verses about Combrit and Treffiagat:

> *Kon-bridiz, traon ha krec'h,*
> *'Zo doganed nemet c'houec'h*
> *Hag ar c'houec'h-ze evez ivez*
> *Panaved resped d'ho gragez.*
>
> [Combrit men, the flatlanders
> and the uplanders,
> Are all of them cuckolds save six,
> And those six would be too
> But for folk's respect for their wives.]
>
> *Treffiagat, pochou laou,*
> *A ia d'ar mor daou-daou,*
> *Da glask tanvez da nea,*
> *Evid ober kerdenn d'ho c'hrouga.*
>
> [Treffiagat, bag of lice,
> Into the sea, two by two
> To fetch the wherewithal
> To plait ropes to be hung by.][6]

* A *litron* is a litre bottle of wine [Translator].

Such counting-rhymes used to be chanted at fishermen – from a safe distance, of course – by the children of farming families. They underline a division into sub-spaces that nevertheless went hand in hand with a social practice of contacts and marriages. They voiced antitheses and gulfs of a symbolic order that did not inhibit concrete relationships of mutual aid and exchange.

The same cannot be said of relations between North and South Bigouden, between which there was and is a real gulf in terms of both speech and usage. The identity of the people of South Bigouden is defined in opposition to that of the people of the northern part of the region.

The people of Saint-Jean-Trolimon speak very readily of what distinguishes them from their northern neighbours:

We don't talk the same. In Tréogat, Peumerit, Plovan, they call each other *vous*: here we say *tu* . . . There's a big difference of outlook. Up there, in Pouldreuzic, Plovan, Peumerit, Plozévet, they have actual pitched battles for the elections between the upper part of the town, which is red, and the lower, which is white . . . They're very proud up there. Down here people don't have the same temperament, they don't get so excited. And the girls there are not as smart as they are here. The boundary is Plonéour . . . Around Creac'h Calvic it's like here.

The speaker alludes to differences of outlook, of speech, even of dress. To the outsider, Bigouden costume looks very much the same in the north as in the south. But to the watchful eye there can be no confusion. Northern women do not use a lace to tie their cap; in Plovan, Tréogat, and Peumerit it is *mod kouz* or old-fashioned. Details like that enable North and South Bigoudens to tell each other apart. The differences of costume stem, we know, from the influence of outside fashion. The area around Plozévet is in fact directly influenced by Quimper fashion, both *Glazick** fashion and urban dress, while the whole of South Bigouden is to some extent shielded from outside influences by Pont-l'Abbé.

To a person seeking to differentiate himself or distance himself from someone else, any sign will serve, whether it be a particular eating habit or a particular way of behaving. The inventiveness of popular speech came up with many different nicknames for the people of North Bigouden. And Plozévet folk were no slackers, either, when it came to making up names for the inhabitants of South Bigouden.

They used to come down from Plozévet to the southern communes to sell their vegetables; they also went to the market held in Pont-l'Abbé

* The traditional blue (Breton *glas*) costume of Quimper [Translator].

every first Thursday in the month to buy and sell piglets. In one person's view it was their eating habits that marked them out: 'they ate their bread piecemeal' (a sign of parsimony), and 'they didn't frequent restaurants'. The fleeting image had struck my informant's memory with sufficient force to make it symptomatic of the difference between 'them' and 'us'. The fact that the people who came down from North Bigouden in their horse-drawn charabancs sold early vegetables earned them the nickname *Ar paotr Carottes Ruz*. They were also called *Paotred ar Balhur*, using a Breton swearword more current in Plogastel, Plozévet, and Pouldreuzic than around Pont-l'Abbé.

This antithesis between North and South Bigouden was not inconsistent with the migration that took place from the former to the latter when the ports and with them the canneries began to develop. The minor industrial revolution brought about by the boom in the canning industry prompted many countryfolk to move to the towns and factories lining the southern coast and to settle there with their families. They came from the inland *bourgs*, where the proliferation of large families was becoming too much for small farms, but they also came from the *palues* lining Audierne Bay, where the soil was very poor. Pouldreuzic, Plovan, and Tréguennec supplied the largest contingents. The men went to sea in the fishing boats, and the women found work in the canneries. The ports of Lesconil, Kérity, Le Guilvinec, and Saint-Guénolé saw their populations swell rapidly. Many a life story illustrates this general southerly migration. A poor farmer scratching a living in the sandy coastal strip becomes a worker in Le Guilvinec; inhabitants of Tréogat and Plovan move to Kérity to help build a trawler.

The frequency of these moves by individuals and households constituted a subjective observation. Studying it objectively meant having recourse to the concept of mobility. Often used by sociologists to relate two generally distinct orders of data in industrial and urban societies, namely family data and occupational data, the concept of mobility also proved a useful tool in the present study of historical ethnology. Employed in its geographical sense, it made it possible to indicate the extent of moves and to show how all social classes were affected by them. Finally, it helped to establish in the firmest possible way the existence of two distinct sub-units within the Bigouden region.

Furthermore, mobility was shown to be a complex phenomenon, the social, occupational, spatial, matrimonial, and familial aspects of which constituted multiple facets of a single whole. Our documented examples of family geographies furnished both cross sections and dynamic views of that phenomenon. They revealed, for example, that when a farm had been left and had fallen to others in succession, it was extremely rare for the third or fourth generation to return there. Gradually the memory of the hamlet of origin faded into oblivion. Mobility forged a mentality of tenant farmers with no symbolic attachment to any particular place.

As an age-old rhythm asserting itself from the eighteenth century down to our own day in the manner of great tidal waves sweeping first from south to north, then from north to south, mobility appeared closely associated with two kinds of data, namely demographic trends and mode of property transmission. These will be examined in the following two chapters.

2

A population explosion

Why a demographic study?

It is not a question of offering up a sacrifice to the traditional historical monograph. Quite apart from their intrinsic interest, demographic data lie at the very heart of our subject. How do fluctuations in the population tie in with the economic and social balance of the region? How does standard of living relate to the variables that condition it, to the system of inheritance, and so on? Clearly natality, fertility, mortality, and marriage rate are of cardinal importance as far as our investigation is concerned.

Social historians and anthropologists are well aware that the application of rules – be they more or less explicit, like those regulating inheritance, or more or less conscious, like those governing marriage – is connected with and itself acts upon population and its constituent elements in complicated systems of interaction that micro-simulation studies seek to grasp.[1] A good, detailed case example is extremely useful as regards giving us a better understanding of the processes at work.

The demography of this corner of Brittany presents some very peculiar features when we compare it with that of other parts of Brittany and France. For instance, its population curves for the nineteenth and early twentieth centuries show a contrasting growth in population. Why, one wonders, did South Bigouden experience a population increase at that time?

To borrow a term beloved of historians (who themselves borrowed it from the cyberneticians), are we dealing here with a 'homeostatic' society – that is to say, one equipped with its own intricate self-regulating apparatus? This hypothesis holds that every society works towards a balance of human and natural resources such as will guarantee its reproduction. However, the twin factors of that balance may change.

Natural resources may suddenly increase with the introduction of new farming methods or an entirely new industry; human resources may change as a result of variations in age at marriage, natural fertility, celibacy, or migration. Inter-relations of this kind are not simple. For example, changes in agriculture have not always been in the direction of a reduction in labour requirements. Take potato-growing, which calls for a high labour input at each stage: sowing, hoeing, earthing up, and harvesting.

The connections between household structure and population movement are likewise complicated and difficult to interpret. Nevertheless, there is no doubt that the two phenomena are inter-related. Even during temporary phases, complex (i.e. extended-family and multiple-family) households may multiply in response to demographic pressure.

If the consequences of population growth are difficult to establish, its causes are possibly even more so.

Demographers are very much more cautious than they were in the 1970s when it comes to explaining the phenomena that they are nowadays, thanks to improved techniques, able to determine with such precision. They take account of all the many social, economic, cultural, and even psychological causes that can influence demographic events. No simple, single correlation is accepted any more, and many more questions are being asked about the reciprocal effects of demographic phenomena – fertility and mortality, for example. Given high infant mortality, do couples tend to exhibit high fertility? Is early marriage associated with high mortality and a high level of natural fecundity? Very particular attention is paid to age at first marriage, the variable that dictates such questions as marriage and fertility rates. Hajnal defined the classic model – late marriage age associated with high celibacy[2] – typical of European peasant societies; does it apply here?

Finally, such non-demographic variables as the structure of domestic groups and notably mode of property transmission also dictate population levels and how they fluctuate. We shall be looking at these problems and their many complex inter-relations with demography in chapter 3.

The demographer needs to exercise a degree of cunning with the materials at his disposal when it comes to analysing different population data. Situations differ, depending on the period under investigation. In the eighteenth century there were few population censuses, and the reliability of civil status data taken from parish registers needs rigorous checking – an art in which demographers are past masters. For the nineteenth century, on the other hand, there is abundant documentation in terms of both population censuses and vital records.

Moreover, it is generally more reliable. So a subject like fertility rate, being the ratio between the number of women of child-bearing age and the number of births, will require different approaches in the two centuries. Where no censuses make it possible to classify the population by age, demographers have to fall back on the method of reconstituting families associated with the name of Louis Henry.[3] The snag about the 'Henry method', as it has come to be called, is that its evidence is gathered from a small number of households when these are relatively mobile. The nineteenth-century measurement, being based on the whole population, gives a better picture of fertility and how it develops. But despite the quality, abundance, and sophistication of demographic analyses, the conversion mechanisms for demographic events have yet to be fully clarified as far as the nineteenth century is concerned. So the reconstitution techniques of the family *fiches* associated with the name of Louis Henry remain relevant even in a century that is amenable to different techniques of demographic analysis. This is the conclusion reached by Etienne Van de Walle, a specialist in the French population in the nineteenth century:

> Our data may simply indicate that a region such as Brittany preserved a high fertility rate coupled with rural poverty and high mortality in the early nineteenth century – in contrast to Normandy, where the data differ in all these fields. That may be interesting as a description, but it does not explain anything.
>
> To explain the timetable and precise causes of the decline in fertility, different approaches would have to be pursued. It would be necessary to analyse the behaviour of couples classified according to a set of economic, cultural, and social criteria. It would be appropriate to follow a series of cohorts right through periods of transition from one type of fertile behaviour to another. Attitudes to fertility would need to be related to those couples' environments.[4]

It is to this kind of discriminating investigation, among other things, that the present chapter is devoted.

The documentation assembled makes it possible to carry out a thorough demographic analysis only for the period 1830–1970 – except, that is, as regards marriage rate. In fact the nucleus of our inquiry concerns the population in the nineteenth century, and vital records were not systematically collected until the first census (1836). Birth and death statistics for the population of Saint-Jean-Trolimon are missing prior to 1830. So our study of population data (level and composition) and data regarding birth, fertility, and mortality essentially concerns the nineteenth and twentieth centuries; the eighteenth century can only form the object of guesswork or sampling. It is different with marriage. Our investigation covered all the marriage certificates of Saint-Jean-Trolimon as well as those gathered in the course of genealogical research. This doubled the number of certificates at the basis of the study, making more than 3,500 in all. So the sample is fairly large and does provide information about nuptiality in the eighteenth century, notably about that key variable of both population structure and social structure: age at marriage.

Our study of households is based on analysis of another documentary source, namely the nominal population lists drawn up every five years from 1836 to the present, of which some use has already been made in relation to mobility (see chapter 1).

Population movement in Bigouden from the eighteenth to the twentieth centuries

There is no proper study of the historical demography of this part of Brittany (Cornouaille) in the eighteenth and nineteenth centuries. The work of historians such as Alain Croix and Jean Meyer relates to earlier periods.[5] According to their findings, general population movement appears to have been characterised by a period of relative growth in the sixteenth and seventeenth centuries followed by a period of stagnation during the eighteenth.

By comparing various statistical indices (parish records, official counts for the purposes of taxation, military conscription, and the like) we can work out that the population of Bigouden remained stagnant until the 1750s, when we observe a gradual increase in population that was again interrupted by mortality crises in the 1780s. After that the whole region recorded a substantial demographic upsurge, with variations as between inland and coastal districts.

Plonéour-Lanvern experienced a population increase of more than 50 per cent between 1800 (2,154 inhabitants) and 1900, with a dip around the middle of the century. Since peaking in 1926 (4,536 inhabitants), the population has been in decline. Tréguennec, with the smallest population in the region, doubled its figure between 1800 (340 inhabitants) and 1900, with fluctuations similar to those observed in Saint-Jean-Trolimon; the drop in its population took place after the 1921 census (755 inhabitants). Plomeur recorded a very sharp rise. Not only did its population double between 1800 (1,132 inhabitants) and 1900; the increase was so marked that in 1881 it became necessary to carve out from it the new commune of Le Guilvinec. Like other rural communes, it has had a falling population since 1911 (2,638 inhabitants). Plobannalec also saw its population rise over the space of a century, and after reaching a peak in 1921 (2,832 inhabitants) it experienced not so much a fall as a stagnation. Loctudy grew by more than one and a half times from 1800 (1,060 inhabitants) to 1900. In fact growth continued there until 1954. Lastly, Combrit virtually doubled its population between 1800 (1,392 inhabitants) and 1911 (2,760 inhabitants), declining slightly thereafter.

The populations of the coastal communes positively exploded during the nineteenth century. That of Penmarc'h increased fivefold, while that of Le Guilvinec, which as we have seen was hived off from Plomeur in 1881, doubled between the date of its foundation (1,968 inhabitants) and 1906. On the other hand Pont-l'Abbé grew very slowly, despite its urban status; in 1800 it was less populous than Plonéour, and it still had only 6,315 inhabitants in 1900. The first half of the twentieth century was marked by demographic stagnation, and the resumption of growth is very recent, dating only from the 1960s.

Within a general movement of marked population growth throughout the region, the story was thus slightly different in each commune. The more northerly and most agricultural communes are those in which average growth was followed by a sharp drop in population. By contrast, the coastal communes exhibited substantial population growth.

Saint-Jean-Trolimon, the second smallest commune in the canton, shows all the characteristics of small rural communes (they are even more pronounced in the case of Tréguennec). What characterised the population of the commune at the heart of our investigation was very marked growth in the first half of the nineteenth century followed by fluctuations in the second (see table 5).

Altogether South Bigouden experienced a substantial increase in population: from 12,270 inhabitants in 1821 to 31,096 inhabitants in 1901. To these it is appropriate to add the population of Plonéour-Lanvern, which by some administrative aberration is included in Plogastel canton.[6]

Investigations carried out in the commune of Plozévet to the north confirm that it experienced the same almost explosive growth in population. And even in comparison with Finistère as a whole, where the rural population underwent a remarkable expansion throughout the nineteenth century, the south coast stood out. Between 1821 and 1901 the nine cantons of the coastal region saw their combined populations shoot up by 70 per cent. Top of the league were Fouesnant, Quimper, and above all Pont-l'Abbé, the only one actually to double its population.[7] As we shall see, this was partly due to record fertility.

A study of the constituents of this demographic movement shows the region to have been a veritable repository of demographic attitudes if we relate them to the general movement of the French population during the same period. Death and birth rates were further marked by

Table 5. *Saint-Jean-Trolimon: population development, showing conglomerate and scattered populations*

	Population	Conglomerate population	Scattered population	As a percentage	Houses
1800	634				
1821	770				
1826	832				
1831	878				
1836	966				
1841	1,022				
1846	1,128				
1851	1,164	65	1,094	93.9	194
1856	1,093	171	922	84.35	187
1861	1,049				181
1866	1,102	67	1,035	93.9	183
1872	956	90	866	90.5	175
1876	978	100	878	89.7	170
1881	986	82	903	91.5	175
1886	998	78	920	92.1	176
1891	1,063	105	958	90.1	175
1896	1,032	104	928	89.9	175
1901	1,045	108	937	89.6	183
1906	1,122	109	1,013	90.2	188
1911	1,124	98	1,026	91.2	194
1921	1,146				
1926	1,082	142	940	86.8	218
1931	1,064	193	871	81.8	237
1936	1,096	237	859	78.3	250
1946	980	260	720	73.4	246
1954	931				
1962	798	241	557	69.7	175
1968	758	268	490	64.6	
1975	666				

features of what demographers refer to as the demographic *ancien régime*, which in many parts of France coincided with the political *ancien régime*. Here it extended into the mid nineteenth century and occasionally beyond.

An archaic death rate

Death and fertility rates have been calculated for the population of Saint-Jean-Trolimon through the nineteenth century on the basis of ten-year periods, which offer the advantage of smoothing out short-term fluctuations (see table 6).

The death rate remains very much higher than that of France as a whole during the first half of the nineteenth century and only really declines after 1900. We need to set aside two particularly murderous

Table 6. *Mortality in Saint-Jean-Trolimon (per thousand)*

Date	Saint-Jean-Trolimon	Plozévet		France
1831–1840	31.2			
1841–1850	28.8		27.9	23.3
1851–1860	30.1			
1861–1870	37.45			
1871–1880	23.26	1876–1885	26.5	22.3
1881–1890	28.02			
1891–1900	19.77			
1901–1910	19.94	1904–1913	15.9	18.6
1911–1920	18.86			
1921–1930	14.81	1926–1935	11.9	16.4
1931–1940	12.12			
1941–1950	10.8			
1951–1960	12.3	1954–1963	14.7	11.6

decades: 1861–1870 and 1881-1890. The culprit in the first case was smallpox, still engraved in the memories of old people; the second is accounted for by a flu epidemic in 1881, which proved particularly deadly throughout Finistère.

Infant mortality, derived from the ratio of the number of infant deaths at less than one year of age to the number of live births over the same period, appears to have remained very high throughout the nineteenth century. It seems even to have been on the increase during the first half of the century, then to have remained remarkably stable from 1860 to 1920, with a slight downward trend. The real drop came only after 1931. While the overall death rate declined from 1880 onwards, we note that it did not affect all age groups equally and that adults tended to benefit most. This may be the reason why, despite a high birth rate, the population did not grow more.

Infant mortality in Saint-Jean-Trolimon was comparable with that in Plozévet for the same periods in the nineteenth century. On the other hand, Plòzévet experienced a big drop in infant mortality at the beginning of the twentieth century, which placed it below the national rate, while Saint-Jean-Trolimon experienced only a much more modest decrease. It was also comparable with the rates for the rural population of Finistère as well as with national rates. A national rate admittedly makes little sense, amalgamating as it does statistics relating to both rural and urban populations as well as to various social classes whose behaviour is very different (see table 7).

Galloping fertility

Birth rate (see table 8) and fertility rate (table 9) serve to measure the explosion of births. The former, calculated by the ratio of the number of

Table 7. *Infant mortality (per thousand)*

Date	Saint-Jean-Trolimon	Plozévet[1]		Finistère[2]	France[3]	
1831–1840	150.2			*195		
1841–1850	188.8		183.8			
1851–1860	129.9			165		
1861–1870	150.4				1864–68	178
1871–1880	133.3	1876–85	167.5		1869–73	187
					1874–78	164
					1879–83	167
1881–1890	152.3				1884–88	169
1891–1900	146.9			140		
1901–1910	110.8	1904–13	127		1909–13	120
1911–1920	141.2					
1921–1930	110	1926–35	85			
1931–1940	82.7				1929–33	80
1941–1950	75.9					
1951–1960	72.5	1954–63	31.5			

*Châteaulin *arrondissement*

Sources: [1]Pierre Kernen, 1967; [2]Daniel Collet, 1982, p. 99; [3]Etienne Van de Walle, 1979, p. 143.

births in a given period to the mean population of Saint-Jean-Trolimon over the same period, remained very high throughout the nineteenth century and indeed up until the First World War. It was subject to fluctuations, with an upward curve towards the second third of the nineteenth century. When it appeared about to drop after 1881, a final burst in the decade prior to the war pushed it up to what was still a respectable level compared with that of Plozévet, though Plozévet itself had a high birth rate. It was also above the figures for Finistère. The record was beaten by Bigouden with its 45 per thousand, whereas the mean rate for Finistère was 37 per thousand between 1823 and 1832.[8] On the map of birth rates by rural cantons, the two cantons of Bigouden stand out. The rate for Finistère and *a fortiori* the rate for Bigouden were well above the national rates, which fell steadily from the beginning of the nineteenth century. With regard to both natality and mortality, the demographic behaviour of the commune was genuinely archaic.

Of greater interest than birth rate, fertility rate is calculated on the basis of the female population of child-bearing age. It relates, for a given period, the number of births to the number of women between fifteen and forty-four years old.

Table 8. *Birth rate (per thousand)*

Date	Saint-Jean-Trolimon	Plozévet[1]		France	Finistère[2]	
1801–1805						37.7
1806–1810						36.6
1811–1815						37
1816–1820						36.4
1821–1825						35.4
1826–1830						37.5
1831–1840	44.03		41.9	27.4	1831–1835	36.9
					1836–1840	33.7
1841–1850	48.2				1841–1845	34.2
					1846–1850	31.5
1851–1860	41.6				1851–1855	31.2
					1856–1860	32.8
1861–1870	43.9				1861–1865	34.1
					1866–1870	34.5
1871–1880	48.08				1871–1875	34.3
		1876–1885	46.6	25	1876–1880	33.8
1881–1890	42.3				1881–1885	32.9
					1886–1890	33.1
1891–1900	40.3				1891–1895	31.7
					1896–1900	31.3
1901–1910	42.4				1901–1905	30.3
		1904–1913	37.4	19.6		
1911–1920	30.8					
1921–1930	29.3					
		1926–1935	20.4	17.4		
1931–1940	24.6					
1941–1950	16.1					
1951–1960	13.3	1954–1963	12.7	18.2		

Sources: [1]Pierre Kernen, 1967; [2]E. Van de Walle, 1974, p. 304.

In Plozévet, it comes out as follows:
1841–1850: 197
1876–1885: 229.3
1904–1913: 292.4
1926–1935: 85.5
1954–1963: 79.1

This calculation brings out the high level of female fertility and shows how it increased during the second half of the nineteenth century, as in Plozévet. Not only is the figure for Saint-Jean-Trolimon higher than that for Finistère; like the latter, it went up in the second half of the century.

So in order to reach a better understanding of the behaviour of families with regard to fertility, a useful approach, even as far as the nineteenth century was concerned, seemed to be Louis Henry's method of reconstituting family *fiches*, a method that has convincingly proved its

Table 9. *Fertility rate according to fertile age classes (legitimate and illegitimate births)*

	1831–1840	1851–1860	1891–1900	1921–1930	1941–1946
Mean no. of births per annum	40.6	47.0	42.2	32.7	15.8
Mean no. of women between 15 and 44	234	274	235	279	249
General fertility rate	173.5	171	179.5	117.2	63

worth. Some 1,400 family *fiches* were drawn up for the period 1830–1974. Not all of them could be used to study fertility. It was the so-called 'complete' families that interested us, the ones where it was possible to follow the woman right through her child-bearing years up until her forty-fifth birthday or, failing that, right through a union lasting more than thirty years. Strict observance of these criteria excluded a large number of reconstituted family record cards for two reasons: firstly, because a high level of mortality broke up unions after only a few years, interrupting the sequence of births; secondly, because a high degree of mobility meant that it was possible to observe households only during the sometimes brief period that they spent in the unit of observation (Saint-Jean-Trolimon).

This is a criticism that demographers often level at the Henry method – that the more mobile the population, the smaller the sample left for observation.[9] How representative, though, is a minority of stable families as against a majority of mobile ones?

Historians rightly stress the drawbacks of the Henry method as far as the seventeenth and eighteenth centuries are concerned because of the deficiencies of parish registers. But there are real problems with regard to the nineteenth century, too, whether in terms of uncertainty about birth dates in the early part of the century or in terms of conjectures with regard to mobility. So it became necessary to adopt certain special conventions in order to preserve a meaningful sample for investigation. Couples were thus included whose marriage had been celebrated elsewhere than in Saint-Jean-Trolimon and whose first child had not been born within the unit of observation but who spent the rest of their lives in the commune. Also marriages were included that had lasted slightly less than thirty years, since that criterion appeared too exacting in a demographic regime characterised by high mortality. On the other hand, couples who had married in Saint-Jean-Trolimon, lived there for

Table 10. *Complete families and the number of their children*

Marriages celebrated	No. of complete families	Mean no. of children	Mean age of woman at marriage	Mean age of woman at last parturition
Before 1830 to 1850	44	8.3	20.34	41.02
1851–1870	42	8.3	21.35	39.8
1871–1890	35	8.5	21.9	39.5
1891–1910	35	6.6	21.2	37.5
1911–1930	37	3.5	21.4	31.6
1931–1940	9	3.1	21.1	32.7

N.B.: The age of women at marriage in this sample is lower than that taken from marriage certificates, because women marrying later were less likely to figure in the sample (risk of 'incomplete' family).

a few years, producing a few children, and then left for a neighbouring commune before returning to finish their lives in the village were excluded when the years spent away covered too long a period of fecundity.

All these conditions for establishing the sample imparted the same slant to it – that of under-estimating the number of births and consequently the fertility of households.

In the end 202 complete families were reconstituted in accordance with the criteria set out here, the total breaking down into six periods:

Date of marriage	No. of complete families
Before 1830–1850	44
1851–1870	42
1871–1890	35
1891–1910	35
1911–1930	37
1931–1940	9
	202

The whole sample was thus fairly well balanced as between the different periods.

The first job was to work out the average number of children per family in each period. In the light of the average age at marriage, it is easy to understand why the fertility rates are so high (see table 10).

Women did in fact marry very young, even remarkably so for the early nineteenth century. We know that, for France as a whole, average age at first marriage was around 26 between 1820 and 1825, 25.8 between 1826 and 1830, 25 between 1831 and 1835, and remained around 24 from 1836 to 1900, subsequently dropping further until 1960.[10]

In Törbel, a village in the Swiss Valais where Robert Netting illustrated in detail the absence of contraception, the number of births per household was very much lower (4.88 for the period 1800–1849, 5.07 for the period 1850–1899) because of the advanced age at marriage of the women (around twenty-eight). The increase in numbers of births there during the nineteenth century was due to a shortening of intergenetic intervals brought about by various socio-cultural factors.[11] By contrast, in the non-contraceptive society of Bigouden it was the early age at marriage that accounted for the fertility of couples – an inverse illustration of Pierre Chaunu's suggestion that the great contraceptive weapon of the seventeenth and eighteenth centuries was precisely a late age at marriage.

We can be quite specific about the chronology of the fertility of couples. It rose gently through the nineteenth century and began to drop before the end. The advent of contraception dated not, as one might have thought, from the First World War but from the late nineteenth century. Two things show this: average number of children per couple and average age of women at their last parturition.

The relative 'youth' of this latter figure may seem surprising. On the one hand it can be linked to age at first parturition, which was relatively low on account of age at first marriage. That would mean that young brides of seventeen who had their first child at eighteen experienced a relatively early menopause, which would account for this average age at last parturition of around forty. On the other hand, since these are averages we are looking at here, it also means that contraception was not wholly unknown to some couples, who did manage to control the number of their children. Van de Walle observed that 'contraception appeared even in regions that maintained high levels of fertility in the nineteenth century, such as Brittany, Alsace, or the Nord [*département*]'.[12] Gérard Delille found traces of it in parts of rural Italy in very much earlier periods.[13] Contraception is clearly indicated in the drop in age at last parturition from 1891 and further confirmed in the final period.

To sum up, these analyses tally with one another in showing a very high level of fertility, despite a relatively low age at last parturition. They also show that, while contraception appeared as early as the end of the nineteenth century, it only affected some couples.

If we accept the hypothesis that the contraception involved was male contraception, how did it become known and how did that knowledge spread?

Taking the couples of our complete families one by one, we find certain couples evincing what appears to have been a different attitude from the others, or at least we notice that with some couples the spacing of births diverged from the generally prevalent model. Even during the earliest period, a small number of couples seems to have adopted a contraceptive mode of behaviour – 'seems' because it is difficult to prove that the spacing of births is the result of a deliberate mode of behaviour and not of some outside cause, whether physiological or social (the husband moving away, for instance).

Such appears to have been the case with Jean-Louis Lagadic and Perrine Run, who married in 1836. She was particularly young (sixteen) at the time of their marriage, and five births between then and age thirty-nine came at successive intervals of twenty-six, thirty-four, seventy-two, thirty-six, and again seventy-two months. The marriage lasted fifty-one years.

Similarly, during the period 1851–1870 ten couples out of forty-two appear to have limited the number of their descendants, though on the basis of a different system. Take, for example, a marriage celebrated in 1853: the wife was twenty at the time, and while the births of her children followed the same rhythm, between the first and the fifth, as those of couples who did not limit the number of their descendants, between the fifth and sixth, her last, fifty-seven months elapsed; her age at last parturition was thirty-four, and the marriage lasted forty-two years. Was this an indication of a deliberate strategy?

Between 1870 and 1890 no couples seem to have practised birth control. Between 1898 and 1910, sixteen couples out of thirty-five practised a form of contraception, the effect of which is reflected in the figures. Those couples had an average of four children. If nowadays that number suggests a large family, in turn-of-the-century Bigouden it was a small one, representing a marked contraction from the families of previous generations with their eight or ten children. Two contraceptive profiles stand out: some couples had large intergenetic intervals; with others, fertility was controlled for good before the end of the woman's fecund period, age at last parturition being twenty-six to twenty-eight, and this time the intergenetic intervals were shorter. So what we have here is a twofold phenomenon:

a classic type of behaviour, between behaviour typical of natural fertility on the one hand and, on the other, a fertility that is controlled but occurs throughout the woman's fecund period, during which intergenetic intervals expand and reduce the number of live births (or there is the hypothesis of systematically induced miscarriages);

a more modern type of behaviour according to which births are grouped in the early years of the couple's life cycle, then, once the desired size of family has been achieved, fertility is cut off (the observed fact of the longer interval between penultimate and last parturitions indicating accidents of birth control).

Were the couples who practised contraception in any way distinguished from other couples, at the social level or in terms of standard of living? Can we identify any ties of alliance, kinship, or neighbourly intercourse among them that might explain the diffusion within this sub-population of the phenomenon that, at a certain moment of social history, marked them out from other couples? And are there family histories of lines practising birth control?

The investigation covered too few couples to give any clear results. Granted, many of those couples were in fact linked by affinal or consanguine ties, though the same could be said of the non-contraceptive couples. Granted, too, some of the former were near neighbours and may have engaged in joint operations on the farm, which meant that they were often together, though the same could also be said of the latter. Nor does anything emerge clearly in terms of social hierarchies. At the end of the nineteenth century these were levelling out, with tenants and owners no longer marked out from each other by their way of life, any more than the inhabitants of the *bourg* (there was a tailor among the contraceptive couples). The hypothesis of birth control stemming from an acculturation of the wealthiest class (or on the contrary the poorest class) finds no support here.

The only characteristic the couples appear to share is the birthdates of the spouses. The couples that did not limit their descendants were older than those that did – particularly the men. Born between 1865 and 1879, they had a mean birthdate of 1870, whereas in the case of the couples that practised birth control the husbands had a mean birthdate of 1878 in a spread from 1870 to 1887. So we have an age-group hypothesis. Knowing that exemption from military service was abolished in 1872, may we not assume that the male technique of *coitus interruptus* was learnt by the country's young national servicemen from that date on? Or, if not learnt (because knowledge of the practice certainly existed in the latent state), applied systematically?

Historians employ the theory of demographic transition to explain the rapid population growth typical of Third World countries and the slowing down of demographic development in the countries of the West. According to this theory, a decline from high levels of mortality heralds a corresponding drop in fertility, and it is the excess of births over deaths during the transition period that generates a population increase. In Saint-Jean-Trolimon the pattern appears to have been different. Right through the nineteenth century – up until 1950, in fact – the high fertility rate was consequent upon (a) a young age at marriage and (b) the absence of birth control. The rise at the end of the nineteenth century was due to a shortening of the intervals between births, which

Table 11. *Saint-Jean-Trolimon: development of numbers of marriages and of marriage rate*

	No. of newly-weds	Population	Marriage rate (per 1,000)
1831–1840	196	1836: 966	20.2
1841–1850	212	1846: 1,128	18.7
1851–1860	160	1856: 1,093	14.6
1861–1870	236	1866: 1,102	21.4
1871–1880	180	1876: 978	18.4
1881–1890	176	1886: 998	17.6
1891–1900	190	1896: 1,032	18.4
1901–1910	202	1906: 1,122	18.0
1911–1920	162	1911: 1,124	14.4
1921–1930	248	1926: 1,082	22.9
1931–1940	186	1936: 1,096	16.9
1941–1950	174	1946: 980	17.7
1951–1960	166	1954: 931	17.8
1961–1970	156	1968: 758	20.0

was in turn attributable to closely inter-related medical, social, cultural, and economic causes.

This was characteristic of Brittany in general and Lower Brittany in particular. After analysing population movements for an earlier period, Alain Croix reached the conclusion that 'the decisive mechanism of demographic growth in Brittany is certainly the early age at marriage'.[14] Our study of the marriage rate fully confirmed this feature of the South Bigouden region.

A brisk marriage rate

Frequent marriages, and a notably low age at marriage: analysing these two demographic features will enable us to understand a number of social phenomena.

Calculated as the ratio of the number of newly-weds to the mean population over the same period, the marriage rate in Saint-Jean-Trolimon was consistently higher than that of France as a whole, which fluctuated around sixteen per thousand in the nineteenth century.[15] This is indicative of brisk matrimonial activity, which suggests that young couples were able to set up house without difficulty, irrespective of social category.

Only in the period 1851–1860 does there appear to have been any difficulty. This was marked by a decline in the rate, which a subsequent recovery made good. A similar slowing-down in the period that included the First World War was also followed by a compensatory rise (see table 11).

Such brisk matrimonial activity was not typical of Brittany as a whole, and spot investigations reveal considerable disparities between regions. In the Léon region of northern Finistère, in particular, marriage appears to have presented greater difficulties.[16]

Two other demographic indices support our finding of a brisk marriage rate in Bigouden: the very low celibacy rate, and the average age at first marriage. For the eighteenth century, age at first marriage (in other words, excluding widows remarrying) was calculated on the basis of the marriage certificates used for the purposes of genealogical reconstruction. It came out around twenty-four for men and twenty for women in the period 1701–1750 and around twenty-three for men and under twenty for women in the following fifty-year period. The ages emerging from these sources were admittedly not as reliable as those worked out on the basis of parish records analysed in accordance with the classic method of the historical demography monograph. Nevertheless, they appear typical of the behaviour of lines whose genealogy has been reconstructed; men and women married young in the eighteenth century – much younger even than in Brittany as a whole a century or to earlier, and much younger, too, than their neighbours in the Léon region.

These averages also disguise the fact that a substantial number of marriages were between very young people – sixteen-year-olds, even. Marriages of youngsters scarcely out of childhood took place particularly at the time of the Revolution, as was observed at Penmarc'h, and may have enabled the young men concerned to evade conscription. On 6 Ventôse, Year III, for example, Jean Le Merdy, aged fifteen, married Jeanne Le Corre, also fifteen; on 9 brumaire, Year IV, Jacques Riou, fifteen, married Marie-Michelle Le Pape, sixteen; on 18 pluviose, Year VIII, Alain Le Corre, fifteen, married Marie-Louise Le Corre, who was only thirteen.

The way in which the sample was made up may have given it a certain slant, for as we shall see later the family genealogies selected were characterised by a certain level of prosperity. The population at the heart of our inquiry, then, married particularly young, but Bigouden as a whole also married relatively young. The local marriage rate in the eighteenth and nineteenth centuries appears to have been much closer to that of the Middle Ages[17] than to that of France as a whole in the same period.

Here again the behaviour of the Bigouden region seems to have been peculiar. Should we describe it as archaic, as if it had simply resumed, after a certain time lag, an older type of behaviour observed elsewhere? Or was it original?

High mobility combined with a labour-intensive type of agricultural production probably contributed towards establishing this model. We find confirmed here the association that has been shown so often between death and marriage. The demographic upheavals of the eighteenth century with its major crises of mortality helped to push young people into contracting an early marriage and starting a family to carry on the farm. The whole population married young, particularly the wealthier section of it, because there were places to be taken up or preserved.

Age at marriage remained low in the nineteenth century, and this time the finding is corroborated by unimpeachable sources. All the Saint-Jean-Trolimon marriage certificates were examined as well as those of Plonéour, Plomeur, and Loctudy for the decade 1851–1860. Age at marriage was very much lower in Saint-Jean-Trolimon, as it was in these other South Bigouden communes, than that recorded in other parts of France and even elsewhere in Finistère.

The period of very early marriages appears to have been over by 1820, after which ages fluctuated between twenty-two and twenty-four for men and eighteen and twenty-two for women. This was still well below age at first marriage in France as a whole, which was around twenty-eight for men and twenty-four to twenty-six for women belonging to the generations born between 1821 and 1900.[18] Between 1851 and 1860 more precise figures for age at marriage in Saint-Jean-Trolimon were 26.4 for men and 22.2 for women, while in Plonéour they were 26 and 22.3 respectively, in Plomeur 25.5 and 22.4, and in Loctudy 25.8 and 21.8 – very similar, that is to say. The average age of women throughout Finistère for the same period was 25.4 (see table 12).

Another striking tendency is a rise in age at marriage in the mid nineteenth century. Here again South Bigouden developed differently from France in general in that it saw a lowering of age at marriage during the nineteenth century. This trend, reflected in all the Bigouden communes, was like a mirror image of the general trend.

Arguing in terms of averages has the inevitable effect of flattening out behaviours specific to certain groups. Better, perhaps, to look at the margins, which are often significant. Similarly, it seemed a useful exercise to analyse ages at marriage within the sibling group. Matrimonial strategies concerned all the children, each strategy depending in part on the number of surviving children suitable for marriage, the combination of sexes represented, and the order of sexes among siblings. Likewise, the age at marriage of each of the children seemed to form part of an overall strategy in which each age depended on the age at which the older sibling had got married and itself influenced the age at

Table 12. *Saint-Jean-Trolimon: mean age at first marriage*

	Number of marriages	Men	Women
1793–1800	18	23.3	23.4
1801–1810	26	24.3	20.1
1811–1820	50	23.6	21.5
1821–1830	59	25.5	21.9
1831–1840	88	25.2	21.7
1841–1850	106	25.3	22.6
1851–1860	80	26.4	22.2
1861–1870	118	25.7	22.3
1871–1880	90	24.1	22.3
1881–1890	88	24.5	21.2
1891–1900	95	25	21.9
1901–1910	101	25.5	20.9
1911–1920	81	24.9	22.2
1921–1930	124	25	21.3
1931–1940	102	25.2	21.5
1941–1950	87	25.5	22
1951–1960	83	26.1	23.2
1961–1970	80	25.9	23.2

which the younger ones would marry in their turn. In other words, the children did not marry in random order, and when the machinery of marriage was put into operation for one of them, all those junior to him or her were affected.

Demography is not the only factor involved, of course; we may also ask to what extent heirs or successors married in special ways (younger or older than the others) or whether the marriages of senior and junior siblings observed distinct rules.

All these questions prompted an analysis, based on the family *fiches*, of the ages at marriage of a number of sibling groups for which more or less complete information was available right throughout the nineteenth century, and an attempt was made to identify regularities among some fifty sibling groups. The rule that emerged from this analysis showed that there was an order of birth to be respected. Girls might marry before their brothers even if they were junior to them, because they married younger; but order of marriage within each sex had to follow order of birth. This rule tied in with demographic circumstances, notably with mortality, which might, in opening up a gap between brothers and sisters, account for the precocity of those marriages that served to fill it.

For the rest, despite many exceptions attributable to individual factors that could not be ascertained, very little variation was found between the ages at marriage of members of a sibling group, with women and men consistently marrying at the same ages, give or take a couple of years. And the answer to our question whether eldest and youngest siblings were distinguished by a specific age was a clear negative. This corroborated our analysis of inheritance practices; there was no successor designate who was predetermined by a marriage that marked him or her out from the other brothers and sisters. The following two examples illustrate these regularities:

Children of Alain Le Berre and Catherine Tanneau

Marie-Catherine 22, b. 17 Jan 1831	married Yves Riou 22	on 14 Apr 1854
Henry 27, b. 30 Jan 1833	Marie Corcuff 23	2 Jul 1855
Sébastien 25, b. 9 May 1838	Françoise Coïc 22	18 June 1863
Jeanne 23, b. 11 Aug 1840	Pierre Le Roux 23	3 Sept 1863
Isidor 28, b. 11 Mar 1842	Marie Cariou 24	13 Feb 1870
Alain 21, b. 5 Sep 1847	Corentine Coïc 20	23 May 1869
Marie-Jeanne 19, b. 11 Apr 1851	Guillaume Berrou 22	8 May 1870

Children of Charles Pape and Marie-Anne Le Pape

Vincent-Charles 29, b. 28 Dec 1838	married Marguerite Pemp	on 12 Jul 1867
Marie-Louise 25, b. 24 Oct 1840	Jean Le Bren	8 Oct 1865
Marie-Catherine 24, b. 21 Feb 1842	Jean Salaun	17 Jun 1866
Pierre 27, b. 25 Aug 1843	Marie-Jeanne Calvez	1870
Marie-Hélène 36, b. 15 Mar 1848	Louis Gloaguen	7 Oct 1884
Marie-Josèphe 27, b. 9 Sep 1849	Nicolas Coupa	17 Jan 1877
Jean-Marie 29, b. 6 Apr 1852	Marie-Louise Buhannic	1880

A study of ages arranged by sibling group revealed that marriage came slightly later for day labourers or small tenant farmers. The wealthier ones appear to have married off their daughters between eighteen and twenty in the nineteenth century, the less well-off two years later. The same time lag was observed for men. This may be accounted for as follows. The poorer tenant farmers kept their children with them longer because they used them as labour. Then the difficulty of finding farms imposed a further delay. The wealthier ones, on the other hand, were not only able to replace family labour with paid labour; they also owned sufficient property to be able to set their children up at an earlier age. Age at marriage emerged as a key variable at the crossroads of demographic problems and questions of inheritance.

Another aspect of the marriage rate, namely widowhood and remarriage, is also of great interest. High mortality made widowhood and remarriage very common. This too was part of the specific character of the Bigouden region, linking it to phenomena that in other parts of France were found more in the seventeenth and eighteenth centuries. Up until 1860 between 20 and just under 30 per cent of marriages were in fact second marriages between a widow and a bachelor, a spinster and a widower, or a widow and a widower. This covered a wide range of situations, depending on whether those concerned were poor day labourers or wealthy tenant farmers, men or women. In general terms, widows had less chance of remarrying than widowers – unless, that is, they were still young and in charge of a sizeable farm. Widowers, whether well-off or indigent, remarried quickly, particularly if they had dependent children. The organisation of work being based on two complementary labour forces, male and female, it was important to reconstitute the unit of production.

A fundamental aspect of matrimonial practice in Bigouden – and one that demographic data fail to reflect – was the succession of remarriages of widowers and widows to persons widowed in previous marriages, creating large numbers of stepchildren. Let us take just one striking example.

Sébastien Le Garrec, a day labourer, was married to Françoise Bescond. She died in March 1864 at the age of fifty-six. Of their four children three were still on the farm when Sébastien married Perrine Le Loeuff, herself the widow of another day labourer by whom she had had ten children, five of whom were still dependent in 1864. Perrine died in childbirth in the following year, aged thirty-nine. Sébastien married a third time, taking to wife Marie Le Quémener, also a widow and the mother of six children, two of whom were still with her. A child was born to Sébastien and Marie in 1867. Think of the number of half-brothers and sisters and even unrelated children who were required to share for a time a home where they had no remaining parental attachments. The instability that must have racked that domestic group!

Our example provides a natural lead-in to a study of domestic groups in nineteenth-century Saint-Jean-Trolimon, the basic family unit in this rural society.

Often complex domestic groups

Individuals were organised, for labour and living purposes, in units having a family structure, namely the households or domestic groups already referred to on several occasions. The household in fact partly overlaps Louis Henry's family *fiche*. It may be confined to the father-mother-children group on which our calculations of fertility rate and inter-genetic intervals were based. Or it may include other persons – relatives, servants, or lodgers. The domestic group may be defined by the criterion of co-residence or shared living-space. That often makes it difficult to pin down in exotic societies in which residential groups are comparatively elastic, changing with the seasons. (Where does one draw the line when a fluctuating number of huts share the same courtyard, which itself forms part of an enclosed group of courtyards?) In our societies the definition appears more clear-cut, since we generally live in permanent constructions and each individual has one residence and one alone (even though he may have a part-time 'second home'). Moreover, the residential unit coincides with the definitions provided of the domestic group, within which particular tasks are carried out. Individuals are thus identified by the activities they perform.

As Jack Goody puts it: 'Domestic groups are those basic units which in preindustrial societies revolve around the hearth and the roof, the bed and the farm, that is, around the processes of production and reproduction, of shelter and consumption.'[19] The domestic context covers two groups of activities: those that belong to social reproduction, to the production and consumption of food, and those that belong to biological reproduction, including the birth, upbringing, and socialisation of children.[20] The analytical usefulness of the concept is increased if domestic groups are studied in a dynamic perspective.

Meyer Fortes stressed that studies of domestic groups are of little value if they fail to take account of the ages of the members of the group, particularly that of the head of the household. The residential criterion that seems essential to qualifying a type of domestic group is never more than a moment in the cycle, and the residential schema is no more than a temporary crystallisation of a process that grows and ungrows as the individuals concerned advance in age. It is these forces of fusion and fission within the domestic group that we need to observe at work.[21]

What we observe in Saint-Jean-Trolimon over a long period is an increase in the number of households coupled with a decrease in their

Table 13. *Household size*

Date	Population	No. of houses	No. of households	Mean size of households
1836	966		155	6.2
1841	1,022		177	5.9
1846	1,128		182	6.1
1851	1,164	194	197	5.9
1856	1,093	187	188	5.8
1861	1,049	181	181	5.7
1866	1,102	183	188	5.8
1872	956	175	175	5.6
1876	978	170	177	5.5
1881	986	175	186	5.3
1886	998	176	184	5.4
1891	1,063	175	183	5.8
1896	1,032	175	186	5.5
1901	1,046	183	197	5.4
1906	1,122	188	207	5.4
1911	1,124	194	212	5.3
1921	1,146		230	4.9
1926	1,082	218	241	4.4
1931	1,064	237	268	3.9
1936	1,096	250	271	4
1946	980	246	279	3.5
1962	798			
1968	758			
1975	666		232	2.8

average size. Following a marked increase between 1836 and 1850, corresponding to the upsurge in the population, the number of households stabilised between 1856 and 1896. From that date there was a second steady increase that initially corresponded to the second population upsurge between 1891 and 1921. Subsequently it parted company with the population curve, the latter dropping again while the number of households continued to rise. Average size of household, having remained stable at a high figure until 1911, dropped away from 1921 onwards. This was a year that from many points of view appears to have brought a break with traditional demographic trends (see table 13).

A look at the number of houses provides a wealth of information. A high figure in 1851 began to decline after 1856, then levelled off until 1896, after which it rose steadily again. It was thus out of step with the increase in the population but traced a curve parallel to that of households. This means that co-residence remained at much the same level, despite the population increase. The relative stability of the number of houses also testifies to the saturation of the area, since in the second half of the nineteenth century each of those houses corresponded (approximately) to a farming lease or sub-lease. Demographic pressure thus came up against the rigidity associated with a particular system of land ownership that limited any partition of farms.

Viewed in a dynamic perspective, the structural criterion of domestic groups is a major indicator of family organisation. Some criticism may be made of its use to draw comparisons between a number of societies apprehended in terms of a count that freezes their household profile at a particular moment in time. But reinstatement of the long-term approach makes it possible to provide a dynamic picture of the family by isolating its structural characteristics or purely cyclical features. Periods of fission and fusion can also be analysed. Despite the criticisms that have been levelled at it, Peter Laslett's typology is still the one that works best.[22] As we know, it distinguishes four main types of household: solitaries, 'no family' (households with no obvious family structure), simple family households (father, mother, children; otherwise known as nuclear domestic groups), and complex ('extended' and 'multiple') family households, in which the couple and their children live together with another related couple (married children, or a brother and sister-in-law).

It is possible, on the basis of the family *fiches*, to establish a dynamic profile of the development of household structures that takes account of a double time dimension: on the one hand, the cycle of each household with the successive transformation phases that it undergoes; on the other hand, the changes that occur in the long term.

Analysis of household structures and of their state of development at five-yearly intervals up until the present day brings out the high incidence of different generations living together – then as now, but with a quite different significance (see table 14).

A widower or widow living together with his or her children or two couples (parents and married children) living together accounted for a high percentage of households throughout the nineteenth century and into the twentieth century. Sometimes a quarter of households were complex, the highest percentages corresponding (with a time lag) to periods of peak population. In 1846, for example, 25.8 per cent of households were complex; in 1871 the figure was 27.5 per cent and in 1911 27.3 per cent. From the beginning of the nineteenth century to the 1850s the proportions of extended and multiple households were more or less equal. After that date, extended households outnumbered multiple ones. Rather than take these figures to the last decimal point, we ought to use them more for the trends they reveal. Not that the statistical base was a narrow one; on the contrary, the investigation bore on more than a thousand households. But household structures are

Table 14. *Saint-Jean-Trolimon: development of household structure*

	1836	1841	1846	1851	1856	1861	1866	1872	1876	1881	1886	1891	1896	1901	1906	1911	1921	1926	1931	1936	1946	1975[1]	1975[2]	1975 (total)	1981[3]
A	1.2	1	1.6	3.6	5.5	3.5	3	3.8	5.1	5.5	5.6	3.3	3.2	3	2.9	6.2	6.9	3.4	9.4	6.7	10.9	9	34.7	24.5	29.05
B	1.8	5		1	0.5	0.5	0.5	1.5	2.2	2.8	1.2	1.1	1	2	1	0.5	1.4	0.8	1	0.4	0.4	3.3	3.8	3.6	
C	76.1	70	72.6	75.6	74.5	76	73	67.2	68.5	72.2	72.5	83.2	77.8	76	70.7	66	67.3	76.2	73	73.6	74.6	79.8	53.3	63.8	67.2
D	4.6	5	11.5	11.6	12	10.5	18.5	16	16	11.2	14	9.3	11.1	11	13.9	18.1	18.6	16.3	14.2	15.3	9.2	5.6	3.8	4.5	3.5
E	16.3	19	14.3	8.2	7.5	9.5	5	11.5	8	8.3	6.7	3.1	6.9	8	11.5	9.2	5.8	3.3	2.4	4	4.9	2.3	4.4	3.6	
F	155	173	182	197	188	181	188	175	177	186	184	183	186	197	207	212	230	241	268	271	279			232	296

A = solitaries; B = households with no obvious family structure; C = simple households; D = extended households; E = multiple households; F = total number of households.

[1] Farmers

[2] Non-farmers and retired people

[3] Figures kindly supplied by the *Mairie* secretary.

linked to the family life cycle, and the composition of a household, as measured at five-yearly intervals, was subject to change as a result of demographic and social factors.

Let us take an example. Maurice Gloaguen and Sébastienne Garrec, aged fifty and forty-eight respectively, settled in Saint-Jean-Trolimon in 1851 with six children. All born in Penmarc'h, these were aged between eighteen years and six months. In 1856 they had two servants helping them. In 1861 they were still on their own with their unmarried children and two servants. In 1866 Jean, the eldest son, was living with his wife in the parental home (and had been since his marriage in 1862). At that time the Gloaguen senior household included, in addition to the young couple and their two-year-old daughter, six other children aged between twenty-two and eight; there were no servants any more, the older children having taken over their work. In 1872, with Jean Gloaguen and his wife installed independently on another farm in Saint-Jean-Trolimon, the Gloaguen parents took in a second son, Nicolas, and his wife (they had married in 1870), together with their son. At this time the place was also home to a further five children aged between twenty-five and fourteen and working on the farm; there were no servants. In 1876 Nicolas and his wife had in turn found their own farm, Maurice Gloaguen was dead, and Sébastienne was housing another couple, Marie-Catherine and her husband. One last son was still living with her, as was her granddaughter Agathe (Agathe's parents were still alive, so was this perhaps to relieve her mother?). Sébastienne then employed a farmhand to help her. Finally, in 1881 the eldest son returned to the farm to succeed his mother.

The domestic group was thus by turns simple and multiple (complex), according to Laslett's classification. Clearly this was a case of co-residence with unstable partners, since the young couple was accommodated only for an interim period. It was a fundamentally different pattern, in other words, from that which prevailed in stem family areas.

This complex household structure was connected with the ways in which property was transmitted, which we shall be analysing in the next chapter. But it is already evident that many young couples (who did indeed, as we have seen, marry very young) co-resided with the parents of one of them until such time as they found independent accommodation. A widower or widow who was still active would thus share the farm with one of his or her married children and spouse, whose residence would be temporary and would give way to a further period of temporary co-residence by a younger brother or sister and spouse.

This kind of arrangement was typical of the better-off families. It virtually never occurred among day labourers, whose households were usually of the 'simple' or nuclear type.

Nowadays these same percentages have taken on a quite different significance. A household that includes two generations is no longer indicative of wealth but on the contrary of straitened economic circumstances. The old parents live with the young couple; often they and their daughter work the farm while the husband has a job in the craft trades or in industry. A home of one's own is now on the contrary a symbol of modern life for farmers or artisans. Finally, a new feature now characterises the population. This is the large number of solitaries – widowers and above all widows who have retired from farming. This is a reflection of the age structure of the population, which in 1984 included a large number of old people (120 out of 800 inhabitants).

Demographic movement in Bigouden, then, is wholly original. Historians generally associate population growth with industrialisation;[23] here it occurred in a purely rural setting. Further peculiar features were a very young age at marriage, which went hand in hand with complex domestic groups living together under the same roof, with generations overlapping rather than succeeding one another. None of the features supposedly typical of domestic groups in northern and western Europe – late age at marriage, independent assumption of the running of the farm by young couples, circulation of young people among households as servants[24] – is observable here. The Bigouden model relates to a world of kinship in which the couple merges with the family network and the socialisation of young children is the province not of parents alone but also of rising and collateral generations. Early marriages led to a remarkable fertility rate that, since there was no emigration from the region, gradually eroded social hierarchies, notably through the medium of the particular local mode of property transmission. In fact there was a highly complex causal relationship between the inheritance system and demographic patterns. In so far as that system governed access to wealth and economic opportunity, it helped to account for the fluctuations in population level as expressed through each domestic group.

3

To each his (or her) share: an egalitarian (partible) system of property transmission

The domestic groups of the Bigouden region were complex in structure. Marriage, often entered into at an early age, did not coincide with setting up an independent household; there was an interim period when the generations lived together. Forces of fission were at work, however, breaking up domestic groups, while throughout the nineteenth century forces of fusion reconstituted them on similar foundations. Movement of people was also a movement of property, though the two were not always simultaneous. To study the ways in which property was handed down from generation to generation is thus to observe both the cycles of evolution of domestic groups and the modes of property transmission that made possible their social reproduction.

For a long time the accepted idea was that the pertinent distinction between modes of property transmission was the principle of partible inheritance as opposed to that of impartible inheritance benefiting one child only (also known as the single-heir system). Attempts have been made to associate each principle with a particular demographic system and a specific household structure. However, such correlations disregard analyses in dynamic terms, comparing census returns that 'photograph' the village at a given date with practices that by their very nature involve the passage of time. We need on the contrary to take account of this essential dimension, which tends to bring together systems that, while distinct in legal terms, nevertheless present certain similarities in practice. As Georges Augustins points out, it is no good confining oneself to examining legal regulations that are constantly being contradicted by the facts. Rules need to be consigned to the realm of ideology, and we have to look elsewhere for an explanation of what happens in practice.[1] particularly since in certain countries – Greece, for example[2] – various codes prescribing divergent arrangements overlap, and usage departs from them all.

Unequal and equal heirs

But does that mean the distinction between impartible and partible inheritance has to be rejected on the grounds that no system exists in which all the children but one actually receive nothing at all and no system that does not tend to benefit one child in order to stop its own ill-effects precluding any kind of social reproduction? Intermediate forms may exist, but specific types do characterise societies that are very different not only at the economic and social levels but also at the symbolic level.

The supposed confusion of the two systems may also stem from a semantic confusion surrounding the word 'dowry'. Itself endowed with many meanings – economic, legal, and emotional – the term nevertheless denotes quite specific modes of inheritance. In the single-heir system, dowry refers to the amount of assets, in money or in kind, given to the child at the time of marriage as the child's definitive share of the family fortune. Dowries are said to exclude all but the first-born. They are often (though this varies from society to society) assigned to the female children. The same term also refers – in the partible system – to the small sum of money or movables (trousseau, livestock) given (or promised) to the bridegroom as well as to the bride on the strength of their marriage. Such property has to be restored to the definitive succession when this ensues on the death of the parents. So the same word serves to describe a quantity of property that in the one case excludes the recipient from any claim to inheritance and in the other constitutes a token or promise of future assets still to come. Strictly female in some instances, in others a dowry is received by both parties to a marriage. Here it serves to complete a principal patrimony in which it is incorporated; there it must be strictly balanced for both parties. In one case it is allotted to a family group, in the other to an individual. One could go on listing such antitheses, particularly since in practice things are often less clear-cut.

The contrast between partible and impartible inheritance also remains pertinent when the concept is related to that of succession. When inheritance is impartible, heir and successor merge into one; the person who inherits the family assets also takes the father's place at the head of the farm – in other words, succeeds him. Partible inheritance, on the other hand, makes designation of the successor very much less clear.

It is still a very relevant exercise to compare the two systems. The one that predominated in South Bigouden in the eighteenth and nineteenth centuries presented such a contrast to the impartible system as almost to caricature it, so it may be useful to characterise the latter system briefly.

Parts of south-west France, the upper Verdon valley (Basses-Alpes), and the Gévaudan region (Lozère), as well as the Spanish Basque country south of the Pyrenees are all examples of peasant societies in which the prevailing system of inheritance distinguishes between a principal heir, a secondary heir, and younger brothers and sisters.[3] The principal heir receives the patrimony, consisting of house and outbuildings, garden, fields, rights to common land, and room in the cemetery. His function is to maintain or enhance the standing of the *maison* ('house' in the familial sense). The rest – secondary heirs and younger brothers and sisters – are dowered and excluded from the patrimony. This system of inheritance ties in closely with the marriage system; heirs marry younger daughters of other houses, and the monetary dowry they receive as a result goes into the collective patrimony, where it will serve, in turn, to dower the excluded brothers and sisters. The system also ties in with the stem-family configuration, where three generations co-reside.

This general model serves as the basis for a number of variants having to do with the sex of the heir (more or less thoroughgoing exclusion of daughters), his place in the order of birth, and the size of the children's respective shares at the time when the property is transmitted. But the main structural and ideological outlines of the system remain the same.

The geography of the impartible system has yet to be drawn. It often tends to be associated with southern Europe. However, it also exists in certain more northerly regions; Robert Cresswell has described how it operated in the West of Ireland in the 1960s.[4]

The impartible system hinges around a house, which carries a high symbolic value, and variants are always contained within the limits of the model. By contrast the partible system opens the door to a broad range of strategies depending on the demographic, economic, and personal situation of the domestic group. The principle, which can be baldly stated as an equal share for each child, is highly flexible in its application. As Lutz Berkner writes, 'family pressure, socialization, or personal choice may lead some potential heirs to accept cash or other settlements rather than claiming their inheritance in land. This will depend on such things as the economic feasibility of dividing the farm, the financial ability to raise the cash settlements, and the accessibility of non-agricultural employment. Partible inheritance systems are basically flexible, and division can almost always be avoided in such a way that the laws are not actually broken and the peasant customs are accommodated.'[5]

The impartible system, which is built around the continuity of the house, succeeds with all its variants in maintaining that continuity. The partible system carries within it the seeds of its own destruction. Each generation faces anew the conflict between ideology, usage, and social reproduction.

Were the ideology of sharing to be observed to the letter, it would soon become impossible, after only a few generations, to keep viable farms going. Each society has to find its own solution to the contradiction inherent in the system: to do justice to the rights of each child, or to sustain a viable business in the context of the local environment, the dominant type of economic activity, and the level of technology available.

Several types of social response are offered. Either the heirs have access only to a right of use rather than one of ownership (as is the case among the inhabitants of Tory Island, off the coast of Donegal) or domestic groups implement a successful policy of contraception and practise matrimonial strategies that make it possible to prevent properties from disintegrating (the system adopted by the large farmers of Minot in Burgundy). At Nussey in the Jura the pseudo-partible system clearly benefits males, whom the deeds eloquently describe as *plus prenants* ('more grasping'). In another place women receive only movables, which makes it possible to leave the land and the means of production intact.[6]

Bigouden society presents a model that is different again from those described here. Studying it as an example and comparing it with other Breton societies together have the effect of strengthening the conviction that there is a wide variety of property-transmission models in peasant societies.

Inheritance in Lower Brittany: a unique model and its variants

A feature of old Breton law as laid down in the *Très Ancienne Coutume* of 1539 is its egalitarianism. It forbade any benefiting of one heir at the expense of another. Instead it upheld the principle of *rapport forcé* whereby any advantage granted to a child during the lifetime of the parents was taken back in the general share-out that took place after their deaths.[7] At the same time it affirmed the advantage of the eldest child, whether of noble or of common birth. Implementation of these two principles eventually produced large numbers of impoverished younger sons and daughters in the Breton nobility.[8] It also led to the countryside becoming dotted with small manors that, once the more disinherited nobles had merged into the mass of the population, passed into the hands of their peasant occupants.

The way in which the *Coutume* and the *rapport forcé* were applied and the preference given to the eldest child in the choice of shares varied enormously between the different Breton regions and even within Bigouden itself. Was this a matter of different models or variants of the same model? It is hard to say, because the sources on which descriptions rely are too diverse.

Alexandre de Brandt, analysing inheritance practices in rural France, observed two models coexisting in Brittany. Except in coastal districts, impartible inheritance predominated. A twofold usage characterised the region: that of sharing out the property while the ancestors were still alive, and that of observing the law of primogeniture. However, closer to the coast it was the eldest child that inherited, regardless of sex. De Brandt further drew attention to the special character of the Côtes-du-Nord region, where (he said somewhat obscurely) impartible inheritance was proportionally the least common.[9] He thus made a distinction of system that coincided with a geographical distinction: impartible inheritance with compensatory payments (*soultes*) to non-inheritors (or celibacy on the part of sisters)/partible inheritance; central Brittany/littoral. The geography is highly imprecise, and the sources of de Brandt's investigations, recalling the secondhand accounts of late eighteenth-century travellers, are extremely flimsy. We are left with a hint of differentiation between the three Breton *départements* of Côtes-du-Nord, Morbihan, and Finistère. De Brandt's insistence on the law of primogeniture is odd. This is another of those tricky terms: here it does not denote anything like the same system as the one in force in regions practising impartible inheritance.

For one thing, the law of primogeniture is not associated with the exclusion of dowered younger brothers and sisters since in principle all children have equal shares in the estate. For another, it poses the question of paternal authority in a rather different way than occurs under the impartible system. Paternal authority generally has the power to 'make' an heir, whether elder or younger, male or female. Here it is on the contrary diminished by the operation of the system, since it is the actual first male or female child who, shortly after his or her marriage, succeeds the father as head of the farm.

Use of the law of primogeniture was not as absolute as de Brandt suggests. In Le Porzay (the region to the north of Quimper), according to Elicio Colin, it was usual for the eldest child to retain all the property on condition that he or she paid a *soulte* or compensatory amount to the other heirs. This was a valued custom in a region of large families numbering – in 1919, for example – up to eleven children capable of inheriting.[10] Nor did Alain Le Grand make any reference to the rule when he looked at the actual mode of transmission of a farm in the Quimper region – that of Robert Huella in Guengat. His study showed that, in the eighteenth century, depending on the demographic circumstances obtaining for each generation, it was sometimes the eldest child who inherited the property, sometimes the youngest.[11]

Farther west, on the Cap Sizun peninsula, a study revealed two different systems of inheritance in adjacent communes – with different marriage systems to match. In Goulien, a society of owner-farmers, the law of primogeniture prevailed; only a few children of the sibling group were married. What mattered was ensuring the permanence of the 'house' and maintaining the line. In the neighbouring commune of Plogoff, where economic activity was divided between fishing and agriculture, land was shared equally among all the children, and most of them married.[12]

Like Goulien, but unlike the South Bigouden communes, Plozévet practised the law of primogeniture. According to Michel Izard's detailed study the first-born – whether male or female – played a clearly preponderant role. When the eldest boy or girl got married, a marriage contract stipulated the date when the newly-weds would take charge of the farm of the parents of the husband/eldest son, in the case of a virilocal marriage, or the wife/eldest daughter, in the case of an uxorilocal marriage. In Plozévet and neighbouring communes, the period that elapsed between the marriage of the eldest child and the couple taking over the family farm was extraordinarily short: one or two years – rarely more than five. The contract appraised the total value of the assets making up the farm. An estimate of money and goods to provide for the retired parents was subtracted, and the rest was divided into as many shares as there were children of both sexes. As soon as he succeeded, the new master had to buy back the shares of such of his brothers and sisters as had left home and be prepared to do the same for those who, whether minors or majors, still lived on the farm and continued to be his responsibility.[13] Though so close geographically, South Bigouden has no law of primogeniture. This may be only a minor feature in a model that resembles Plozévet's, but it is important enough in native eyes to make it a criterion of distinction between North and South Bigouden, along with the way in which the *coiffe* is worn and the use of *vous* and *tu*.

A pronounced characteristic of the whole Bigouden region is the genuinely equal status of men and women in the matter of inheritance. The system does not play hanky-panky with daughters as it does in the Jura; here, provided they marry, they are full heirs. The husband very often comes to 'act son-in-law' (*faire gendre*) with his father-in-law. Moreover, no unmarried child would be chosen to succeed; he or she would retain his or her theoretical rights to the land, but that share would never materialise in the form of a viable farm – only as a *soulte*. So there is no difference in the way male and female children inherit land, leases, and movables. The husband administers his wife's assets, but she retains ownership of them. On dissolution of the joint estate, their respective separate properties are carefully inventoried and valued.

Equality of the sexes in the matter of inheritance forms the basis of the profoundly cognatic system of filiation. It is because children inherit from their father and their mother and because women hand down assets in their own right that the two lines are balanced in importance. In South Bigouden no distinction of birth order or sex decides which of the children will succeed; they are all *heritourien*.

Oral research and legal documents: two complementary sources

We have seen that it is no good confining our attention to the rule as formulated in the law books. So how should the practice of property transmission be studied?

The first method is by direct inquiry. The farmer reveals that he bought his farm thirty years ago or that 'the family has been here for 600 years'. Given such essential information, the researcher is able to start from the present and work backwards in time. Occasionally, some old documents still kept in a drawer of the dresser will throw light on the history of a particular farm. A contact with a landowner will make it possible to undertake the same analysis on the other side of the tenancy relationship. All this is inadequate and piecemeal. Several times a farmer showed astonishment at the present researcher's puzzling interest in old papers that he had only recently burned. All that mattered to him was the immediate past and above all the uncertain future.

How, then, was one to study the mode of inheritance over a long period? An initial, rudimentary way of doing so was to trace lines of inheritance through residence data. The number of children in each household was known, as was the number of surviving children and their marriages. One therefore observed which child succeeded the father or father-in-law on the farm. It might be a matter of inheritance (a lease transferred, a field or some cows given); it might be only a provisional arrangement indirectly linked to the final hereditary settlement. In other words, observing residential succession was useful but still not adequate.

So there was no alternative to consulting notarised documents if one wanted to find out how things were done over a fairly long period. But how to approach such a mass of documentation? Nineteenth-century

Pont-l'Abbé kept four or five notaries extremely busy (today two practices survive). There are mountains of files on deposit in the departmental archives in Quimper but no indexes enabling the researcher to identify particular locations. Deeds relating to successions, marriage contracts, wills, donations, and post-mortem inventories are lost among a mass of deeds of a purely economic nature: farm leases, mortgages, loans, and so on. How was one to find the ones that related more particularly to the farms in which one was interested?

It was necessary to take something of a long way round. This consisted in combing the registry tables for the deeds of post-mortem changes of ownership relating to natives of Saint-Jean-Trolimon. The government levied a tax on successions, which laid an obligation on the deceased's family to declare the rightful claimants and to value the total amount of movables and real estate left to the heirs. For a village population that numbered around 900 throughout the nineteenth century, only 500 such deeds were found between 1830 and about 1920. In other words, a large proportion of the population died without leaving anything that might have constituted evidence of a tax liability. These deeds also referred to the involvement of notaries who had drawn up marriage contracts or post-mortem inventories. This offered a way of locating the deeds relating to the households under observation. The deeds once extracted from the notarial files, a corpus was established of some 150 deeds relating to various households between 1830 and 1920; to these were added about twenty wills, marriage contracts, partitions, and exchanges of real estate. Why so small a number? First because of the irregularity of the notarial records, secondly because such deeds – notably the post-mortem inventories – reflect some family crisis. No deeds does not mean no succession; this may perfectly well take place over several generations by amicable settlement within the family, with no notary being involved.

This study of practices relating to succession in South Bigouden is thus based on a combination of oral and archival research. So how does the partible system organise the access of the generations to assets and property?

Patrimony is not an undifferentiated, universal entity; its content varies according to the type of economic activity and mode of land ownership involved. This last point is crucial. A partible system of property transmission may conflict with the need to keep a landholding in one piece, but it suits a tenancy system. It is easier to make one's heirs equal when one has only movables, livestock, or money to pass on to them. Granted, the successor in a tenancy enjoys an advantage over his brothers and sisters, who will have to find tenancies of their own. But the ability to keep a lease in the family cannot always be taken for granted; it is the will of the landowner that is paramount.

South Bigouden resembles other parts of Brittany in being characterised by an ancient mode of land ownership that, so far as the tenant farmer is concerned, represents a combination of tenancy and ownership. As was pointed out in the Introduction, the *domaine congéable* makes for two owners: one of the land and one of the 'reparative rights, edifices, and surfaces' (sometimes referred to as the *convenant*). The owner of the latter is called the *domanier*. Not all farmers were *domaniers*. Some – very few – owned their own land; others were simple tenants, enjoying a lease in writing or agreed verbally.

So what a man or woman might receive from parents varied according to the legal system under which they worked their farm and according to the level of parental wealth, which was itself linked to the size of the farm worked.

What was it, then, that circulated? The answer is: freehold or reparative rights; a right at the discretion of the owner regarding transfer of the tenancy from father to son or son-in-law by sub-rogation of the lease; and rights in respect of movables, ready cash, livestock, agricultural plant, furniture, and household linen.

Certain rights in respect of common lands such as heaths, *étangs*, and peat bogs that provided animal fodder and domestic fuel were attached to a farm and transmitted with it but were not individually transmissible to each of the children.

The family life cycle and property transmission

It was usually marriage that marked the beginning of the younger generation's rights in respect of the patrimony held by the older generation. In discovering that what was handed down to the children at marriage was relatively insignificant in comparison with the parents' entire patrimony, we establish right from the outset that in this society marriage did not constitute the critical moment so far as succession was concerned; that was pushed back to the end of the parents' family life cycle or even to a point after their deaths.

It was relatively rare for marriage contracts to be drawn up. Only about thirty were found for couples who died in Saint-Jean-Trolimon, and on most of the deeds of post-mortem changes of ownership there is no mention of marriage contracts. In fact, some deeds state expressly that no marriage contract was drawn up: such arrangements as were made at the time of the marriage were organised verbally. Nevertheless, a certain number of couples did prefer to seal the clauses of their matrimonial association in writing in order to protect the separate properties of each of the spouses. Contracts are mainly found among well-to-do farmers, though one also comes across them between relatively poor people, where husband and wife contributed only modest estates.

The ancient Breton legal institution of the joint matrimonial estate, as provided for by the civil code, was adopted without difficulty by couples entering upon marriage. The first article of the marriage contract laid down that the spouses-to-be should hold their property in common from the day of their marriage before the registrar. It was sometimes only acquests that accrued to the joint estate in that the estates brought in by the two spouses were constituted as separate property that must return to the donors in the event of the donee dying childless. Some contracts stipulated that, the dowry aside, every asset falling in succession should remain separate property. If certain assets of one spouse were sold during the marriage without reinvestment to his or her advantage, he or she might recover the amount against the assets of the joint estate before partition. The distinction between separate property and property held jointly was aimed primarily at protecting the children of different unions in a situation in which widows and widowers frequently remarried. It underlined the strong position of the woman, who though married still retained possession of her property, even if it was her husband who administered it.

Dowries varied in both kind and value. The commonest references are much as follows:

> In consideration of this marriage Louis Le Corre and wife do constitute as dotal settlement of portion by anticipation, to be charged initially to the succession of the first of them to die and subsequently, if need be, to that of the survivor, and do give to their son Henri Le Corre, who accepts it, the sum of twelve hundred francs, which the said Le Corre couple [Louis and his wife] undertake jointly and severally to pay and supply to the [young] couple.[14]

Poorer peasants organised a payment plan for the dowry, spread over a longer or shorter period. Here is part of the marriage contract of the daughter of some small farmers who held a lease near the coast and a tailor living near the *bourg* of Saint-Jean-Trolimon:

> The bride's father, Bertrand Boderez, does by this present deed *inter vivos* . . . make over to his said daughter, who accepts it, the sum of six hundred francs to be deducted before any partition and after his demise from the surest part of his estate.[15]

However, the sum was not to be paid until three years had elapsed.

Occasionally, when a joint household was organised between parents and children, provision was made for the dowry to be paid to the married couple only when they ceased to co-reside with the parent couple. For instance:

The said Péron couple declare that they are tying up for the benefit of their son Jean-Louis Péron, who accepts it, the sum of six hundred francs, which they undertake severally to pay at such a time as it shall please him to go and live on his own together with legal interest payable annually.

And as security for the said sum, principal and interest, the Péron couple do particularly assign and mortgage all their real-estate rights to the locality of Le Castellou in the commune of Saint-Jean-Trolimon.[16]

Mention is made here of the common practice of paying interest on the promised dowry in the event of deferred settlement.

Actual cash payment of the dowry was no problem for well-to-do farmers. In fact, among the notarised deeds the occasional dowry receipt stipulates that the sum has been paid 'in metal coin' (*en espèces métalliques*). Often payment was in fact deferred, complicating the succession when the donee died before it had been completed.

For example, when the post-mortem inventory of Marie-Perrine Le Sevignon, who died in June 1842, was drawn up at the request of Hervé Lucas, to whom she had been married since 1839, it contained the following reference:

Under a marriage contract authorised by Maître Kernilis, notary at Plonéour, which cannot be produced because the relevant fees have not yet been paid, Pierre Le Sevignon settled on the deceased, who was his daughter, the sum of nine hundred francs, three hundred francs of which are still owed by the afore-mentioned grantor.[17]

Dowries were not necessarily paid in cash or made in the form of promises of sums of ready money. Instead there might be an as yet unsettled credit transfer or inheritance account, or possibly a sum that the husband or wife already owned as separate property, either as a product of his or her labour or as a settled inheritance account. In this way Anne Le Lay, whose father was dead, settled on herself, as dowry, a simple expectation of inheritance:

Anne Le Lay declares herself to be the owner of about eleven or twelve hundred francs owed to her by the above-named Jean Bargain and wife in completion of her father's inheritance; although this account has not yet been settled, she nevertheless intends to tie up for her benefit and to exclude from her future joint estate the total amount of this guardianship account, whatever it may be once discharged.[18]

Less well-to-do farmers settled a dowry of livestock or furniture that, while it might well represent a drain on their business, was one they were better able to tolerate than taking out a large sum of money. Here is an example from some small tenant farmers.

who settle as dowry on their daughter, who accepts it: 1) the sum of six hundred francs stipulated as payable in future as soon as they wed before the registrar, 2) a wardrobe and a cow that are together valued at one hundred and fifty francs and will be delivered to Louis Gloaguen. The bride-to-be will contribute as her marriage portion the capital sum of nine hundred and fifty francs, which has come down to her from the estates of her afore-mentioned deceased father and mother.[19]

Sometimes a single heifer was donated – for example, to swell the livestock of a farm that was already a going concern when a girl married a widowed leaseholder. Or it might be some furniture and some young stock.

Take Guy Le Pape, who was living with his widowed mother. She farmed at Tronoën, on the edge of the sandy coastal strip where the parents of his bride-to-be, Catherine Le Loeuff, lived. On the authority of his deceased father, he declared himself the owner of a share of the surfaces and reparative rights of a holding at Plomeur. In addition his mother gave him, over and above his share:

A box bed complete with bedding of two feather mattresses, two sheets, and a bolster, together valued at twenty-one francs, and a pair of two-year-old bullocks, one black haired, the other piebald, valued at forty-eight francs, making a total of sixty-nine francs. Corentin Le Loeuff further makes over to his daughter, who accepts it, an irrevocable donation *inter vivos*, on account against her future inheritance, of a two-door pine wardrobe or sideboard, valued at twelve francs, and a three-year-old piebald cow with head markings, value thirty-six francs, or forty-eight francs all told.[20]

What shines through this example is an evident desire to balance the contributions from both sides, to make them complementary, and to establish a nucleus of furniture and livestock such as would enable a young couple to set up home on their own at a later date.

All these examples relating to dowries, not just as they appear in marriage contracts but also as they must have operated independently of notarised deeds, show evidence of what Jack Goody calls 'contrary pulls':

On the one hand, to get a spouse of the right standing for one's offspring there has to be (at least as a promise) a specific settlement from the estate. Moreover, to retain the new couple on the estate and to give them the right motivation for work, one may have to hand over control. On the other hand, this process of pre-mortem transmission weakens the control of the senior generation over their very livelihood,[21]

which of course rendered their retirement more precarious. The conflict was reduced to the minumum in Bigouden society in that marriage portions did not enable the young couple to set up on their own. Marriage did not coincide with inheritance.

A sum of money that depended on the parents' level of wealth and the odd gift of furniture or livestock certainly did not allow the young couple to take charge of a farm. Such an undertaking required furniture, farm implements, livestock, seed, and all the rest of it.

Where, then, did the young couple live? Depending on position in the sibling group, or because of accidents of demography, there might be co-residence, either temporary or permanent (succession), or the young couple might leave to set up their own home – *en leur particulier*, in the phrase used by notaries in marriage contracts. The contract sometimes fixed the details of what at the time of the marriage was regarded as a temporary arrangement. Here, for example, is a deal made between a parent couple and their son, the eldest of ten children, who married in 1854 (by the 1856 census the young couple had already left both the parental home and the commune of Saint-Jean-Trolimon):

> The husband and wife named here, Jean Péron and Jeanne Le Corre, do in support of the proposed marriage undertake to have the husband and wife-to-be in their home for two years. There, in return for the work they undertake to do on the farm, they shall receive two hundred and forty kilograms of wheat and two hundred of barley annually in produce of fair market quality. For the duration of their stay with their parents they shall have the use of twelve ares of land [1 are = 100 m^2] suitable for planting potatoes and six ares of land suitable for planting hemp, which to provide a basis for the collection of duty is valued at the sum of fifty francs, all this being accepted by the husband and wife-to-be.[22]

In this contract payment is provided for in kind. Farm rents were still paid in the same way at the time. Here it was a sort of wage granted in acknowledgement of the farm labour that the young couple would supply in place of servants. The parents also gave up some of their land for the young couple to use during their temporary stay. A cow and calf might be thrown in too. With the money earned from the sale of cereals and the produce of their bit of land, the young couple could put together a nucleus of capital that would enable them to leave the paternal farm.

So the marriage contract, though seldom used, served essentially to protect the patrimony. For the most cautious families it was also a way of guarding against the risk of an engagement being broken off. Two references were found that would come under this heading:

> It has been expressly agreed between the parties, Widow Le Loc'h standing surety for her grand-daughter and the man Le Corre and his wife answering for their son, that, in the event of one or the other of the future spouses wishing to go back on the projected marriage, the one who shall refuse to comply with the

said undertaking shall within eight days of such refusal being made known legally pay to the other the sum of three hundred francs as indemnity for expenses incurred on account of the marriage.[23]

And this one:

Should either of the future spouses renounce the projected marriage, he or she shall pay the other the sum of thirteen hundred francs as soon as such renunciation is established.[24]

(Both the above marriages did in fact take place.)

Most of the time dowries were settled without there being any need to involve the notary. The young couple began their married life within the terms of purely verbal agreements between the families concerned. Co-residential arrangements on a single farm and sub-leases granted by leaseholding parents were matters on which one gave one's word. Only when a transaction between tenant and proprietor was involved – subrogation of a lease from father to son, tenancy agreements, *domaine congéable* leases – did the notary draw up a deed.

It often happened that the family life cycle was interrupted by the death of husband or wife. Concern to safeguard the property of the children of the marriage and to determine what would come to them in their own right from the inheritance of their deceased father or mother in the event of the survivor remarrying led to the drawing up of a post-mortem inventory. Such documents did not state which child would inherit which share of land or rights. They simply established the total amount of property that the heirs, spouses, children, and grandchildren would have to share, distinguishing between assets held individually and assets held jointly and sorting out the rights of recovery and the rewards due to the succession. In capturing domestic cycles at the various moments when death interrupted them, they reveal how the capital held jointly by husband and wife was put together and divided up.

After listing the rightful claimants, who are usually present or represented by a guardian or surrogate guardian, the notary states his intention of proceeding to a 'faithful inventory and precise description' of the movables, farm implements, livestock, crops, cash, papers, and particulars appertaining to the joint estate (or inheritance), all found in the house and outbuildings of the farm concerned.

In order to arrive at the most accurate valuation of the said objects, the notary enlisted the help of neighbouring farmers or relatives. These actually came to the house for the purpose, promising to give advice to the best of their knowledge and belief and to 'conduct [themselves] loyally and well' in the execution of their duty.

There follows a detailed description and valuation of every object or animal located in its place in the house or outbuilding in which it was used or kept. This of course adds enormously to the interest of such descriptions, giving us a picture of the objects of everyday life. And when, as in our study, the number of persons occupying the dwelling and running the farm is known, that interest is even greater. On the other hand certain questions need to be asked about gaps and omissions. There is rarely any mention of cots, for example, which is surprising in so prolific a society. Among the animals, poultry are hardly ever recorded. Also, how much was concealed in terms of undeclared money or securities, and how much was undervalued? One is always coming across 'supplementary statements of inventory'.

Furniture, heating and cooking utensils, stocks of meat, crops in the granary or in the field, carts, farm implements, livestock listed by age and colour, linen, and yarn were all valued and together constituted the assets of the deceased, to which were added ready cash and monies owed. The deceased's personal belongings were always valued separately and did not form part of the joint estate; they were regarded as the separate property of the deceased. Set against these assets are various debit items such as the expenses of the deceased's final illness, the funeral expenses and cost of the coffin, and the notary's fee – in other words, all the things that death inevitably involves. These items are relatively more expensive, the smaller the inheritance. For example, in the post-mortem inventory carried out at Kergonan on 8 November 1842 at the request of the children, following the death of their father, the notary's fees incurred three years earlier on the occasion of the post-mortem inventory after the mother's death appear on the debit side among monies owing. The same inventory itemises nine francs as the cost of the coffin, thirty-one francs for the deceased's funeral expenses, and twenty-one francs for prayers in his memory.[25] The debit side also lists loans contracted with relatives, friends, or the notary, an estimate of the amount of wages still owing to farmhands, rent owing on the farm, and as yet unsettled guardianship accounts, on which it is often difficult to put a precise figure.

The inventory goes on to specify the claims that both spouses are entitled to make against the joint estate when assets owned separately by either have been sold and the amount not reinvested to the original owner's advantage. The document thus lists the various rights and lands that have been sold and estimates their total value. Conversely, it itemises and values the assets purchased by the joint estate for which it ought to receive compensation.

If the balance was in credit, the real work in relation to the estate now began – the job of sharing it out among the children. Two distinct situations presented themselves: either the parents had already, during their lifetimes, organised the succession that was to take place following their deaths, or the heirs had to exercise certain choices and a process of circulation of assets within the sibling group began. The parental will, whether or not it was organised in notarised testamentary arrangements (it rarely was, in fact), sought to reconcile two opposing tendencies: equality among heirs and minimum partition of the estate.

The tenant farmer had only his lease, while he owned the movables of the farm. But he would often try to transfer the lease to one of his children. Either the lease remained in effect and there was subrogation from father to son or son-in-law, or the lease expired. The movables of the farm were appraised, and the child who eventually took up the lease compensated his or her brothers and sisters. Successions of tenant farmers seem in fact to have been few and far between. Landlords did not hesitate to get rid of tenants who had difficulty in paying the rent if the harvest failed. For their part, tenant farmers would often seek to take on the lease of a larger farm as their children increased in number. Occasionally (though this was unusual) a tenant might persuade his landlord to divide a farm in two in order to accommodate his children. Some farms were in fact fairly large (twenty-five – thirty hectares). They could stand being split in two and still remain viable, given the technological level of nineteenth-century agriculture. Ultimately what circulated among the children of tenant farmers were only sums of money and pieces of furniture.

The *domanier*, as we have seen, owned an asset that was difficult to divide up in practice, namely the reparative rights, edifices, and surfaces attached in a complex manner to the actual land of the farm. Every child had a claim on these, but only the one who had the use of them might own them. The rule of succession was here upheld by a legal condition. For a *domanier* to enter into possession of the lease, he must be the owner of the reparative rights; if a son succeeded his father in the lease, with the landowner's authorisation, he must first purchase those rights from his father or from his brothers and sisters. In contrast to the position of the simple tenant farmer, who had very little scope at all, there was room here for a paternal strategy. It was for the father to put a value on the reparative rights he intended to surrender; if he undervalued them, he benefited the child who took his place and harmed those who had left home, because their share was diminished accordingly.

A brisk operation of simultaneous lateral repurchase and resale came into play when a couple decided, following the deaths of both parents, to stay on the farm where they were living. If a farmer was taking over his father-in-law's farm, his and his wife's joint estate would buy back from his wife's brothers and sisters their shares in the reparative rights, while the husband would sell his shares in the family farm from which he had come (always supposing that his parents, too, were *domaniers*) to whichever of his brothers and sisters took up that lease.

Take the example of Marie-Louise Le D., who married Isidore M. on 17 May 1879. The young couple lived with Marie-Louise's parents until 1882, then took the lease of a farm at Kersine. Marie-Louise lost one of her parents in 1881 and the other in 1884. The young couple then moved back to Castellou, where they proceeded to buy up all the reparative rights of the other two heirs (only three of five children having survived). At the same time Isidore sold his own shares back to his brothers and sisters, for his future did not lie on the parental farm, where a brother or brother-in-law was in occupancy.[26]

When looking at *domaniers*, we need to distinguish between those who owned one lease or *convenant* and those who owned several. A number of households had farms where the profits were such that they were able to put together a bit of capital to compensate those of their children who would not inherit a lease. Hervé T. and Catherine G. held four leases and had five children. Two of the farms concerned were at Lescoulouarn, one at Keréon, and one at Kerfiat. They comprised some eighteen hectares each. The four daughters each received a lease and the son a sum of money. The son preferring to have a farm, the eldest daughter Catherine gave her brother Lescoulouarn and used the money to buy the hôtel de la Croix Verte in Pont-L'Abbé (which has since become a large garage).

When parents owned a number of farms, there was no danger of them being broken up and perfect equality could be applied, as in the following instance:

Auguste C. owned four leases covering sixty hectares in all. Of the eight children born to him, only four (three sons and a daughter) were still living at the time (*c.* 1920) when he arranged his succession. Two of the sons got farms of eighteen and sixteen hectares respectively; the others received farms of only thirteen hectares. To make the partition strictly equal, the father stipulated that two and three hectares of the larger farms should belong to the heirs who received the smaller farms on the express condition that they resold those lots to the brothers who inherited the farms from which they had been detached. Two concerns – that no child should be favoured above any other, and that farms should not be broken up – were reconciled on payment of a compensatory *soulte* between brothers.

Where there was only one farm, the owner might adopt a variety of strategies ranging from setting up a number of smaller but implicitly still viable farms to maintaining the integrity of the unit of production by compensating the other children who did not remain on the farm. Take the example of the succession organised in their lifetime by the man P. and his wife, who in 1900 acquired a farm of some twenty-one hectares at Gorré-Beuzec. Seven of their children reached maturity, with a twenty-year gap between the eldest and the youngest. At a time of intense population pressure, with the demand for land reaching a peak, the three oldest were married off to tenant or owner-farmers working farms of a similar size. Of the four who were left, two more got married while the other two remained unmarried and went on living on the parental farm. The parents disposed of their estate as follows: they divided their farm into three holdings of seven hectares each, two for the married children and one for the unmarried ones, and gave the older children a money share strictly equivalent to the value of each of the three holdings.

If he lived long enough, the father's wish might even jump a generation. Take the example of Le Steud, where the head of household appointed his grandson to succeed him:

Jean D. and Marie-Louise C. had six surviving children. The three sons became schoolmasters, two of their sisters took the veil, and the eldest daughter and her husband left to rent a farm elsewhere, returning only in the 1920s just as Jean D. and his wife were settling their succession. Taking the view that he had given his children equal shares – the sons in terms of their education, the daughters in portions for the convent – old Jean D. sold his farm to his grandson at a price well below its value, leaving it for the young man to compensate his uncles and aunts, all of whom had left the district. In this case the paternal strategy expressed a concern to safeguard the integrity of the working capital – that is to say, the farm considered as a means of livelihood.

Juxtaposing these different examples brings out the variety of the solutions adopted, the flexibility of the system within the context of a general rule, and, contrary to all expectations, the force of the father's wish.

Another aspect to emerge from this examination of testamentary arrangements is the concern of certain testators to protect children who might be incapable of defending their rights against rapacious siblings. Here is part of a deed of gift made on 12 March 1848 by Marie-Anne Le Berre, widow of Louis Le Palud. The eldest of her four daughters, Marie-Jeanne (b. 1815), is listed as 'imbecile' in the 1841 census:

> The said Marie-Anne Le Berre, being advanced in age and wishing to forestall any arguments that may arise among her children and grandchildren on

distribution of her estate, has decided to share among her children and grandchildren all the property that would belong to them following her death, which property comprises the place known as Keryoret bihan, both land and reparative rights, situate in the said commune of Saint-Jean-Trolimon, entirely belongs to the said Marie-Anne Le Berre, provides a net annual income of ninety francs, and has a capital value of two thousand francs.

Four lots are described in detail in the deed, which goes on to specify:

Whereas the principal dwelling of the place is wholly contained in the fourth lot and whereas for that reason the said fourth lot is worth more than the first and third lots, which have no dwelling, the owner to whom the fourth lot falls shall be obliged to pay to the owners of the said first and third lots, giving half to each, the sum of six hundred francs by way of balance [*soulte*], which sum or balance shall fall due without interest one month after the decease of the said Marie-Anne Le Berre.[27]

In this case the widow organised her succession during her lifetime by, on the one hand, reserving to herself the tenure and usufruct of all her property and on the other hand requiring her children to assume all her debts. She further took care to safeguard the operation of the farms by requiring the joint heirs to grant one another way-leaves along the cart tracks.

Another will, which is the original betrays a touching confusion in the transcription of Breton into French, well illustrates the tensions between brothers and sisters inevitably provoked by a system that delayed settlement of the estate indefinitely:

Marguerite Le P., widow of Vincent Le P., a farmer living at Kerveltré in Saint-Jean-Trolimon, being diseased in body but of sound mind, dictated her will in Breton to the undersigned notary as follows: anxious to treat all my seven children equally and fearful of being unable in my lifetime to give to each what I have given to a number of them, I desire and formally intend that those who have not yet received anything from me should after my death and before any partition of my property receive an advantage equal to that which I have given the others. I declare accordingly that I have settled upon my son Pierre by anticipation the sum of four hundred francs and upon my daughter Marie-Louise, since deceased . . . the sum of four hundred francs. If following my death one of them or their husbands or heirs should dispute this settlement of portion by anticipation that I declare on oath having made upon them, I give and bequeath to my other four children, Vincent, Pierre Jean, Corentine, and Marguerite Le P., further to their regular share, the sum of four hundred francs each, on the understanding, however, that this bequest shall have effect only in the event of one of the others not acknowledging voluntarily and when bound to tell the truth the settlement of portion by anticipation made upon him.

This will was dictated in Breton by the testatrix in the presence of witnesses to the undersigned notary, who translated it into French in full and in his own hand as it was dictated to him. It was subsequently read back, first in French, then in Breton, to the testatrix, who stated that she fully understood it and wished to abide by its provisions, which are a true expression of her last will and testament.[28]

The old woman was afraid that the children whom she had already endowed would forget the cardinal rule of Breton succession, namely that heirs must state, at the time when a succession was settled definitively, what they had already received.

Heirs and successors

Our examination of property transmission practices reveals the variety of the solutions available to parents or abruptly imposed by death. Having made this analytical digression, we come back to the question: who eventually succeeds to the farm, since clearly, if all children are heirs, not all are successors? And is there – second question – a liquidation of assets in the hands of the older or the younger children or in those of sons or sons-in-law? To answer these questions, tables of succession were drawn up for fifty-two farms – those, namely, for which genealogies had been reconstructed and in respect of each of which we had ascertained, for each generation, the number of children, the number of children still living at the time when the succession was settled, and which of them finally succeeded. A study of these tables furnishes a certain amount of information about successional usage.

To begin with, it appears to have been common practice for several households of brothers and sisters to run the parental farm for a number of years before setting up on their own elsewhere. This phenomenon is distinct from that of co-residence, which occurred during the parents' lifetime. Short-term verbal letting agreements enabled parents to make available to their children a small house or a few fields on which the young couple could live or at least survive until something better turned up. When the parents died, several of their married children occupied the family farm one after another.

These 'successive successions' were mainly observed on the farms of the hamlet of Kerbascol, which lies on the edge of a *méjou* or open-field area. Given the peculiar nature of the terrain, the farmer was able to extend himself somewhat by carving out a few lots from the land that tended to be farmed collectively. We find the same thing, however, on enclosed-field holdings at Kergonan, Tréganné, Méout, Tronoën, and also Steud bras, where evidence of divisions and successive occupations can still be seen in the houses built in a row.

Let us take the example of Kerbascol farm, occupied by Louis Loch and his wife Anne Lay from 1836 to 1872. Possessed of a large fortune, they died without direct heirs, which revealed the extent of their kindred. The farm came from Louis's father's side of the family, but

Louis's brother-in-law moved in there and the latter's wife stayed on, despite being widowed several times. Anne Lay lived at Kerbascol from 1856 to 1872. When she was widowed and remarried, she left the farm to go and live at Kermathéano. Her younger brother Louis took her place from 1866 to 1872, when he was replaced by another sister, Marie-Jeanne, and her husband Jean Draoulec, who had in fact co-resided on the farm for a while at the beginning of their married life. They can be seen as the true successors, since it was the children of this couple who followed on (see fig. 4).

Fig. 4. Example of 'successive successions'. The Kerbascol case.

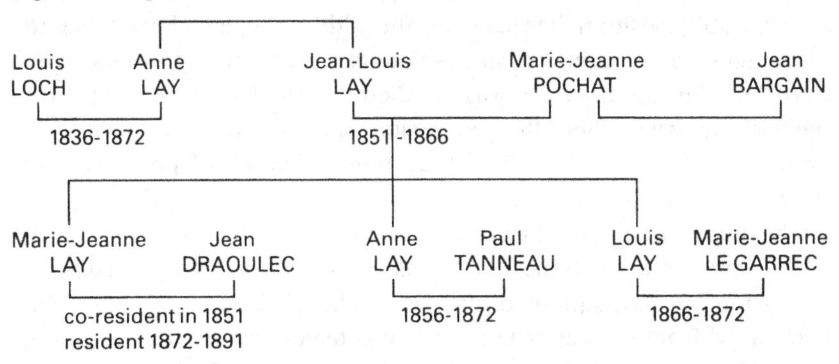

One consequence of these successive successions was to leave in abeyance the transmissions in connection with which it was difficult to define where they began and where they left off.

The following generation saw similar overlapping successions for the three children of Jean Draoulec, made possible by the existence of several dwellings.

At Méout the female fourth child of a group of eleven surviving children took up residence. She and her husband lived there from 1861 to 1881. But none of her children followed on; it was her youngest brother, born twenty-one years after her, who succeeded (see fig. 5).

Fig. 5. Example of 'successive successions'. The Méout case.

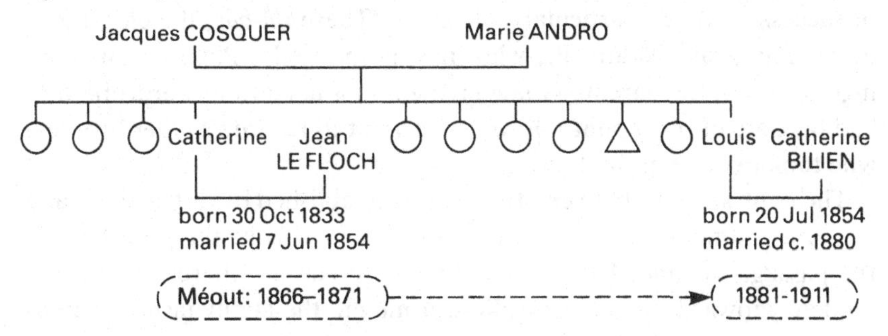

Is it possible, working from the tables of succession, to find out which child eventually succeeded its parents? When sex and order of birth were established for all the successors to the farms examined, certain tendencies emerged. Sons succeeded their parents more often than daughters and sons-in-law (seventy-five sons against forty-eight daughters). And eldest children succeeded more often than younger children (thirty-four eldest sons against twenty-one elder daughters). This interesting result might suggest that, as in North Bigouden, a form of primogeniture survived here. It did not. Often the eldest child took the farm over again after spells on other farms; he was not succeeding so much as returning. The most that can be said is that he found himself in a favourable position because, as the eldest, he had been able to accumulate more capital than his younger brothers and sisters, who were less far on than he was in their family life cycle. This was particularly true when the succession was effected by *vente avec licitation* – that is to say, sale by auction in one lot of property held *indivisum* (see table 15).

The illusion of a right of primogeniture may also stem from the way in which the table of transmissions is constituted. All the dates of succession appear squashed up, as if the situation had remained unchanged from the beginning of the nineteenth century right up until the 1930s. Actually, in the mean time the population first shot up, then families started to become smaller; a flight from the country set in, which had the effect of restating the problems of inheritance in fresh terms.

The people interviewed (and it is probably their testimony that provided the safest data) had had very different experiences of succession. 'The youngest is in the best position', they would say. 'It's the youngest child who succeeds.' And that seems logical in so far as the final settlement is not made until the parents are either very old or dead, by which time the older children are already established on farms elsewhere.

The contrast between North and South Bigouden with regard to mode of succession is felt particularly strongly. 'There [*là-bas*] the eldest gets everything', said Isidore P., who, having married a Plozévet girl, had high hopes of his parents-in-law giving him a handsome wardrobe that had his date of birth on it. However, it went to his wife's elder brother, who took over the farm 'there'.

The contradiction between the facts as established by our analysis and the oral testimony of those interviewed also has something to do with the passage of time. During the nineteenth century there was a slight tendency towards the eldest child returning to the family farm, whereas

Table 15. *Succession by rank in sibling group*

	Boys										Girls									
Rank	1	2	3	4	5	6	7	8	9	10	1	2	3	4	5	6	7	8	9	10
Number	22	12	8	7	4	8	2	4	5	3	9	12	5	10	3	1	1	3	0	4
	34										21									

over the past fifty years or so it has usually been the younger ones who remained there. That is the whole problem of studying changing modes of succession over a long period.

Tensions among siblings

Being the successor had its advantages, but it also carried obligations with it. The child who succeeded must compensate his brothers and sisters, and if his parents were still alive at the time of the transfer he had to take care of them until their death. These responsibilities went with his position.

The payment of lateral *soultes* or compensatory balances among brothers and sisters created tensions within sibling groups of which the notarised documents furnish only a feeble echo. Such rivalries were exacerbated when the children of more than one marriage found themselves competing for the succession. Where one of the children of a second marriage succeeded, and where he or she was a stranger to the farm and to the assets attached to it, a row would break out among the step-brothers and step-sisters.

Family rows provide excellent evidence of the complexity of a flexible system that was heavily dependent on the random influence of demography. Take Kerameil farm, the lands of which were purchased by its occupier around 1850.

Henri T. lived there until his death, when his youngest son took over

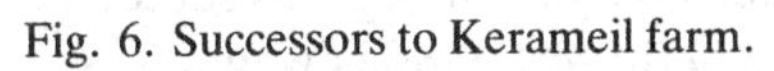

Fig. 6. Successors to Kerameil farm.

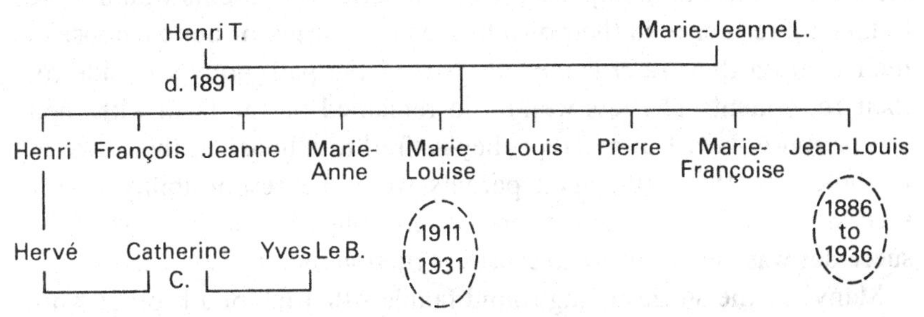

the lease. The son's marriage remaining childless, however, part of the farm was sub-let to his nephew and godson Hervé, eighth child of Henri T. Jr. Hervé died in 1910, and his widow Catherine got married again to a man who had been related by marriage to her first husband. She remained on the farm with her new husband. However, her uncle Jean-Louis himself became a widower and remarried, this new, late marriage giving him four children – future successors. A quarrel ensued, leading to a definitive rupture between Jean-Louis's children and Hervé's, who wrongly felt that they were entitled to succeed.

All systems of inheritance generate rows of this kind. Apart from such references as can be gleaned from notarised documents, it is impossible to pin them down except through direct inquiry. The ethnologist finds himself tackling a sore subject here – painful for his interviewees and perilous for himself. In asking questions about the transmission of family estates and reviving old quarrels, he touches the quick of family sensibilities and finds himself invading the most private realms of identity. Even in a good relationship of long standing, the interviewee may clam up, refuse to talk, and actually throw you out, shouting, 'You're worse than the Gestapo!' The only lesson to be learnt from this kind of bitter experience is that the subject is still taboo and carries a powerful emotional charge.

Most family rows concern the respective amounts of shares and what is considered to be wrongful occupation of a tenancy. They are of a very different kind from those that Alain Collomp observed in the upper Verdon valley in Haute-Provence, where a variant of the 'house' system operates with systematic co-residence of two generations. There the rows occur between father and son or son-in-law or between brothers over matters arising out of co-residential arrangements.[29] A different motive underlay the family quarrels of the *heritourien* of Brittany, and it had to do with the fate of the aged parents. It was agreed that the successor, who was implicitly regarded as possessing an advantage, would look after his disabled brothers and sisters and his aged father or mother until their death. This was in exchange for the right to succeed. At this point in the family life cycle, the better-off parents would leave to take up residence in their own house in the *bourg* or build a house of their own on their *réservation*, the part of the patrimony set aside for their retirement. The less well-to-do remained on the farm with their son or son-in-law. But whether they retired to a house of their own or remained on the farm, aged parents were the responsibility of the successor couple. Non-observance of this obligation on the part of the successor was one of the main causes of ructions among siblings.

Many are the stories going round families that tell of a brother who, having inherited a farm (seriously undervalued in the opinion of his brothers and sisters; they had accepted the low valuation, but it had meant small monetary shares for them), forsakes a dying father and makes a sister living ten kilometres away come and fetch him by charabanc and take him home with her. With that all the fissile forces latent within the sibling group are released; no one now speaks to the brother or the sister who failed to comply with the fundamental moral obligation to help one's aged father or mother to die at home.

Where none of the children had succeeded their parents, the upkeep of the old people was attended to collectively once a year during what was known as the *Fricot Mamm Gouz* 'grandmother's meal'. All the children came together for this and paid out an equal sum each for the parents' upkeep. When the parents died, their furniture was auctioned among all the children and equal shares made of the sums raised.

The absence of a prompt appointment of a successor had considerable influence on the way in which children become socialised and on their matrimonial fate. None of them was brought up in any special way, and none was married for the particular reason that he would be the one to continue the family farm. The social fates of the children in principle remained uniform, since their marriages were all required to be made with homogamous partners. That is why our interviewees' accounts of the excellent harmony reigning among brothers and sisters should not be thought to be designed purely for the ears of the ethnologist with the object of concealing the true nature of family relationships. They reveal usages quite as genuine as those that generated conflict.

Partible transmission and erosion of patrimony

Bigouden society was a hierarchical society. At the end of the eighteenth century, certain members of it were distinctly well-off while others were extremely poor (a detailed analysis of this situation will be found in chapter 7). The wealthiest members accumulated several *convenants*, which they sub-let. They possessed gold and credits (monies owing to them) with the notaries; the problem of keeping the farm in one piece did not exist for them, because they (part-) owned several farms, which in the next generation they distributed among their children. These then became owners of only one farm, over which a number of children had rights. As for the gold and credits, they were rapidly dispersed. Several factors combined to lead to a steady impoverishment of lines that at the beginning of the nineteenth century had possessed large amounts of capital: for example, the peculiar mode of land ownership, which made real partition of farms impossible, and the size of families.

It was only the prosperity of the region that, thanks to considerable expenditure of human energy, made it possible to combat the impoverishing effects of the egalitarian system. A hardworking tenant farmer, provided he enjoyed a long life and good health, could reconstitute the domanial property of the agricultural establishment to which he had succeeded. It would be hard for him to accumulate any capital. So his children would be poorer than he was.

South Bigouden does not present the sorts of phenomenon that Marie-Claude Pingaud found in Minot in Burgundy. Despite a partible inheritance system, certain Minot farmers were able to make a social breakthrough and accumulate patrimony over and above the shares they had inherited. Admittedly, the Burgundians discovered contraception sooner than the people of Bigouden.[30] In Bigouden the only brake on the gradual frittering away of patrimonies was the institution of the *domaine congéable*. When farmers started to become their own proprietors (owning both land and reparative rights) in the late nineteenth and particularly the early twentieth centuries, there was no further limit on the division of holdings. All farms became so parcelled out as to make it quite impossible to run the individual holdings viably. The partible system thus prevented farmers from acquiring estates of any size. It also explains why couples from the same sibling group, who all received approximately equal shares and contracted socially identical marriages, might nevertheless experience very different destinies.

Beginning in the 1850s for the poorest section of the population and from 1870 for everyone, certain choices became unavoidable if a person wished to stay in the region. If he could not find a farm he followed the call of the sea, as it were; he became a fisherman or a cannery worker. As the fishing industry expanded, canneries were set up and other associated trades moved in. All our genealogies contain examples of children born and brought up on farms who subsequently took up non-agricultural trades.

This was one consequence of the partible system, and it distorted its effects. As long as hierarchies were sustained, the kindred network was relatively close-meshed and socially homogeneous; as patrimony became dispersed, hierarchies weakened and kindreds broadened out until eventually they embraced both the wealthiest and the poorest. Economic success was a fragile thing. A run of bad harvests, an illness, and the tenant farmer had no further capital to fall back on. Socially, he went into decline. The whole social level dropped, in fact, as matrimonial strategies changed. There being no further need to safeguard and consolidate capital, kindreds were able to open themselves up to other alliances.

In a system that tended to impoverish heirs as the generations went by, one demographic factor had an important compensating effect. This was a high death rate, which had the effect of settling on a small number an inheritance that ought to have been scattered among a larger number. On the other hand, when death removed the young the patrimony leaped a generation. When it came to settling the estate, the share that could be claimed was so small that a compensatory money payment removed all right to an asset that in consequence became reconstituted in the hands of the principal heir. There is no doubt that patrimonies would have been broken up very much sooner had mortality not done its work.

The system of property transmission helped to keep the population where it belonged. It discouraged permanent emigration and furthered intra-regional mobility. Moreover, it had an undeniable effect on access to wealth, and it made households poorer in a context of population growth.

The study of how it worked in practice relates to a particular family life cycle and to a specific organisation of the kinship group. Following an early marriage, the young couple often spent an interim period co-residing with parents or parents-in-law. The older generation did not relinquish their farm until the end of their family life cycle. In doing so, they reduced the contradiction inherent in the dowry system, which obliged them to deprive themselves in order to provide for their children. Their authority remained unchallenged until their death. While seeking to maximise their own material and symbolic capital, they continued to play the equality game. Parents might try to resist the dividing up of their farm (when they owned it), yet they would share their movable estate among all their children. The system was profoundly egalitarian in that sons and daughters were equal with respect to inheritance and equal with respect to marriage. Bachelors were rare here, and there was no matrimonial strategy giving an advantage to one child in particular.

Yves Le Gallo expressed surprise at the low level of religious fervour in the Bigouden region as measured by the number of priests it produced, notably in comparison with the Cap Sizun region or with Douarnenez-Ploaré to the north. If we reduce Bigouden to some twenty parishes, he said, it seems that the number of priests it produced in the nineteenth century barely exceeded sixty, and of those about half came from Pont-l'Abbé and Loctudy. By contrast, the four parishes of the extreme cape produced eighty-eight, Cléden-Cap-Sizun accounting for thirty-six, Plogoff for sixteen, Goulien for twenty-five, and Primelin for eleven. Peumerit, Tréogat, Tréméoc, and Saint-Jean-Trolimon provided none at all and Plovan and Combrit only one.[31]

But was it really a question of religious fervour? Did it not have more to do with the inheritance system, which structurally rejected non-heirs, instead absorbing all the children? The latter hypothesis seems the more likely. All children had a 'right' to marriage, as they had a right to a share of the inheritance. Furthermore, they all had a right to the same kind of marriage. Since the successor was not designated in advance, no special marriage strategy was suggested to one child in particular; all were expected to make good marriages. The consequence of this egalitarian attitude was to disperse the sibling group. Mode of land ownership, inheritance system, matrimonial model, and mobility are structurally inter-related. If one had to define this society in a nutshell, it would be as the diametrically opposite system to that of the 'house': no patrimonial domain constituting an entity, no single heir, no symbolic attachment to a place with which a lengthy genealogy could be associated.

This characteristic system of property transmission also relates to the social organisation of kinship beyond the domestic group. There was no unilineal descent, as in exotic societies, and there was no very deep-rooted cognatic descent through which a person might claim rights of ownership or use. In South Bigouden, rights to land might go back one, two, or at most three generations. The idea of rights coming down from a remote common ancestor was totally alien to this society. A person might be aware of having been born on a particular farm, but once he had left it he had no hope of ever returning. From the legal standpoint, if descent is regarded as a body of rights capable of being activated, descent in Bigouden had a short span. On the other hand the domestic group was situated within a vast network of consanguine and affinal ties with which there was no patrimonial interaction but via which essential information regarding tenancies and marriages would circulate. A genealogical memory that was brief in relation to previous generations but vast at the level of collaterality and alliance – those are the twin features of family organisation as delineated by the mode of property transmission. In a word, the system was distinguished by its horizontality.

The way in which dowry operated had to do with this peculiar structuration of kinship. As Jack Goody points out, the dowry system is essentially bilateral because it distributes rights that do not tie ownership to sex. It is associated with a type of diverging devolution that enables property to be transmitted outside the unilineal descent group and weakens the structuration of corporate groups. By way of comparison, in African societies that do not have dowry but where bridewealth circulates between clans and lines, such bridewealth reinforces the structuration of those groups.[32] In South Bigouden the sole function of the dowry, however small it might be, was to establish a joint matrimonial estate. It might be the sum used to pay compensatory *soultes* to brothers and sisters, but it did not link together the marriages of siblings. On the contrary, it individualised them, and it was these new domestic groups that launched the process of genealogical fission that was to culminate three generations later.

However, there was nothing fixed about this mode of property transmission. On the contrary, it evolved when population pressure increased but resources, given the mode of land tenure and the level of technology, hit a ceiling. But was this in fact a case of evolution or simply one of a fresh response, within an unchanged mode, to a new situation? Did South Bigouden observe primogeniture, like its northern neighbour, in the period prior to the one investigated, and did population pressure substitute the youngest for the eldest child? It is difficult to say. One would have to push one's researches further back in time, combining a study of the domestic life cycle with that of the generations, which falls into a more extended rhythm. The present state of historical documentation is ill-suited to such an inquiry.

We can, on the other hand, say without hesitation that the partible system of inheritance had the effect of making people poorer. One often falls into the habit of thinking of the social evolution of human groups as tending towards permanent improvement. People aged seventy and over are well aware that their grandparents were better-off than their parents. They have seen gold coins and heard tell of the many estates that the old folks owned and sub-leased. Marie T. tells how, on the occasion of her uncle's wedding, her grandfather entirely filled with gold coins the wooden box (the *ar hastell*) in which the ball of bread dough was carried to be cooked in the oven.

The nineteenth century may have marked the apogee of a rural civilisation, judging by its costumes, its furniture, and its folklore; in economic terms it was a century of gradual individual impoverishment that tended to level out social categories. This had direct consequences as regarded the marriage system, which was not fixed in time but changed in response to economic, social, and demographic conditions.

4

Regular relinking through affinal marriage

In *sociétés complexes* (those that, according to the distinction introduced by Claude Lévi-Strauss and taken up by Françoise Héritier)[1] do not prescribe marriage partners – there are evident difficulties in the way of observing matrimonial regularities in the sense of exchanges of marriage partners recurring regularly down the generations. In South Bigouden, which is characterised by the mobility of its domestic groups, by its egalitarian system of inheritance, and by the small number of peasants actually owning the land they farmed, the chances of observing such regularities seem even slimmer. There are no motives for such strategies. Investigation of supposed regularities is complicated by Bigouden's having no identifiable stable units of exchange. The region lacks the clans or lineages of exotic societies. It does not even have the *oustas*, *ostals*, and other 'houses' of those parts of France that acknowledge a principal heir.[2]

Does that mean we must be content with the standard observation of consanguine marriages, the incidence of which is usually calculated on the basis of dispensations granted by the church, taking – it follows – no account of marriages outside the prohibited degrees?

Bilateral genealogies

As far as looking for regularities is concerned, the usual sources used by ethnologists and historians are unsuitable. Marriage certificates make it possible to delimit the area of matrimonial mobility or socio-professional endogamy by comparing the professions of the spouses or their fathers. They may also serve to identify unions dubbed 'remarkable' – double brother–sister marriages – in the mistaken sense of 'unusual'. To find out whether marriage occurs among kin – and what sort of kin – we need genealogies. Moreover, those genealogies must cover many generations, and they must be fully bilateral. That is to say, in each generation they must take account of both the paternal and maternal lines, given the full equality of men and women with regard to the transmission of property.

It is whether they rely on oral genealogy or on written genealogy that distinguishes studies relating to elementary or semi-complex societies from those devoted to complex societies. The former have no choice, in fact, since the genealogies available are exclusively oral.

Oral genealogies display several characteristics. They are often not located in time, which renders demographic events imprecise. Those who recite the genealogies have selective memories, depending on the rules of kinship that obtain. In a patrilineal society, for example, it is the female ancestors who get forgotten. Some genealogies go back no farther than three or four generations. Finally, biological lines of descent are not the same as social lines of descent, though this is of greater concern to the geneticist than to the historian.

The ethnologist dealing with a historical society can choose to reconstitute oral and/or written genealogies. It would be more appropriate, in fact, to speak of genealogies based on interviews and genealogies based on documents, since both are eventually committed to writing and occasionally represented graphically. The ethnologist thus gathers from the lips of his informants genealogies that are like living trees of consanguines and affines traced through the spoken word with the occasional help of a diagram to convey the position of each relative.

In South Bigouden, genealogical memories do not go back very far. Often one grandparental patronymic in four has been forgotten, and in the case of great-grandparents the proportion is even higher. The genealogies reconstituted here orally, in interviews, are shallow in depth and soon peter out. People not only do not know; they attach no importance to knowing. Probably what underlies this attitude is an awareness that one's family has belonged to the same territory since time immemorial. How it contrasts with the meticulous concern to probe as far into the past as possible that motivates the people who can be found 'doing the family tree' in the reading rooms of archive libraries up and down the country! People who are busy living their family tree and are conscious of how it has shaped them feel no need to reconstitute it.

But if the genealogical depth of the rural memory is slight, its knowledge of collaterality is extremely detailed. When it comes to uncles and aunts (even great-uncles and aunts, very often), cousins, siblings, and nephews and great-nephews, people know everything: how many there are of them, whether they are still alive or have passed on (in which case the date of death, a chronological landmark to which great importance is attached, will be known to the month and sometimes to the day), whom they married, where they live, and whether they are tenants or own their own land. Oral genealogy amasses precise data – demographic, economic, and social. Ego locates himself within a kinship system that is invariably very concrete and never disembodied from locality and employment. And he will furnish similarly precise information about his spouse and her relatives. The relevant field is the kindred, the whole body of consanguines and affines, together with the consanguines of affines and vice versa, of which Ego will also have extensive knowledge.

So what is one to make of this web of links forged between 1840 and 1970, spreading out into collaterality and affinity and running from one individual to another, one household to another? What is one to do with the metres of tracing paper recording up to five or six generations that cover the office walls with a splendid maze of multi-coloured lines? Often the overwhelming emotion experienced by the researcher when faced with what Lévi-Strauss so aptly termed the *turbulence** of the matrimonial field is despair. Certainly the frequent assertion that 'We're all kin here' will be seen to be the literal truth if we take a cross-section of the population at any given date. This has to do with village size, with demographic conditions, and with the fact that, in this region of high geographical mobility, movement nevertheless remained very local.

But while oral genealogy provides fascinating insights into the workings of family memory,[3] the experience of kinship, and the choices and personal events that colour social development, it is inadequate as far as looking for matrimonial regularities is concerned. Consequently other genealogies had to be reconstituted with the aid of vital records (for the nineteenth century) and parish registers (for the eighteenth century). Given the size of the population (around 1,000 on average, in the nineteenth century) and its mobility, there was no possibility of reconstituting them all. A decision was therefore taken to trace only a few lines that offered the observer sufficient links in the commune at the heart of our inquiry, namely Saint-Jean-Trolimon. How, in fact, was one to trace the descendants of those who had left the commune to take up residence elsewhere? Granted, such couples stayed within the confines of South Bigouden, but there was no knowing in which commune they had settled, and locating them in any systematic way would have been impossible without assembling all the vital records for the entire region. The same objection applied to the option of working from a stock of

* It means both 'turbulence' and 'unruliness' [Translator].

ancestors and following their descent. Why choose one parish rather than another, and again: how was one to set about tracing the descendants? Another solution would have been to take the population as a whole in the 1970s, which is when research for the present project was begun, and trace the genealogies of those residents. This too seemed an unsatisfactory solution if one was trying to understand matrimonial regularities in conjunction with modes of property transmission.

In the end a decision was made to select from among the wealth of lines those that looked fairly stable. Each unit of residence (farm or *penn-ty*) was examined at five-yearly intervals (governed by the census dates) throughout the nineteenth century to see whether or not the successive occupiers were identical – that is to say, whether they were mobile or stable – and, where the occupiers had changed, whether the new ones were related to the old. In other words, all the farms where there was an observable genealogical continuity from father to son or son-in-law were identified. It was in those cases that genealogies were reconstituted – genealogies not so much of individuals as of successors to farms. In this first phase the small farmer who occupied a farm for the nine-year period of a lease, or for an even shorter period if it was a question of a verbal sub-tenancy, and who then left to continue his wanderings in neighbouring communes would be eliminated. Nevertheless, these weaker lines would be picked up again through the ascending or descending genealogies. The principle of selection was to choose lines of which the descendants were still living in Saint-Jean-Trolimon and in connection with which archive documentation could be gathered, providing economic and social data about their way of life. Using that principle, from around 160 farms listed for Saint-Jean-Trolimon in 1836, fifty were isolated that exhibited genealogical continuity over at least three generations.

For those fifty farms, genealogies were reconstituted as follows. The couples successively residing on the farm were traced from 1836 to the present day. A line of successors was followed, and the genealogies of the successive spouses was added to it. Furthermore, lest the argument be based solely on the marriage of the child who succeeded, the genealogies of the spouses of the other members of the sibling group were reconstituted, the brothers and sisters who would not stay on the farm but had been born to the resident parent couple. For each generation, it was possible to examine the complete stock of marriages of members of the sibling group in order to establish whether successors married any differently from non-successors. The genealogies of those successors and their spouses, together with those of their sibling group

Fig. 7. Commune of Saint-Jean-Trolimon, showing the fifty farms where genealogies were reconstituted.

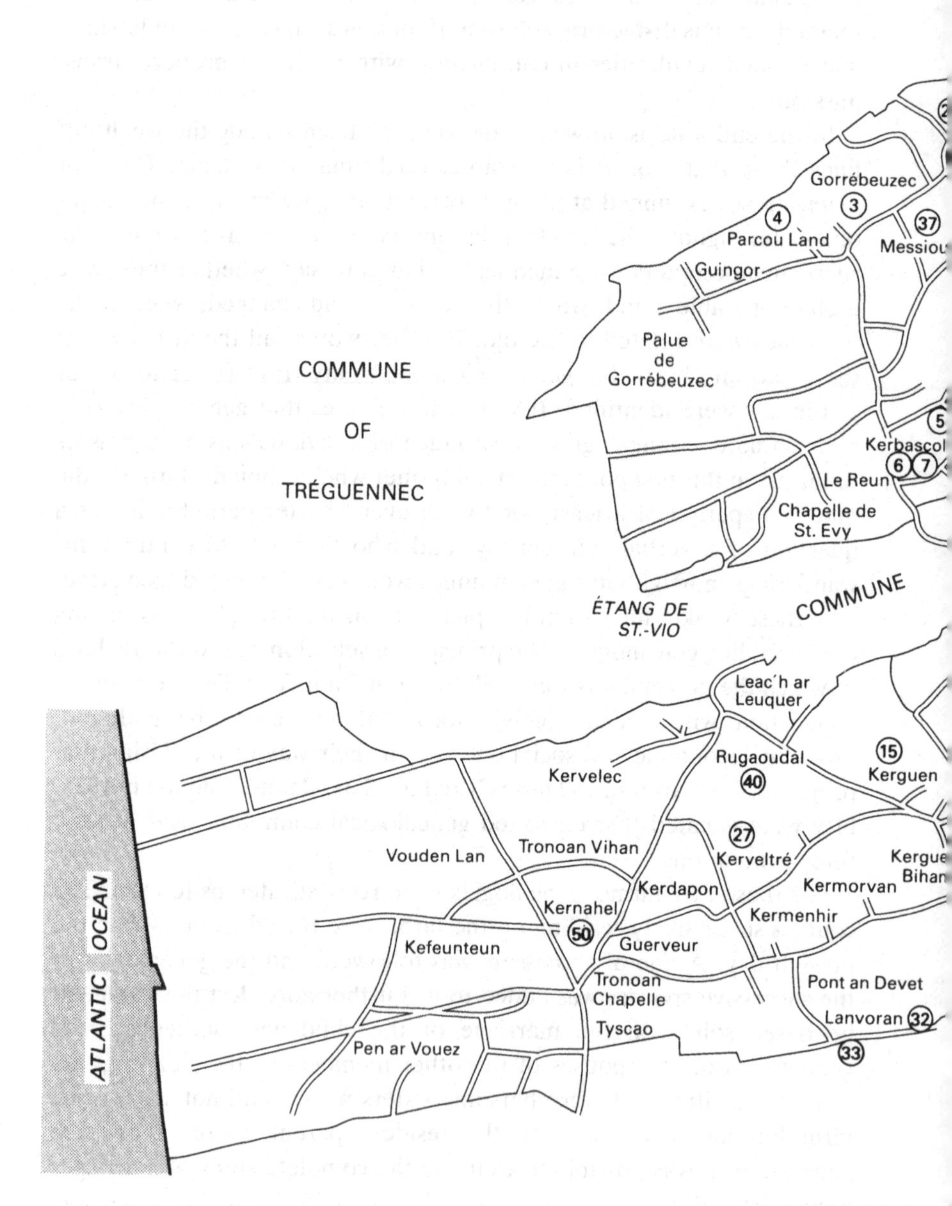

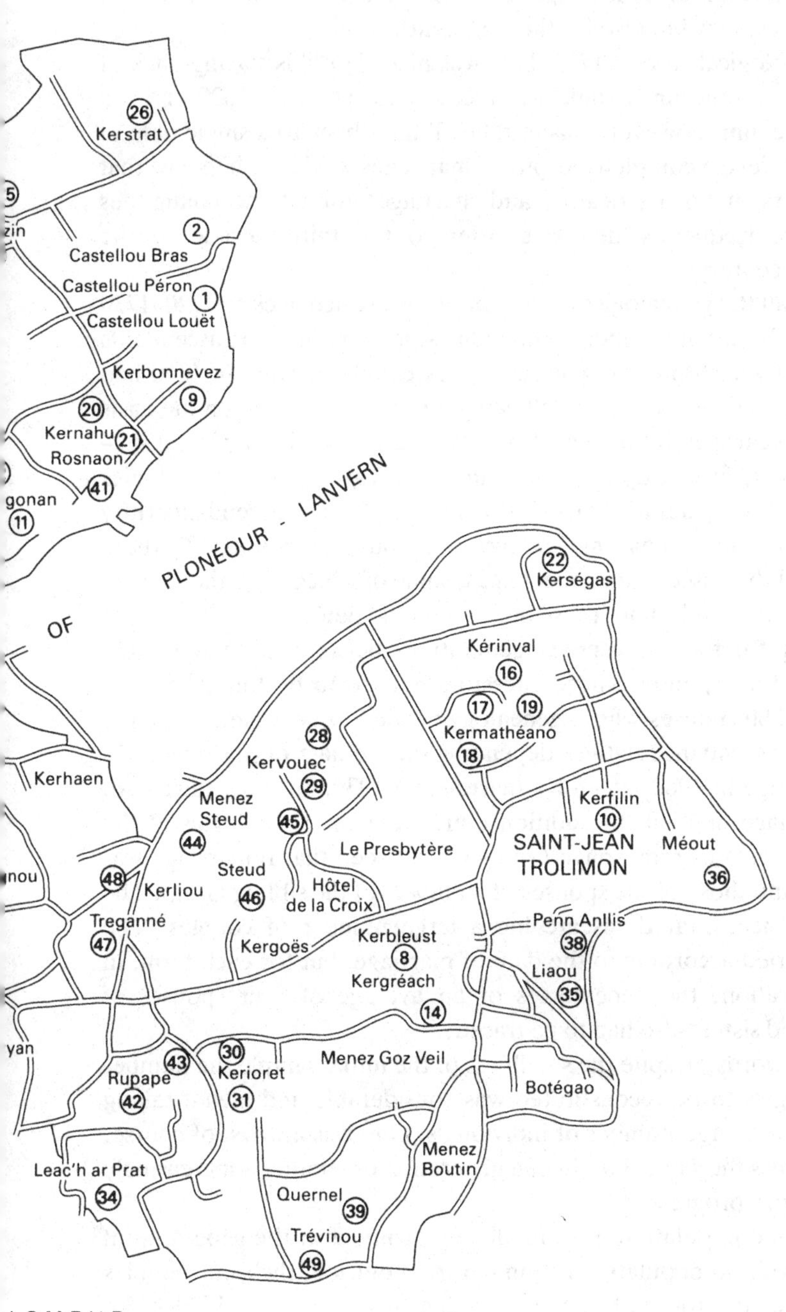
26
Kerstrat
5
zin
2
Castellou Bras
Castellou Péron
1
Castellou Louët
Kerbonnevez
9
20
Kernahu
21
Rosnaon
41
gonan
11
PLONÉOUR - LANVERN
OF
22
Kerségas
Kérinval
16
17
19
Kermathéano
18
28
Kervouec
29
Kerhaen
Menez
Steud
45
44
Le Presbytère
Kerfilin
10
SAINT-JEAN
TROLIMON
Méout
36
nou
48
Kerliou
Steud
46
Hôtel
de la Croix
Treganné
47
Kergoës
Kerbleust
8
Kergréach
Penn Anllis
38
Liaou
35
14
yan
43
Rupape
42
30
Kerioret
31
Menez Goz Veil
Botégao
Menez
Boutin
Leac'h ar Prat
34
Quernel
39
Trévinou
49
LOMEUR

and *their* spouses, were traced back up all lines of descent, wherever they were located in South Bigouden. This was done not only for the nineteenth century but also for the eighteenth.

A chronological stop of 1720–1730 was placed on this tracing-back of genealogies. Some lines could be traced back as far as 1630 or even earlier. The aim, however, was to take all lines back to a single date in order to achieve a complete corpus of marriages, and it so happens that the registers of births, deaths, and marriages for all the communes studied are frequently defective prior to the third decade of the eighteenth century.

Tracing all the genealogies of the sample as defined back to 1720–1730 took in a substantial number of individuals, if we include the ascendants needed to reconstitute the generations. A couple married in 1960 had 512 ascendant couples in 1720, allowing for ten generations on the basis of a thirty-year generation gap. It was decided to work on fifty farms – that is to say, fifty couples present in 1830. Taking their genealogies back about four generations to 1720 gave a total of 800 ascendants (fifty couples × the sixteen parents involved over four generations). To these were added the descendant genealogies, some of which were the same in each generation and some (those of the spouses) new.

Allowing for four descendant generations between 1830 and 1960, that made four spouses whose ancestors needed to be found, or fifty farms × 4,200 spouses whose genealogy needed to be traced back, the number of ascendant relatives depending on the date of marriage: 512 for a marriage in 1960, 256 for a marriage in 1930, but only thirty-two for a marriage in 1840. In addition there were the genealogies of the spouses in cases of remarriage (and we have seen that remarriage was frequent) and those of the spouses of members of the sibling group. The latter also necessitated researching a large number of couples. The number varied according to the date of marriage, but for each farm, in each generation, the genealogies of an average of four spouses of brothers and sisters also had to be traced.

In other words, despite the smallness of the initial sample the number of genealogies to be reconstituted was considerable and meant taking into account a large number of individuals. One was always, of course, coming across the same lines in one generation or another – increasingly so as the work progressed.

In a closed population, profoundly endogamous at the geographical level and with no population coming in from outside, the same couples are found at the top end of the ascendant genealogies in 1720–1730. These are the 'ancestors' or 'founder ancestors', so-called because their parents are unknown. There are something over a thousand of them (see table 16). This was ascertained during manual searches and suggested that the corpus possessed genuine coherence. It also prompted the use of data processing to gain a clearer view of information so complex as to be beyond the capacity of one person to analyse. The data-processing method used and the results obtained after computer analysis of the corpus of genealogies, which comprised more than five thousand individuals genealogically interlinked over fifteen and in some cases as many as eighteen generations, have been developed elsewhere.[4]

More than 80 per cent of couples are connected by relinking through affinal marriage, a practice first named by French ethnologists studying the Burgundian village of Minot twenty years ago.[5] It denotes the fact that two couples have affinal connections with two pairs of ancestors in common. Conversely, it may be defined as two ancestor couples exchanging husbands and wives over several generations in accordance with the following pattern:

Fig. 8. The principle of relinking through affinal marriage.

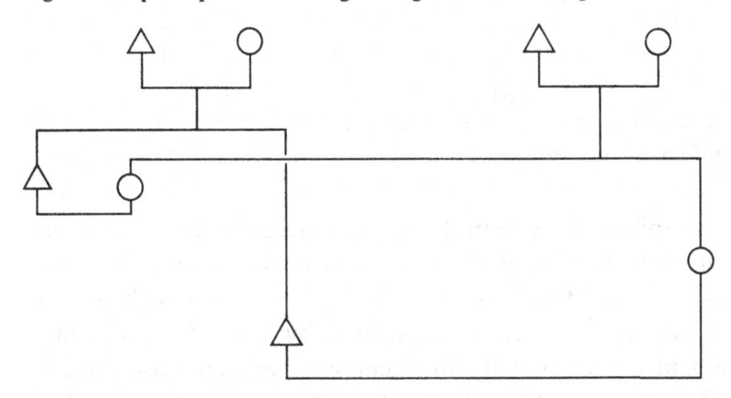

Computer analysis brought out the very high incidence of relinking through affinal marriage particularly in the years 1820–1850. It revealed the existence of kindreds that were particularly integrated by numerous links within the lines they encompassed.

Relinking may be described as a thoroughly bilateral phenomenon bringing both paternal and maternal lines of descent into play in such a way that no preference can be identified at this level. Furthermore, bilateral relinking was practised preferentially among close kin. This was confirmed on all the data for which it was possible to compare the total ancestry.

The computer analysis produced figures, drew chronological curves, and reinforced anthropological hypotheses. The computer's great

Table 16. *Distribution of founder ancestors by generations*

Dates	Generation	Individuals without parents and without children (spouses of widows)	Founder ancestors	Total
1610–1630	1	2	79	81
1630–1650	2	2	257	259
1650–1670	3	4	186	190
1670–1690	4	13	146	159
1690–1710	5	18	152	170
1710–1730	6	15	156	171
1730–1750	7	18	214	232
1750–1770	8	12	182	194/ 1,190*
1770–1790	9	18	121	139
1790–1810	10	11	86	97
1810–1830	11	12	60	72
1830–1850	12	11	54	65
1850–1870	13	12	26	38
1870–1890	14	4	7	11
1890–1910	15	9	6	15
1910–1930	16	3	0	3
1930–1950	17	0	0	0
1950–1970	18	0	0	0
		164	1,732	1,896

*Total no. of founder ancestors

strength is its ability to deal with large quantitites of data very quickly and come up with totally reliable results. However, it produces only what it has been asked to produce. Nor do these results exhaust the problem of relinking, which may proceed by way of collaterality, remarriage, and a succession of affinal and blood relationships. But it is hard to see how they could have been taken into account otherwise than by using complex, time-consuming programmes.

Widows and widowers who frequently remarry

The practice of relinking is peculiarly reinforced if unions contracted through remarriage following widowhood are taken into account.

In the eighteenth century, large numbers of widows and widowers remarried, their successive unions connecting one with another to form impressive chains. Was there a particular rule of remarriage? The number of dispensations for remarriage between affines does not suggest that affinal remarriage was genuinely preferred in earlier times, unlike in the nineteenth century. On the other hand remarriage within kindreds does seem to have been preferred.

The widow or widower, where he or she was relatively young (and notably where the woman was still of child-bearing age), was in a different situation from a first-time spouse. He or she was in place on a farm where labour was based on the complementary nature of the sexes. As well as requiring a father or mother to raise the orphaned children, the establishment was short of a male or female worker. The father–maidservant or mother-manservant solution was unacceptable in so far as farms were in short supply and demand was always high. We find that either the widow/widower married a bachelor/spinster, which found a home for a child, or two widowed people married each other, in which case, since one necessarily moved in with the other, the one who was quitting a farm left a place free for others to take up. Either way, remarriage represented a solution at the individual level (for two people whose lives had been shattered by the deaths of their partners) and at the collective level (in the light of the demand for farms). At the individual level it recreated an operational working and living cell; at the collective level it put a vacant unit back on the farm market.

Remarriage followed two patterns that appear to have been in contrast but were in fact two facets of the same tactic. It might join together two widowed persons not residing in the same commune. (To find a spouse of the right age and social standing for the widowed person, it was often necessary to go looking in another parish, in which case a go-between would identify possible partners and bring them together.) Or it might take place within the kindred, the widowed person marrying an affine of his or her deceased spouse. The church raised no objection here, in contrast to its attitude to marriage with a consanguine of the deceased.

All combinations of affinal remarriages seem to have been possible, but a fairly common case was the widowed partners of two siblings of opposite sexes marrying.

In the eighteenth and early nineteenth centuries the commonest patterns were cascade marriages, with widowers marrying the widows of deceased widowers, etc. Since these remarriages took place within a confined geographical area, cases of looping are found. Such loops are actually more like spirals, closing one genealogical rung lower down, so to speak, with the last in the series sometimes marrying a descendant of the first in the series.

Without there having been any question of a rule-induced regularity or of conscious behaviour, it is possible to see these chain marriages as a manifestation of the thing that underlies the whole system of kinship and marriage in the region, namely the enormous stability of kindreds and the high level of people's knowledge about their matrimonial and social situations.

Take five marriages celebrated between 1747 and 1779. Rather than five men and five women we find only three men and three women, woman no. 5 making her second marriage with the son of man no. 2. The son, in other words, married his stepfather's widow.

There were other kinds of loop that were not consanguine. These involved marriages of remarrying widows' and widowers' offspring from previous unions.

Widowers and widows who moved from one farm to another were always accompanied by such of their children as were still unmarried. These then occasionally found themselves in the care of a stepfather or

Fig. 9. Example of relinking involving widows and widowers.

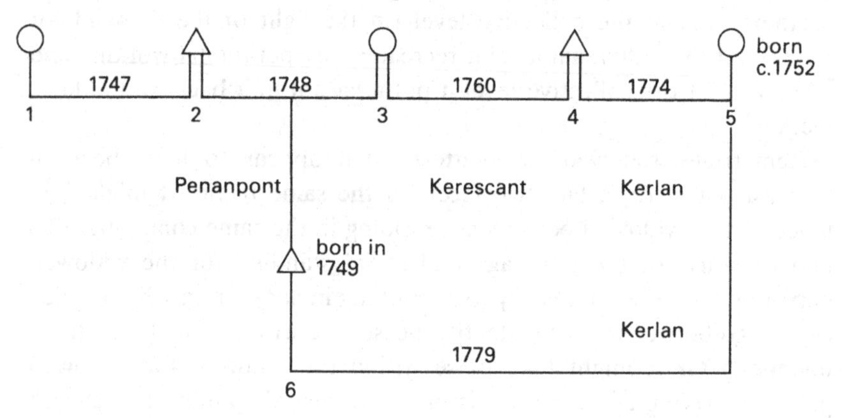

Fig. 10–16. Examples of relinking involving widows and widowers over a number of generations.

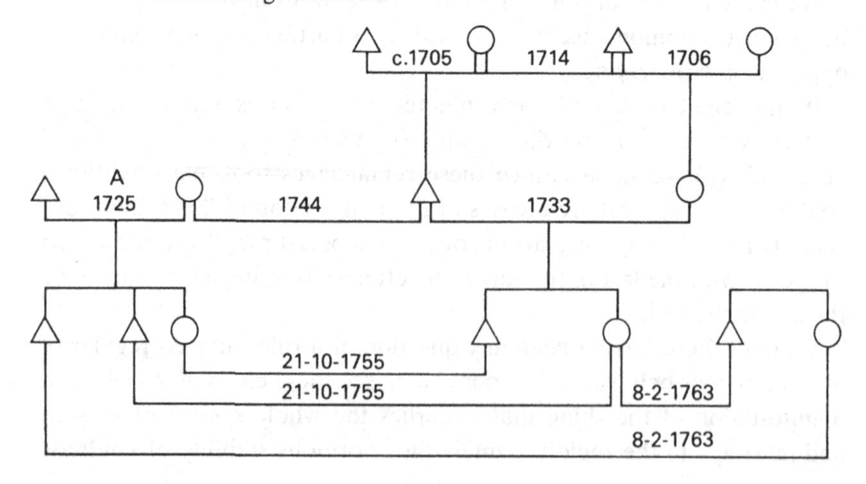

stepmother when their surviving parent died. This explains how, in the succession just mentioned, an unrelated man and woman of compatible ages (the woman younger than the man) found themselves face to face on the same farm. Their marrying each other made it possible to settle some tricky succession problems, since the children's shares could be used by their stepfather. It was also a way of restoring the heir's rights. Usually the remarrying parents married off their children of previous unions together. The weddings were celebrated simultaneously: between brothers and sisters of corresponding ages and/or between the widowed parents who were remarrying and their children.

Here is a typical example, involving a double marriage in 1755 between brothers and sisters born of different unions, followed in 1763 by the simultaneous marriages of a third child of couple A with a girl whose brother at the same time married the woman who had become the widow of one of his brothers, married back in 1755.

This double principle was operated in a limited number of variants.

For instance, a father and his son married two sisters. It was the father's second marriage, the son's first, and the two were celebrated on the same day:

Fig. 11.

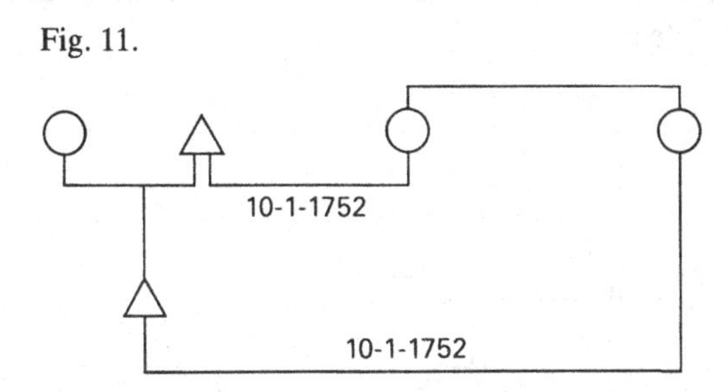

Occasionally the marriages of children of widowed persons added several remarriage links, as in this example:

Fig. 12.

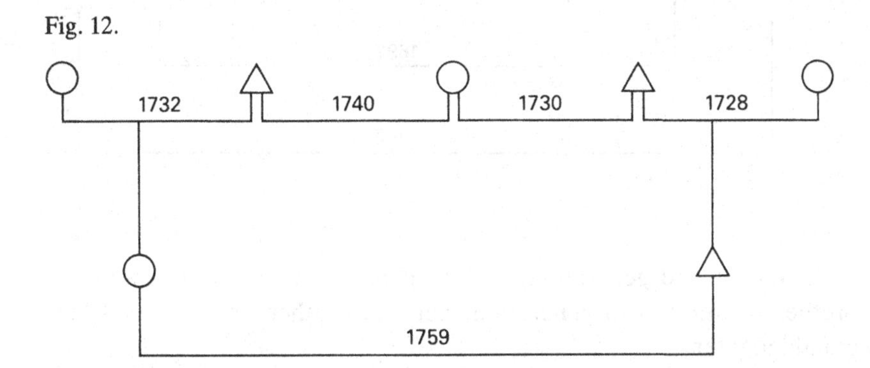

Another variant related to the next generation rather than to the intermediate links of widowhood. A man married a girl whose mother had made a second marriage to his half-brother:

Fig. 13.

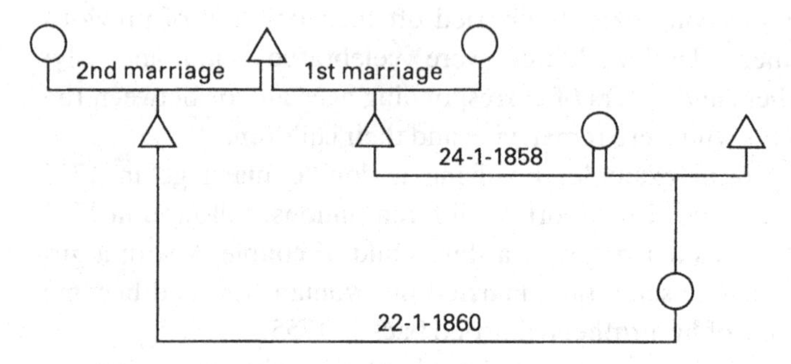

Extending these diagrams of marriages and remarriages, we very soon find that they intermingle in generation after generation, as in the following examples:

Fig. 14.

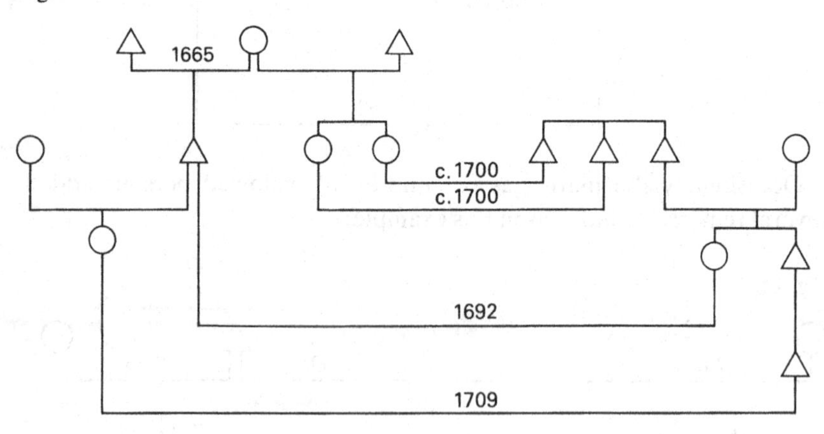

In the second generation, A's daughter married her stepmother's brother; in the fourth generation, her half-brother's son married her granddaughter.

Fig. 15.

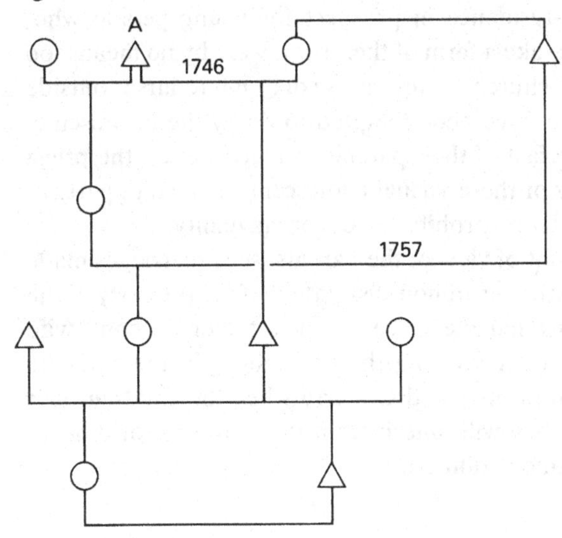

In this case, 1 and 2 married a woman and her niece, then 3 married his stepmother's niece:

Fig. 16.

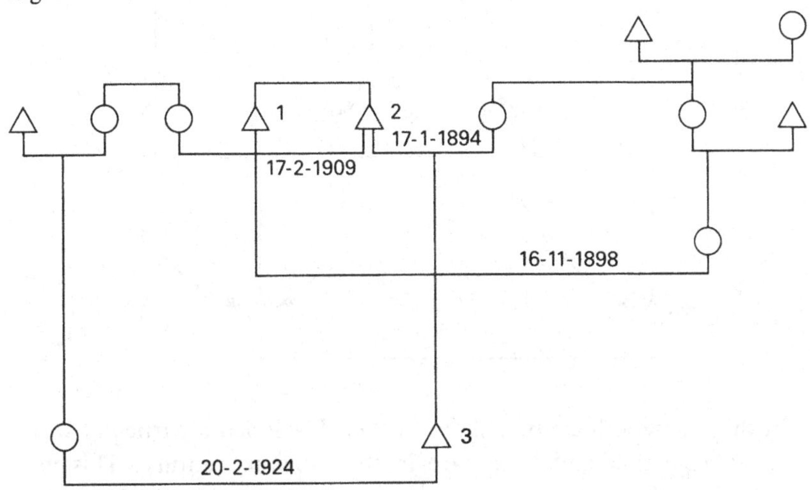

The practice of marrying off the children of remarrying widows and widowers ties in closely with the youth of some brides and bridegrooms in the eighteenth and early nineteenth centuries. Among factors that might account for such marriages, two suggest themselves.

First, the average Bigouden farmhouse is very small. The church, which wielded great authority in this rural society, even in what we should nowadays consider to be the sphere of private life, took a dim view of the kind of co-residence in prospect for young people who, though still too young to take a farm of their own, were by no means too young to infringe the church's ban on sexual intercourse outside marriage – as they might have been tempted to do by the promiscuity imposed on them by the fact of their parents' remarriage. So the priest did not stand in the way of these virtual adolescents marrying, particularly since they came under no prohibition of consanguinity.

Secondly, from the point of view of the parents, such marriages made it possible to mobilise a strategy of non-dissipation of the property of the four lines involved. Marrying the niece of one's father's second wife could be seen as a way of recovering the patrimony. In the example below, if the grandchildren of A and C marry when the children of B and D have remarried, they will inherit from the four lines of descent rather than see that patrimony diluted:

Fig. 17. Inheritance and relinking.

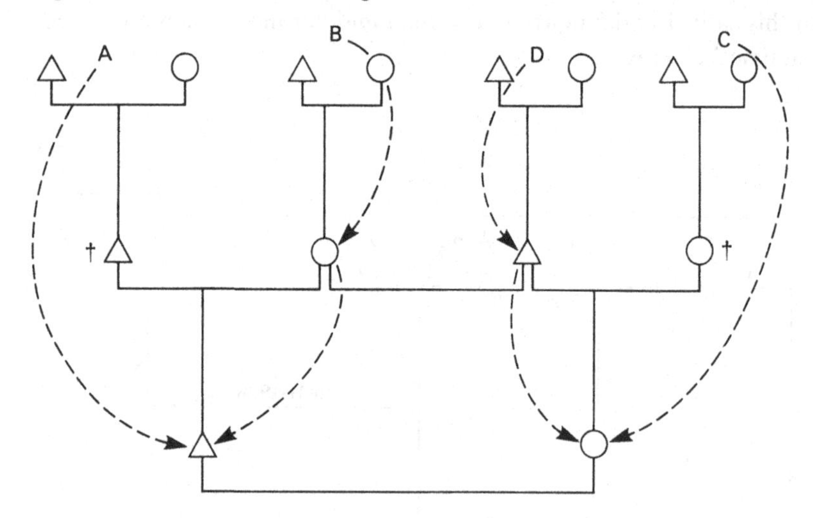

In the nineteenth century there further developed a particular form of remarriage that had been rare in the previous century. This was affinal remarriage of the levirate or sororate type. Of thirty-nine dispensations granted in Saint-Jean-Trolimon between 1849 and 1974, eight concerned affinal ties: seven widowers marrying the wife's sister and one more distant relationship (4–3). In cases of marriages involving another commune, six dispensations concerned widowed persons remarrying relatives (brothers or sisters) of the deceased spouse and two concerned 2–2 remarriages (to first cousins).

The following example was elucidated by a female descendant of marriage A, without whose help it would have been difficult to interpret. The couple marked with an asterisk required a dispensation of 4–4 consanguinity (children of third cousins) and 1–1 affinity in order to remarry. Subsequently their daughter married the two sons of her mother's first husband, one after the other.

Fig. 18. Evidence of a nineteenth-century levirate marriage.

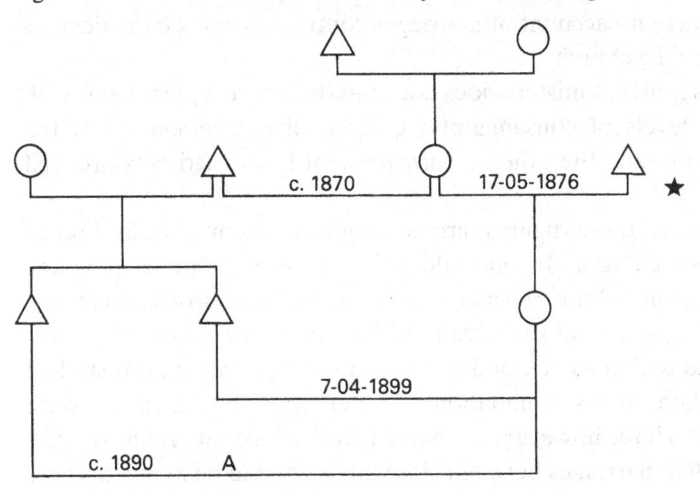

The development of the levirate can be explained by the impoverishment of lines through the nineteenth century. Marrying one's brother's widow meant filling a place that had fallen vacant, a place so precious that there was no time to let it grow cold.

So to the relinking studied on the basis of the genealogical file as run through the computer we have to add that which occurred through widowhood. These were two variants of one and the same practice, a practice that tied in with the other type of dominant behaviour, namely consanguine marriage.

Low incidence of consanguinity

It is commonly said that Brittany is a region where marriage between cousins is of frequent occurrence. Bigouden in particular is often symbolised by a limping farmer's wife suffering from congenital dislocation of the hip as a result of in-breeding. Just how frequent were such marriages, in fact? And were they a product of demographic chance or of a deliberate strategy?

Jean Sutter and Léon Tabah drew up maps showing the geographical distribution and frequency of consanguine marriage in France. We need to bear in mind that their calculations were based on the ratio between the number of Roman Catholic religious marriages contracted and the number of dispensations granted in the years 1926–1945.[6] The period studied was relatively late, since consanguine marriages were at their most frequent in the second half of the nineteenth century. Moreover, after 1917 the church stopped demanding a dispensation for marriages between grandchildren of first cousins (known as 4–4 marriages). So their statistics fail to reflect actual practice, particularly since they necessarily take no account of marriages contracted outside the degrees prohibited by the church.

Be that as it may, Finistère does not head the list of *départements* with the highest levels of consanguinity; Corsica, the *départements* of the Massif Central, and the other *départements* of Lower Brittany are well ahead of it.[7]

Furthermore, these figures are a long way from certain French Canadian percentages. In one village on Lake St John consanguine marriages accounted for 19 per cent of all marriages contracted between 1900 and 1909, while between 1950 and 1967 the figure was 13 per cent.[8] In the isolate of l'Isle-aux-Coudres (also in Quebec), which was studied from the date of its foundation, 90 per cent of marriages were consanguine. Here, however, in contrast to the Lake St. John study, it was found that marriages between close kin tended to be avoided; there were very few marriages between first cousins but a larger number of 4–4 unions. In a completely closed population, which had grown from thirty founder ancestors in 1728 to almost 2,000 inhabitants in 1967, demographic constraints appear to have determined choice of spouse. One imagines a conscious rule may have operated as well, but the study was not conducted along such lines.[9]

Certain parts of France also had very much higher rates than those recorded for other periods by Sutter and Tabah. In Brière (Loire-Inférieure) between a third and a half of the marriages contracted between 1859 and 1879 required dispensations. Probably the overall rate was even higher (approaching the l'Isle-aux-Coudres figures), for the population had grown from a stock of ancestors unregenerated by spouses from outside the region.[10]

It is interesting that, contrary to what one might imagine, in-breeding is not a necessary phenomenon, arising out of demographic constraints removing all freedom of choice as far as spouses are concerned. For example, the work done by Jacqueline Vu Tien Khang and André Sevin in four villages in a valley in the Pyrenees showed that actual consanguinity was consistently lower than it should have been under a panmictic system – that is to say, if choice of spouse had occurred wholly

Table 17. *Comparative consanguinity in the Bigouden and Cap Sizun regions*

Marriages	Plozévet[1] 1860–1960	Goulien[2] 1859–1960	Plogoff[2] 1903–1904	Saint-Jean-Trolimon 1831–1970
2–2	76	29	28	3
2–3	30	8	7	3
3–2	1			
3–3	116	20	46	11
3–4	24	9	5	5
4–3				1
4–4	18	14		8
Total	265	80	87	34
Marriages	3,925 6.7%	937* 8.5%	973 8.9%	1,234 2.7%

*937 marriages between 1843 and 1973.
Sources: [1]Izard, 1963; [2] Delroeux, 1979.

at random.[11] The difference was appreciable as far as close cousins were concerned, less so for second cousins or cousins at a genealogical remove.

Even within the same *département*, consanguine marriage rates often differ widely from commune to commune. Let us compare the figures for Saint-Jean-Trolimon with those for other communes in the region (see table 17).

Goulien and Plogoff appear very much more in-bred than Plozévet, which is in turn very much more so than Saint-Jean-Trolimon. Two things limit the significance of these observations: the periods compared are not strictly identical, and consanguine marriages, as Michel Izard discovered, are frequently under-registered. It is more interesting to study the degree of relationship. In Goulien there appears to have been a preference for close relatives – first cousins (2–2) or cousins of unequal degrees – while Bigouden seems to have preferred marriages between more distant kin – second cousins (3–3) or beyond.

The present study examined consanguinity in South Bigouden in the standard manner, using the dispensations relating to Saint-Jean-Trolimon on the one hand and a data-processing programme based on the corpus of reconstituted genealogies on the other. Given the period covered by the genealogies it would have been desirable to gauge the incidence of consanguine marriages in the eighteenth century. Unfortunately the official registers that would have made this possible have disappeared. It is true that in-breeding increased substantially during the nineteenth century (particularly the second half) and in the early part of the twentieth century. But it was by no means absent in the eighteenth century; consanguine marriages may be found among ancestors in our reconstituted genealogies. Occasionally the church required rehabilitation when the bride and groom had failed to point out that they were related and to ask for a dispensation – unless of course, they had acted in ignorance. It happened to Henry Peron and Maria Cariou at Loctudy in 1772. The marriage having been consummated, the church 'rehabilitated' it. The couple were both 3–3 and 4–4, so the priest required them to say rosaries by way of penance.

Assessing the frequency of such unions is a hazardous undertaking. Among the 300 certificates collected for the eighteenth century, eighteen dispensations for consanguinity were found, one for affinity, and one for *parenté spirituelle* (the relationship between godparent and godchild), putting the rate of consanguine marriages around 6 per cent. That is a very high figure for the period, but it can be explained by the fact that it relates to a batch of selected lines of descent, as noted above. One point emerges strongly: those dispensations were granted for distant degrees of kinship. Of eighteen dispensations, eleven concerned 4–4 relationships, six concerned 3–4 and 4–3 relationships, two concerned 3–3 relationships, and one concerned a 3–2 relationship. In effect that made twenty dispensations altogether, because we have to add two remarriages in which the spouses were linked by a double tie of kinship, as in the case cited above.

In Penmarc'h nine dispensations were found for the period 1720–1790, eight of them for 4-4 marriages and one for a 3–3 marriage.[12] The figures are on the low side, but there is no doubt that marriage between distant cousins was an ancient custom – at any rate in the social group of the farmers covered by the genealogies.

For the nineteenth century it was important to determine the sum total of consanguine marriages contracted by the inhabitants of Saint-Jean-Trolimon and to compare this rate of consanguinity with that for Bigouden as a whole.

As regards the total of consanguine marriages in Saint-Jean-Trolimon itself, the second half of the nineteenth century saw a steady increase. A surge after the First World War was followed by a decline (see table 18).

The mobility of the population suggested that the scope of analysis should be widened. A consanguine marriage was not necessarily one made within the village. Kindreds were scattered throughout South Bigouden, so it was necessary to look for consanguine marriages contracted with relatives not living in Saint-Jean-Trolimon. To give an overall view of consanguinity as far as Saint-Jean-Trolimon was concerned, dispensations granted for marriages celebrated in other com-

Table 18. *Consanguinity in Saint-Jean-Trolimon, 1831–1970*

No. of marriages	No. of consanguine marriages	
88	1831–1840 – 0	0%
106	1841–1850 – 1	0%
80	1851–1860 – 3	3.7%
118	1861–1870 – 3	2.5%
90	1871–1880 – 4	4.4%
88	1881–1890 – 5	5.6%
95	1891–1900 – 4	4.2%
101	1901–1910 – 1	0%
81	1911–1920 – 8	10%
124	1921–1930 – 3	2.4%
102	1931–1940 – 1	0%
87	1941–1950 – 0	0%
83	1951–1960 – 1	0%
80	1961–1970 – 0	0%

munes with former inhabitants of Saint-Jean-Trolimon from the early nineteenth century up to 1970 were also looked at. Kinship recognising no administrative boundaries, it was found that the number of consanguine marriages contracted by folk from Saint-Jean-Trolimon doubled (see table 19).

Chronologically speaking the same curve emerged as was found for the Saint-Jean-Trolimon dispensations, with a surge towards the end of the nineteenth century followed by another in the decade that included the First World War. As was to be expected, the largest number of consanguine marriages was contracted with inhabitants of the communes of Plonéour and Plomeur. The degrees of kinship uniting the marriage partners, both in the study of consanguinity in Saint-Jean-Trolimon itself and in that of the dispensations granted in the communes of Plonéour, Plomeur, Tréguennec, Loctudy, and Plobannelec in the first half of the nineteenth century,[13] indicated that marriage between first cousins was uncommon, as it had been in the eighteenth century. On the other hand, the numbers of marriages between second cousins and between third cousins were quite similar, assuming that the number of 4–4 unions was an under-estimate (remember that the ban on this degree was removed in 1917, during a period that saw a comparative surge in marriages between kin). There is no doubt that marriage between distant cousins was preferred to marriage between close cousins (see table 20). Furthermore, when marriages were contracted between unequal degrees, the husband always belonged to an earlier generation than his wife. One found 3–4 marriages and 2–3 marriages: the reverse was very exceptional.

Table 19. *Number of consanguine marriages celebrated with one spouse born or resident in Saint-Jean-Trolimon*

	Saint-Jean	Plomeur	Plonéour	Pont-l'Abbe	Tréguennec	Plobannalec	Combrit	Tréméoc	Loctudy	Penmarc'h	Total
1831–1840	0										
1841–1850	1										1
1851–1860	3										3
1861–1870	3	1	3					1			8
1871–1880	4	1	3	1				1		1	11
1881–1890	5				1						6
1891–1900	4		3			1		1			9
1901–1910	1	3	1							1	6
1911–1920	8	2		1							11
1921–1930	3	2		1	2						8
1931–1940	1				1		1			1	4
1941–1950	0	1							1		2
1951–1960	1										1
1961–1970	0										
Total	34	10	10	3	4	1	1	3	1	3	70

Table 20. *Frequency and nature of dispensations of marriage granted in the first half of the nineteenth century*

	Consanguinity						Affinity							
	2–2	2–3 and 3–2	3–3	3–4	4–3	4–4	1–1	2–2	2–3	3–3	3–4	4–3	4–4	Total
Plonéour 1806–1858			4	2		5			1	1				13
Plomeur 1802–1858		2	4	10	1	7		2		3			1	30
Tréguennec 1824–1853		1	1			4								6
Loctudy 1810–1858			5	6		7		2		1	1		1	23
Plobannalec 1803–1858	1		5	6		8		2	1					23
Nature of dispensations for consanguine marriages in Saint-Jean-Trolimon granted in the nineteenth century														
	8	3	24	5	0	19	9	2				1		71*

Note: * One more than in Table 19 because one marriage was both consanguine and affinal.

Affinal marriages seem to have traced a certain development. Marriages of the levirate or sororate type (1–1 unions with the brother or sister of a deceased spouse), very unusual in the eighteenth century and the first half of the nineteenth, appear to have become more frequent in the late nineteenth and early twentieth centuries.

One cannot help wondering about the role of the church in connection with the frequency and nature of the dispensations granted and the procedure for granting them. In the diocese of Quimper the *recteur* or parish priest conducted a canonical examination when the engaged couple came to ask for a dispensation, seeing each party separately. He used the model described in great detail in the diocesan statutes, Published in 1710, these were reprinted in 1786 and remained in force until 1851. The results were then forwarded to the bishop, who if he held an indult for minor cases granted the dispensation. Consanguine marriages other than those involving close degrees generally fell into the category of minor cases; dispensations for 2–2 unions were granted by the Pope.

The examination was conducted with each partner separately in order to assess, from the church's point of view, whether he or she was a free agent or was responding to parental pressure. The form of words used was *enquis si c'est de son ordre qu'il ou elle sollicite ladite dispense* ('find out whether it was his or her idea to ask for the said dispensation'). The priest went on to investigate the motive as well as the income and expectation of income of the spouse-to-be. It was on this basis that the *componende* the couple must pay for the dispensation was calculated. On top of that came the sealing fee, which was usually three francs. The amount of the *componende* might vary considerably, as can be seen from the registers of alms and sealing fees.[14] It appears to have done so according to the wealth of the couple concerned. It ranged from nine to more than fifty francs, which at the beginning of the nineteenth century represented approximately the value of a horse. Finally, the common link of kinship was reconstituted.

It is difficult to assess the dissuasive effect of dispensations, given the impossibility of comparing the respective requests and refusals. As regards close degrees of consanguinity (2–2), are we to assume that the people of Bigouden exercised a kind of self-censorship, knowing the complications of a papal dispensation? Or was their rejection of marriage between first cousins tied up with their notion of kinship and their perception of incest? How are we to interpret the absence of 1–1 remarriages before the end of the nineteenth century and the subsequent high incidence of them? Was there a connection between a repugnance on the part of the church and a repugnance on the part of the local peole? Did obtaining a dispensation constitute an obstacle difficult to surmount, or was it merely an inconvenience that put the wedding celebrations back a few weeks? And did the position of the church not change in this regard?

Unfortunately it was not possible to study the parish priests' canonical examinations because these archives have never been classified. They would in fact help to answer our questions. It would be important, moreover, to examine the nature of the motives dissected in them. One file dating from the eighteenth century revealed that the bride-to-be was an orphan, that her assets were managed by a great-uncle who was also her guardian and that marrying the latter's son would make it possible to settle matters of inheritance. What was the incidence of this type of motive, of which one can see the inner logic? How closely was consanguine marriage connected with the phenomena of mortality? All these questions must remain unanswered for the moment.

Let us compare the consanguinity figures obtained from a manual search with those produced by running our reconstituted genealogies through the computer. The programme was designed firstly to test the consanguinity of the sample in relation to the general data for consanguinity in the commune and the region. In the second place, it sought to go beyond the ordinary limits of studies of consanguinity to take account of consanguine marriages contracted outside the degrees of kinship prohibited by the church. It therefore compared the ancestries of both spouses in 2,590 couples, looking for possible common forbears as far as the fifth generation back – that is to say among thirty-two pairs of ancestors. As well as revealing the sheer power of the computer in all its simplicity, this gave a more complete view of consanguinity than that provided by the study of dispensations alone. The results were as follows:

Consanguinity in the sample of genealogies

Degrees of kinship	No. of consanguine marriages
2–2	7
2–3	9
3–2	5
3–3	20
3–4	10
4–3	10
4–4	34
Sub-total of marriages within the prohibited degrees	95
4–5	10
5–4	6
5–5	21
Sub-total	37
Total	132

If we take only the first sub-total, the genealogies of our sample present a consanguinity of the order of 3.6 per cent, based on the 2,590 couples included. The chronological distribution of consanguine marriages up to the 4–4 degree fell within the period 1810–1910, with a surge between 1850 and 1910, which is comparable with the general trend. Furthermore the preference for marriages between distant relatives comes out again.

If we now look at *all* consanguine marriages, including 5–5 unions, the figure increases appreciably to reach 5 per cent of the total. There is a technical reason for this. As genealogical depth known increases, it is logical that more marriages of this type should appear. Also they were at their most frequent in the years 1850–1890. In relation to the total number of consanguine marriages, 5–5 unions represented 28 per cent. Are we to conclude that the relative frequency of 5–5 marriages supports Françoise Héritier's hypothesis, according to which people married the nearest relative available beyond the prohibited degrees?[15] There is insufficient data to allow a satisfactory answer to that question. How is one to distinguish between the effects of the demographic structure of the population and parental wishes? It would have been necessary to work on simulations of population with highly complex hypotheses of fecundity.

The problem is complicated by the fact that one's genealogical data are never complete, remember that in 5–5 marriages we need to know thirty-two ancestors for each partner. That is a considerable number, invariably prompting one to ask how far the percentage of marriages of this type is distorted by lack of data.

To test this, a study was made of the frequency of 4–5 and 5–5 marriages among individuals whose genealogies were known in full up to thirty-two ancestors and beyond. Such individuals numbered 246. They contracted 112 consanguine marriages, 29 of which were 4–5 or 5–5. The proportion of marriages of this type to the total number of consanguine marriages amounted to 25.8 per cent, which is lower than the figure for all consanguine marriages.

On the basis of this sample alone, the proportion of consanguine marriages was considerable but the number of unions contracted outside the prohibited degrees showed no appreciable increase. When the same test was applied to 5–6 and 6–6 marriages, 118 individuals were found whose known ancestry comprised sixty-four or more forbears. They contracted forty-seven consanguine marriages of all degrees, including eleven at 5–5 and 6–6. Intrinsically, the proportion of consanguine marriages was considerable – almost 40 per cent – but with respect to the total of consanguine marriages the proportion of 5–5 and 6–6 marriages was even lower than in the previous sample, amounting to 23.4 per cent. Although the absence of other figures rules out comparisons, we are forced to conclude that the increase in consanguinity in the degrees not prohibited by the church is small in extent – the more so if we relate it to the theoretical number of distant cousins that each individual could possess in those generations.

The data-processing programme having drawn up the list of those ancestors in respect of whom relinking occurred most frequently, the list was found to overlap that of common forbears in consanguine marriages. There is no doubt, therefore, that a large proportion of marriages between kin must be attributed to the better-off kindreds. Consanguine marriage appears to have characterised the poorer elements. We can speak only of tendencies here, given the small number of dispensations on the one hand and the highly unstable nature of the economic and social situation of the less well-off on the other hand, but it would be a mistake to see consanguinity purely as something practised by the wealthiest farmers. It often tends to be associated in people's minds with the regrouping of land dispersed through inheritance. However, no such correlation can be established for the non-land-owning peasants of Bigouden. The chief characteristic of consanguinity in Bigouden, where the incidence of it is in any case relatively low,[16] is the genealogical distance observed between marriage partners. Local practice tends to provide a good illustration of the Breton proverb:

Bugale an gueff ny anded
Goaczaff querend a zo en bed
Ha guellef ma veent demezet[17]

Cousins five times removed
Are the worst kin in the world
And the best people to marry.

Quoted without geographical reference and in an archaic etymological form, this dictum appears to represent as the norm conduct that, to judge from our statistics, was actually quite marginal. Note, however, that, whereas consanguinity and closeness of kinship generally go together, here the opposite is the case. When you marry a first cousin, your aunt or uncle becomes your mother-in-law or father-in-law; there is a tightening of the kinship network as a result of combining in one person what are usually distinct blood and affinal relationships. Here the consanguine unions contracted between very distant relatives merely mean attributing seven pairs of ancestors to the couple in place of eight in the case of marriages between second cousins and fifteen instead of sixteen in the case of marriages between third cousins. What difference does that make, especially since all the ancestors concerned are quite obviously long dead?

Lastly, consanguinity was only one of the special forms of a much wider phenomenon more typical of matrimonial behaviour in South Bigouden, namely relinking – particularly since many consanguine marriages tied in with marriages that relinked lines formerly joined by marriage.

Take the case of Jean Garrec, who was mayor of Saint-Jean-Trolimon in the middle of the nineteenth century. A native of Tréguennec, he owned the land of his farm of Kerstrat. Four of his seven children married kin: one daughter married a 3–3 cousin and the others 4–4 cousins. But the distant cousins wed also happened to be brothers, and that fact may have counted for more in the conclusion of the marriages than the fact that the couples were distantly related.

Highly integrated kindreds

Relinking was usually practised among close affines, consanguine marriage among remote kin. Assuming that both represented conscious behaviour – that is to say, that these forms of marriage proceeded from deliberate choice rather than having been imposed by the demographic structure of the population – short relinking does seem to have taken precedence of a blood relationship that was remote in terms of genealogical space. Indeed, horizontality permitted the circulation of information about the point in the family life cycle that each domestic group had reached and about its situation in terms of patrimony. Short relinkings, irrespective of the sex of the common affines, thus appear to have been the rule in this rural society. Did they help to close kinship groups back on themselves, or did they ultimately integrate the whole or part of the community within which they occurred?

Among the relinkings identified by computer, a certain number of couples were found to be characterised by their frequent practice of this type of behaviour. There emerged a kinship structure with a tendency to close up on itself, though not in a watertight manner. How many couples did that structure involve? Was it unique or could it be found for other sets of lines that were also fairly self-contained, though not so much so as not to be open to others? In other words, did the marriage market consist of relatively closed kindreds that eventually covered the whole geographical field as well as the whole social field?

Matrimonial space might be represented as a set of kindreds effecting a number of relinkings, the number increasing with the degree of affluence of the class of peasants concerned.

As an example, a set of lines was isolated that practised relinking actively and of which the domestic groups resided in Tréguennec, Plomeur, Plonéour, and Saint-Jean-Tolimon between 1730 and 1830. To keep the diagram legible, a selection was made that showed only marriages contracted within the network. This shows how the kindred opened up to take in new lines of descent, certain members of which were phagocytised, so to speak, by the lines weaving the network of alliances. An alternating double respiration stands out; one respiration absorbing the line that provided a new spouse, followed by a second respiration that closed up, incorporating another spouse from the new line in an already existent network as if to enhance its integration (see figure 19).

The occurrence of relinking through affinal marriage within a fairly closed kindred suggests that we should look for the cause in the system of inheritance. Take a set of twenty couples with 100 units of wealth each in generation 1. The same 2,000 units will be found distributed among the married children one or two generations later but grouped differently; allowing for demographic factors, it is likely that the 2,000 will be distributed not among twenty couples but among thirty couples belonging to the original lines of the kindred. Where there is widowhood and remarriage, the same strategy operates at a more individual level, working to keep the patrimony within the kindred that owns it.

The patrimonial line of argument helps to explain the density of relinking. This is greater, the larger the patrimony to be preserved; the number of lines involved in the networks of relinkings increases as the patrimony concerned diminishes in size. The small tenant farmers and day labourers for whom manual searches disclosed identical matrimonial configurations practised them in a less systematic way. In their case it is hardly possible to talk in terms of closing the kindred.

This accounts for the very much lower relinkage rate among groups of full siblings than at Plozévet. In North Bigouden this practice made it possible to regulate transmission of a patrimony over a relatively short period, namely one generation while in South Bigouden it was a question of several generations seeking to protect their patrimony. The same circumstance may also, in a related way, explain distant consanguine marriage. The latter practice did not make it possible to reunite a patrimony that might have been broken up by the rule of partible transmission; on the contrary, close relinking introduced distant consanguinity by keeping common ancestors within the same field of kinship.

Relinkings were not effected as a way of getting round the prohibitions of the church. They constituted an indigenous practice with a logic

Fig. 19. Relinking among seven lines between 1730 and 1830.

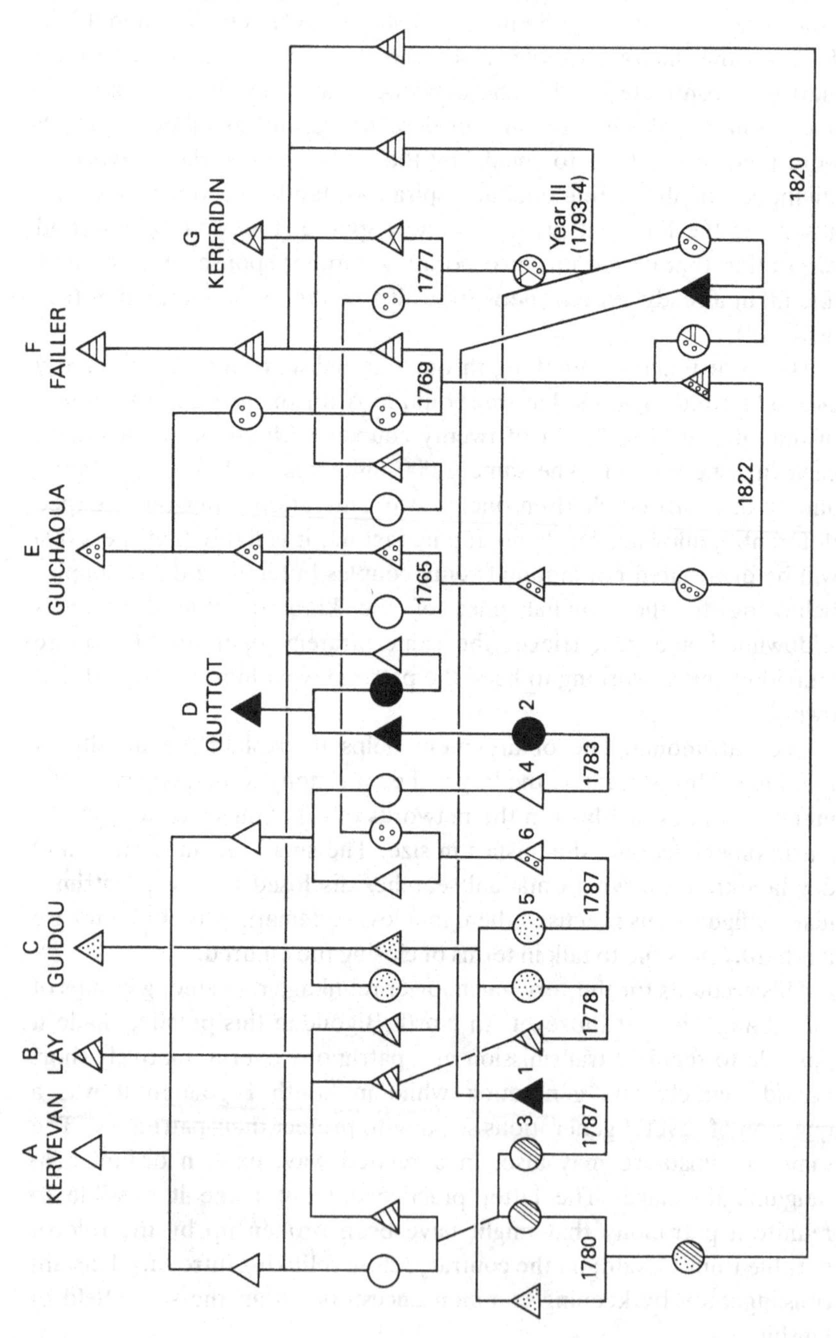

of its own that led eventually to consanguine marriages between remote kin. As a result, we are led to consider the problem of consanguine marriage in this particular context rather differently than usual as in some sense a by-product of relinking through affinal marriage within relatively closed kindreds. Jack Goody's hypotheses regarding the permanence of matrimonial practices having to do with 'the hidden economy of kinship' are thus confirmed.[18]

However, the frequencies calculated by computer show that from the 1860s their practice of relinking underwent a change. Relinkings remained very numerous, but they were effected less often with regard to the same ancestral couples and within blocks of couples who were themselves very active practitioners. There was an opening-up, a loosening of kindreds. People made fewer attempts to *rester entre soi*, to 'keep it in the family'. Parents became less determined to repeat marriages within the same stock of affines. The reason was economic and social in nature and had to do with the effects of the system of inheritance.

Let us go back to our example of twenty original couples preserving for their descendants (who now formed thirty couples), the same patrimony totalling 2,000 units and representing what was still a significant sum in terms of maintaining a hierarchy among the population. When in the next generation the 2,000 had to be shared among sixty or seventy couples, obviously each one was going to be very much poorer than the twenty founder couples had been. What was the good of continuing to marry *entre soi*? Other lines of descent would do just as well. On the other hand, parents' desire to go on planning marriages came out in the continued practice of relinking and in the new prevalence of a particular type of remarriage, namely remarriage between affines. People were less keen to marry a widowed person occupying a farm in another part of Bigouden than to slip straight into the place left vacant by the death of a brother or sister.

An archetypal example perfectly illustrates these modes of behaviour and how they evolved over the years, with the inexorable effect of impoverishing the better-off.

One genealogy in particular stood out. It was that of the R. family, who came originally from Loctudy but because of their fertility spread throughout South Bigouden. One Pierre R., who lived around the time of the Revolution, appears as the ancestor of the line, all bearers of this patronymic being descended from him. Perfectly identifiable on the male side, the line was at that time characterised by its exceptional wealth. Local historians such as Alain Signor interested themselves in it,[19] while the collective memory still preserves an echo of its splendid past. The matrimonial behaviour of the R. family makes an interesting study in that it offers an extreme version of configurations and practices that can be observed in a more attenuated form elsewhere.

So the descendant genealogy of Pierre R. was reconstituted. For some branches this could be done right down to the present day; for others the line stopped at the fourth generation. In the first generation Pierre R., husband of Tudyne P., partitioned his estate in 1790, leaving 13,000 *livres* to his three children (see diagram, p.120). How to marry them into the same social class? In the second generation, two out of three marriages united a brother and sister to a sister and brother. Of ten marriages in the third generation, seven were relinkings and three admitted other lines. In the fourth generation twenty-eight marriages were contracted, including ten consanguine ones, thirteen relinkings (with no double brother–sister marriages), and ten admitting other lines. In the fifth generation only forty-two marriages descending from ten couples of the previous generation were served. Four of them were relinkings (including two brother–sister doubles), thirteen were consanguine, and twenty-eight were contracted with partners belonging to fresh lines.

In chronological terms there was a certain development. Relinkings were more numerous in the third and fourth generations than in the following generation. Certain lines systematically practised marriage to consanguines or affines, while others admitted outsiders.

Relinking occurs when lines of the same social and economic level identify one another and exchange marriage partners. People marry their equals, which is a form of opening, but at the same time they reject those who are not of those lines. Such a policy (*rester entre soi*) loses its justification when the inheritance system combines with demographic pressure to impoverish lines – some much more than others. Whether or not your name was R., henceforth you could marry someone chosen on the basis of geographical proximity or economic affinity, for you were no longer different from them. The increase in consanguine marriages had nothing whatever to do with the possibility of reconstituting a farmstead broken up by inheritance but rather with a body of information circulating within the kindred about such and such a farm that had fallen vacant or was about to do so. However, some lines were more successful in maintaining their social status and symbolic prestige than others, which sank into the anonymous mass of small farmers. Note the line of mayors, who practised a powerful homogamous matrimonial policy throughout the whole of South Bigouden.

The desire to *rester entre soi* resulted after three generations in over-qualifying individuals as far as kinship and marriage were concerned. Two of the four children of Louis R., who had wed Jeanne R., daughter of the first *adjoint* or deputy mayor of Plonéour, married the mayors of Loctudy and Plonéour. François G., broadly integrated in the R. kindred, successively wed two R. cousins (they even had the same first name), the first being a sister of the mayor, the second his daughter; the children born of those marriages further reinforced the integration of a kindred already so dense that the generations were confused and each individual stood in a multiple consanguine and affinal relationship to every other. Take the marriage of François G. and Marie C. Her mother was the sister of his father's first wife; the couple were also second cousins. The half-sister of François, Marie-Louise, wed Pierre R., who was her second cousin at the same time as being, through the G. branch, her first cousin. The children of Louis R. and Pierre D. also made close consanguine marriages.

Relinkings and consanguine marriages mounted up in the branches where there were patrimony or prestige to be preserved. But they were neither as systematic nor as dense once social and economic differences had found their way into the lines of descent. The fact that relinkings were particularly dense in the second and third generations prompts one to identify a certain logic in the exchanges involved: they occur between five principal lines and two secondary lines (as remarriages).

The archetypal example of the R. strongly underlines the main features of the marriage system observed in every kindred. A combination of (mostly distant) consanguine marriages and close affinal relinkings has the effect of over-qualifying the members of the kindred, each one occupying a plurality of positions in the network of affinal and consanguine relationships instead of those roles being distributed among a large number of kinsfolk. The system actually impoverishes kinship, for when your aunt is also your mother-in-law you have one kinswoman where you might have had two. The policy of consanguine and relinking marriages was pursued more systematically by the wealthier lines or those that were better placed politically.

Relinking through affinal marriage thus appears as a constant practice among the people of South Bigouden. Its incidence was in proportion to the prosperity of the section of the population involved. However, to state that there existed exchanging kindreds would seem to border on the banal. Does not every village have its lines that interchange spouses? When it comes to defining the geographical space over which those lines extend, we find that it goes well beyond the boundaries of a single commune to take in villages – in the Breton sense of *hameaux* – scattered right throughout South Bigouden. The greater the number of well-to-do lines a kindred incorporates the larger the area over which it

Fig. 20. Relinking and local political power in nineteenth-century Bigouden.

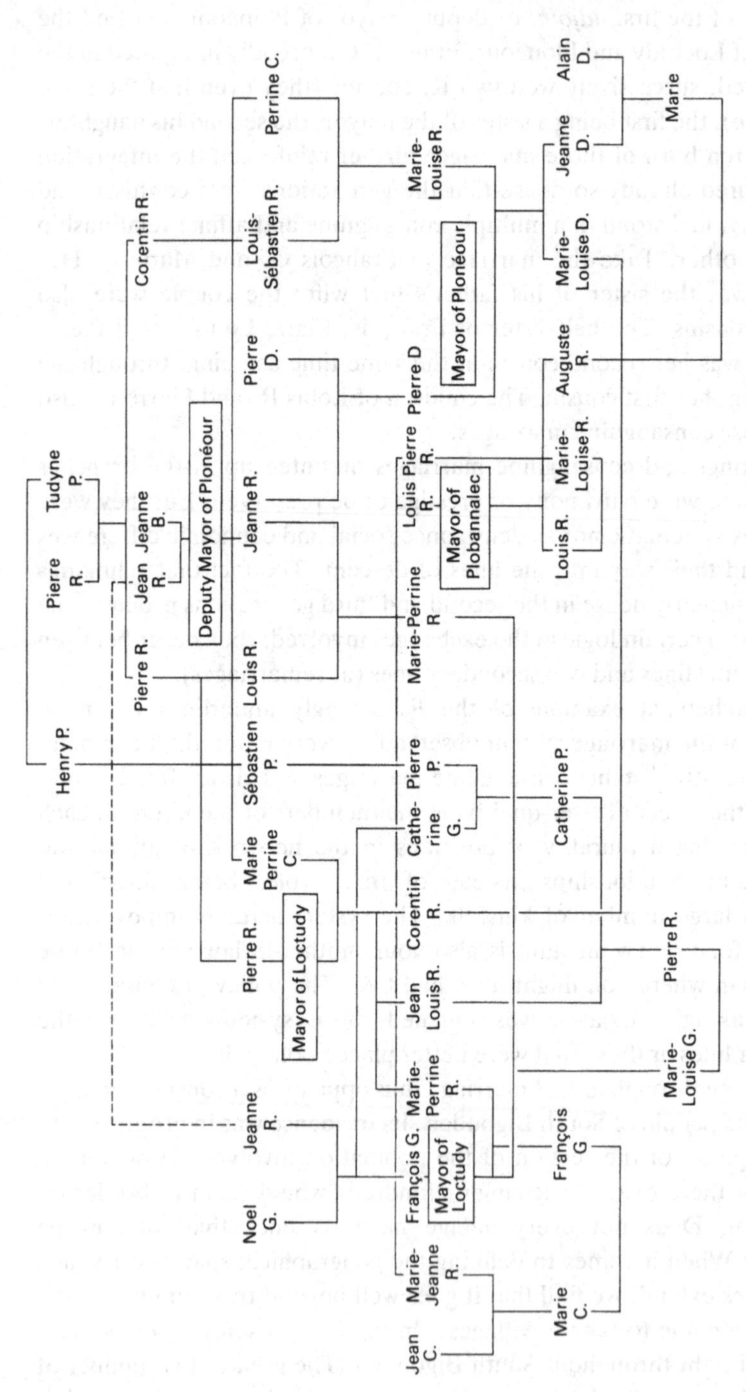

extends. This is a well-known phenomenon; in the interests of social homogamy, the higher a person's place on the social ladder, the wider the area over which he looks for his like. The kinship networks of the local bourgeoisie are very extensive, for example, those of the minor Breton nobility even more so.

A variant of generalised exchange?

Relinking, an extremely ancient practice of which it is hard to date the origin but that continued into the 1950s, appears as the typical matrimonial regularity of South Bigouden, varying in density according to the social group concerned and according to period. The effect of such relinkings was twofold. First, they brought dispersed parts of the patrimony back into the one line. Given the mode of inheritance, we cannot speak of re-concentrating it in the hands of a small number of heirs. Nevertheless, such practices did avoid extreme dispersal and made it possible to sustain a relative hierarchy over the short period of two or three generations. The other effect was at the level of kinship. Relinking led to a considerable diminution in the number of kinsfolk as pluralities of consanguine and affinal relationships became concentrated in the same individuals. However, the kindreds that did the most exchanging were not unlimited in number; they can be pictured as nets placed one on top of another: the closer the mesh, the better-off the lines concerned.

This model of matrimonial regularities, though it may appear at first sight to have been original, can in fact be related to a typological framework that places this peasant society in the continuum of human societies.

> Generalised exchange [wrote Claude Lévi-Strauss] establishes a system of operations conducted 'on credit'. A surrenders a daughter or a sister to B, who surrenders one to A; this is its simplest formula. Consequently, generalised exchange always contains an element of trust (more especially when the cycle requires more intermediaries, and when secondary cycles are added to the principle cycle). There must be a confidence that the cycle will close again, and that after a period of time the woman will eventually be received in compensation for the woman initially surrendered. The belief is the basis of trust, and confidence opens up credit.[20]

The very spirit of the Bigouden system appears to be *a priori* opposed to that of generalised exchange. There were no groups of givers and no groups of receivers of spouses; there were no exchanging groups akin to the *maison* of southern and central France. The strictest bilaterality and absence of differentiation between the sexes made it impossible to identify the lines that would serve as the framework for the system of circulation of spouses. Nor was there any exchange in the way that Lévi–Strauss defined it, because Bigouden had a dowry system and not a system of matrimonial exchanges with a bridewealth, a body of assets permanently set aside for marriages, which circulated in the opposite direction to the women given in marriage. Exchanges were not directed, so that the notion of the cycle or closed loop was foreign to the Breton system.

What gave that system its structure were those very stable kindreds that counter-balanced the individual mobility of domestic groups. But does that mean we must reject the whole question of exchange? Within themselves, kindreds organised an active policy of exchange that was not without a certain spirit of reciprocity. If it was the logic of the struggle against dissipation of the inheritance that led to this marriage policy, each couple that had surrendered a spouse would expect to receive one, be it after a delay, be it through the medium of other couples in such a way that the patrimony did not become detached from the set of lines with which it was associated. Never mind whether the spouse was male or female. Never mind the fact that that patrimony circulated physically without being tied to any one farm. Reciprocity must be observed for a time.

Furthermore, a number of formal factors relate relinking even more closely to generalised exchange. As Lévi-Strauss also wrote: 'Widening of the circle of affines and polygamy are thus corollaries of generalised exchange,'[21]

Did not such relinkings open up to other lines, which in a manner of speaking they phagocytised? Could not remarriage be likened to a form of successive polygamy? This opening-up process was necessary to avoid falling into the trap of consanguine marriage while still remaining among people one knew. Similarly, when Lévi-Strauss lays down 'that generalised exchange leads, almost unavoidably, to anisogamy, i.e. to marriages between spouses of different status; that this must appear all the more clearly when the cycles of exchange are multiplied or widened; but that at the same time it is at variance with the system (in that the latter presupposes equality) and must therefore lead to its downfall',[22] we are bound to admit that the same phenomena were at work in Bigouden society.

Relinkings would tend to isolate a set of kindreds, to close lines back on themselves hugely increasing for each individual the number of identifying references to the other through the superimposition of consanguine and affinal relationships. At the same time with the inclusion of other lines kindreds opened up and the beautiful regularity of the system became diluted. As time went on and the generations became poorer, marriage became detached for the kinship matrix. The economic criterion that had prompted marriage between affines now led to its being abandoned. As a result of the general impoverishment of the population and of the fact that people no longer sought to register their specificity by marrying their peers, who were always the same people, picked out from all over South Bigouden, immediate neighbours or acquaintances now made perfectly acceptable spouses. As a principle, relinking indeed contained the seed of its own demise.

The R. example provides a perfect illustration of these mechanisms. Kinship and patrimony went together to the extent that there is some doubt about whether the latter constituted a true variable apart from the former. Consequently, marriage was used to relink lines that, having identified each other as homogamous were reluctant to let each other go. All marriages were arranged by the older generation. A temporarily closed group had to open up – and do so fast because its assets were melting away. Marriages between cousins were a form of belated response to the impossibility of relinking. Unable to marry an affine, one married a consanguine.

The break-up of the kindreds did not put a stop to the use of relinking, which continued into the 1950s. This revealed the multi-functional nature of the practice. Where affinal marriage had served for a time to combat the dissipation of patrimonies, it continued even in the absence of patrimonial considerations to renew contact between two lines already linked by marriage. It was then a purely symbolic manifestation, marking a desire to *rester entre soi*.

We must stop seeing matrimonial practices purely in terms of highly functional correctives of an inheritance system. Symbolic practices were just as important, and these were very much taken into account in the wedding ceremony.

5

The Bigouden wedding ceremony

What we have called relinking is simply a category designed to meet the requirements of an analysis that seeks to be scientific, global, and thus temporarily detached from the whole affective and symbolic aspect of the subject. But what did the people of Bigouden think of such marriages? What was their experience of them through the medium of the wedding ceremony? So let's hear you, bombardons and bagpipes,* and let the Bigouden bride and groom come forth at last, resplendent, in their costumes!

Kin and yet not kin

In the foregoing analyses relinking was defined as marriage between affines, people who were not complete strangers to each other and who were already related as a result of a marriage (usually recent, as we have seen) prior to their own. The people of Bigouden have their own way of describing that relationship. In these bilingual times, the people I spoke to talked of three classes of relative.

The first category, 'close kin' (*parents proches* or *tud ker tost*), comprises father and mother (*tad* and *mamm*), together called *mad zud*, 'my parents'), brother and sister (*breur* and *c'hoar*), brothers-in-law and sisters-in-law – that is to say, brothers' and sisters' spouses (known as *breur kaer*) – aunts and uncles (*moereb* and *eontr*) on both sides, including both Ego's and his spouse's, and 'proper' cousins (*kendirvi*) – that is to say, first cousins. Note that there is no expression for someone who joins the household of his or her father-in-law (*mab kaer*).

The second category comprises those who are said to be 'slightly connected' (*un peu parents*); these are *cousined*, a word derived from

* The *bombardon* or base tuba and the *biniou*, the Breton bagpipes, traditionally provided the musical accompaniment to a Bigouden wedding [Translator].

the French *cousin* to refer to the children of first cousins. These relatives acknowledge one another as such, particularly when they have remained in the same locality, any slack in the tie of kinship being taken up by the neighbourly relationship. Members of the 'slightly connected' category regard themselves as being of the same stock (*memes gouen*).

The third and last category covers those who are said to be 'no kin at all' (*pas parents du tout*). These are the more distant relatives in connection with whom people speak of 'going outside the family' (*ed er meaz*).

So the Bigouden family is not as extended as people have tended to claim. When several ties of consanguinity and affinity become superimposed in the course of relinkings, the closest always prevail. Michel Izard noted that in Plozévet, too, kinship was restricted to very close relatives.[1] Sylvie Postel-Vinay observed the same phenomenon in Minihy-Tréguier, where people also spoke of having 'gone outside' kinship.[2]

Whether or not a person considers himself related to another, how aware he is of his kinship and the kinds of dealing with others that are implicit in his recognition of such ties are key factors in the negative or preferential choice of a spouse.

To acknowledge a kinsman is to possess a whole body of demographic, social, and residential information about him or her. However, there is a certain ambiguity about the depth of people's genealogical memory.

On the one hand, many people said they did not know the names of their grandparents. 'My grandfather was a Garrec', one person told me, 'but my grandmother, I don't know, I never knew her.' Another confessed: 'We didn't talk about that sort of thing much . . .' When asked about the names of great-grandparents, many people said that that was 'going back a long way', and they swung a hand limply from the wrist as if to indicate an eternity.

This is why a marriage between two partners whose grandmothers were cousins was not regarded as consanguine. That was 'going back a long way', but understandably it was relatively common because the parents who put the marriage together retained the memory of this kinship link, which marked out a set of descendants as so many potential spouses.

'The old people used to know their kin better', Marie-Louise L. told me. 'My mother knew all her male and female cousins and used to say "Good morning, *kerentro*!" to them. That was what they were called in those days.' On the whole people 'called cousins' with each other less nowadays, whereas a generation ago links between first cousins were strong. However, the observation that family memories had deteriorated was constantly challenged by remarks like these: 'The old mayor's father was a cousin to Corentin's grandfather', or as Jeanne P. said of her husband, 'Our two grandmothers were cousins, but we weren't acquainted'. That was a case of a 4–4 marriage that had not required a dispensation, having been celebrated in the 1920s. The bride and bridegroom, who were third cousins, had not known they were related. Identification of the relationship had prompted no particular reaction, however, genealogical distance having abolished the tie of kinship.

Beyond the fourth degree, kin solidarity and even recognition of kinship cease. Two men may bear the same patronymic, but not knowing their precise relationship they consider themselves no longer related. Social mobility, too, plays its part. Where one branch has become poorer and another richer, genealogical distance is reinforced by social distance. Whether or not folk recognise each other as kin will then depend on circumstances. A man may wish to preserve a family connection that flatters him while passing over one that does not. Someone may no longer wish to be identified with the 'stuck-up' R. family even though he bears the name. As a matter of fact, with the exception of this patronymic, all bearers of which know for certain that they are related, a high incidence of homonymy permits a certain amount of speculation on the range of possible identities. 'That's not the same Tanneau family', one might hear, or on the contrary, 'We're the same stock, we come from Kerliou' (one of the hamlets of Saint-Jean-Trolimon).

At the same time, genealogical knowledge extends over a great distance collaterally, particularly among old people. There thus emerges, alongside kinsfolk recognised as such, a somewhat ambiguous category comprising those who are not kin, who have 'gone outside' kinship, but who are still not strangers.

These are the descendants of great-grandparents and beyond, the affines of consanguines and the consanguines of affines. Of the latter, who provide the basis for relinking, people will declare vigorously: 'We're no kin at all.'

The fact of being kin implies obligations to help one another, emotional ties, and also ties of interest. However, there is a difference between 'close kin' and those who are 'slightly connected'. Among the former a competitive relationship exists with regard to the family patrimony; every member of the family is entitled to a share of this asset, and that title needs to be asserted. Ties of kinship, restricted to very close relatives, are thus characterised by affectivity and competitiveness.

Kin of that sort are avoided in marriage. With those who are 'slightly connected', however, relationships of patrimonial competitiveness are usually rare, except where comparatively remote kin end up after a series of deaths, taking charge of children. Dealings with those who are 'slightly connected' and those who are 'no kin at all' tend to be marked by a certain coolness and distance, and marriage with them is prescribed. For instance, during an interview conducted at the house of Pierre-Jean R., it emerged that his wife is sister-in-law to my landlady's daughter. Although they are neighbours, the two women see each other only rarely – much like any neighbours who might bump into each other at the corner shop. The kinship tie was mentioned only when the identity of surname prompted a question about it.

To judge from indigenous statements, then, there seems to have been a conscious policy of not marrying relatives where there were interests at stake. It further appeared that those with whom one's emotional and economic ties were slack would be most sought-after as partners: very distant cousins or close consanguines of affines and *vice versa*.

The contrasting high incidence of marriages between first cousins in the Valencia region of Spain, for example, can be explained by operating the same principles in the diametrically opposite manner. There the nuclear family predominates, and the children of brothers may share no affective ties. That accounts for the number of marriages between first cousins – that and a partible inheritance system that in each generation divides up patrimonies that a consanguine marriage may help to put back together.[3] In Bigouden first cousins saw themselves as being closer than was the case in the Spanish society studied by Joan Mira. Many of them were orphans brought up by uncles and aunts who were also their godfathers and godmothers. Moreover, there were no reasons at the patrimonial level for contracting such close unions. The important thing, the thing that governed the whole approach to marriage was finding a farm. In this respect the kindred constituted a high-quality information network covering a huge area.

Kindreds were not groups of kin sharing a collective feeling of belonging, owning assets in common, and having identical interests to defend. They were nameless groups comprising those individuals who were 'no kin' but were nevertheless not total strangers. It was members of his kindred that a person must marry, without any feeling of incest being involved – first because of the coolness of affinal relationships and secondly because the church did not prohibit such unions. A member of one's kindred was preferred as a spouse because his or her own kinship (and therefore also economic and social) situation was already known as a result of a previous marriage. And there could be no relationship of competitiveness with such a person over a shared patrimony. That may explain the native logic of repeated relinkings within the same kindreds as a deliberate strategy on the part of all concerned.

The deliberate practice of a nameless type of marriage
Yet there is no special name for the practice of relinking.

Fig. 21–23. Contemporary examples of relinking, as described by informants.

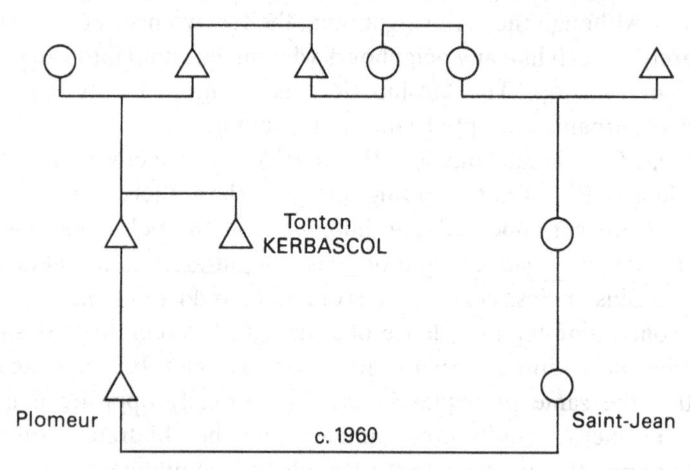

Jean-Louis L. remarked of his marriage: 'I often visited Tonton ["uncle"] Kerbascol [the nickname derived from the farm the man lived on] as a lad on the threshing team. That's how I got to know Saint-Jean-[-Trolimon] so well and met Marie-Anne' (whose blacksmith father was well-known in the district). Their marriage repeated an earlier union two generations back.

People say in connection with such relinkings: 'This marriage is not in the family.'

Fig. 22.

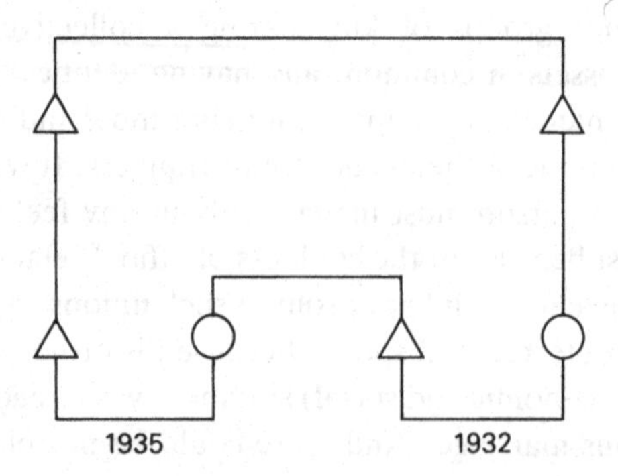

Here the first marriage was celebrated in 1932, the second in 1935. The two young people concerned, a sister of the bridegroom and a cousin of the bride, had 'been made bridesmaid and escort at a wedding: they were fifteen and seventeen. They waited for each other for five years.'

Everyone is aware of the relatively high incidence of marriages between kinsfolk, and since nowadays the practice is seen as belonging to the past and is at the same time regarded with some embarrassment, people state all the more vehemently that 'we're related on both sides . . . but we're not kin at all' – as in the above example, in connection with which I was told: 'A [female] cousin of my mother's married a [male] cousin of my husband's mother.' The recent marriage was represented as the fruit of chance.

Fig. 23.

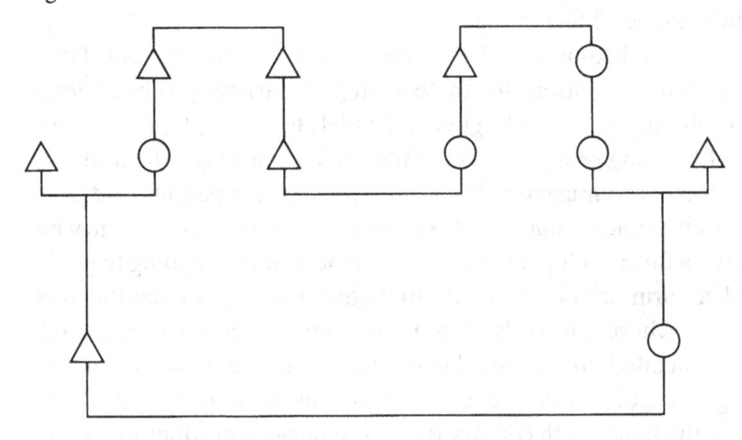

Relinkings and marriages between remote kin were the result of parental wishes. Everyone was agreed on this point. 'There was a time when marriages were arranged by families . . .' 'There was a farm to be had and they jumped at it . . .' 'Finding a place' was the recurrent leitmotiv accounting for these marriage patterns.

Between well-to-do families the paramount concern was for an arrangement that should satisfy both parties. 'You virtually had to weigh everything to get married. Wealth is wealth. People used to say: "How much are you giving your son and your daughter?" If it wasn't the same amount, no chance of contracting the marriage. If there was property, wealth was measured in gold. Parents married off their children like that and the youngsters, no one asked them, they just had to go along with it.' Another account recalled bygone days: 'It was the parents that married off the youngsters. It was more like a sale . . . You needed a pair of scales.' The parental will was decisive: 'Marie-Jeanne K. wanted to marry Corentin L.; the uncle (who was the girl's guardian) had the marriage broken off because I wasn't good enough.' So marriage was dictated by a strong awareness of economic rank. Males and females were treated identically. Dowry was something that had to be equivalent on both sides. In such conditions the young people themselves did not often have much say.

To start with, the respect due to parents was almost sacred. They were obeyed and their decisions accepted without discussion. As confirmation of this subservience to parental authority, take the example of Henri D., who arrived at his future father-in-law's farm by appointment and was told: 'There are two daughters – take your pick.' Jeanne Cossec was told by her mother: 'You should marry that one' (Hervé Calvez), Jeanne's comment was: 'I was in no hurry to get married . . . Later I married his brother Jacques.' Freedom of choice was confined to the sibling group.

For the less well-to-do – and the closer we get to the present day – there is a shift in the reasons for these arranged marriages. It becomes a question of finding a farm. Marguerite Daniel, for example, is the third daughter of a sibling group of seven. 'We couldn't all stay', she told me, 'and when the *koritour* came to suggest Pierre Coïc my parents said yes.' Marriage went hand-in-hand with settling on a farm. 'A person who doesn't have a farm can't get married . . . Where are they going to go?'

To find a farm it was necessry to know where a succession was imminent or where an early death had left a place vacant. Such information needed to be circulated. That was the function of the marriage go-between, a very common figure in Brittany known as the *koritour*. In the nineteenth century they were beggars or itinerant tailors who travelled from farm to farm and had a good knowledge of all the families and of the point each had reached in its life cycle. In the early years of the twentieth century it was the woman who kept the café–restaurant in Saint-Jean-Trolimon who often organised marriages; she was the *koritourez*.

Even consanguine unions or relinkings required the intervention of a go-between. Marie B. said of her marriage: 'Our two grandmothers were cousins, but we weren't acquainted. We call it *kendro pell* [remote cousinhood]. We'd seen each other from a distance at fairs or on our bicycles – that sort of thing. Stephan from Lesbervet was the *koritour*.' The hamlet of Lesbervet lies half-way between Gorré-Beuzec, where the girl lived, and Keregard, the young man's home.

Apart from professional go-betweens, there was always one member of a kindred who kept a look-out for potential spouses among relatives –

such as the woman who was to become a cousin of the bride and bridegroom, who wrote to Joseph H. (then in Canada) to tell him that Catherine N. was preparing to take over her father's farm. Rolland Daniel explained: 'It was my father-in-law who told me: go and propose your sister-in-law to Michel' (Rolland's first cousin) a relinking marriage that did in fact occur.

A story recounted by Canon Le Floc'h concerning the circumstances of a particular marriage well illustrates both the part played by parents in the choice of spouse and the scope of the search for what seemed the best-matched partner. There was a not particularly wide-awake boy, an orphan brought up by his grandparents, who had not succeeded in finding a bride at a wedding dance or one held on the occasion of a *pardon*. His grandmother and her sister were talking outside church one day about a possible wife for him. 'Who could we find for that one, then?' they wondered. They searched and searched among all the girls in the family who were only 'slightly connected', and eventually, in the dim recesses of memory, they hit upon the one who was to become the young man's wife (the bride and groom were grandchildren of first cousins). The discussion ended with the words: 'We couldn't see her although she was family!' As in most peasant societies, the men settled the details of the agreement while the women did the actual prospecting for a marriage partner.

Did people have to have forgotten about their relatives in order to marry them? Certainly not. On the contrary, they needed to know them well in order to make sure that the genealogical distance was sufficient for them to regard each other as only 'slightly connected' or 'not kin at all' – that is to say, an affine of a consanguine or a consanguine of an affine.

Establishing that it was families that made marriages is not necessarily a self-evident conclusion, certain historians, such as Jean-Louis Flandrin,[4] consider that it was the young people who did the choosing, having interiorised the constraints with regard to choice of spouse, and who, on introducing their intended to their parents almost invariably obtained their consent. Here a different model operated; marriages were made by the parents. Their concern, of course, was to find for their children a marriage partner whose relatives possessed a fortune equivalent to their own and/or to find a place in which to settle that child. The two concerns often overlapped, though not necessarily. It was the older generations that were aware of the matrimonial chessboard: 'Youngsters never got about', they say. 'On feast days the girls had to be home before dark.' A high degree of social reserve also prevented young people of different sexes from coming into contact with one another, not to mention the influence of the clergy, who did everything in their power to keep them apart. So young people were obliged to look to their parents for a possible marriage partner. And parents had to be constantly on the alert for the twin eventualities of a spouse of the right age and social standing and a farm about to fall vacant.

Parents certainly did get about. They attended the weekly fair in Pont-l'Abbé as well as local *pardons* and weddings; they worked in the fields, and they formed mutual-aid teams that went from farm to farm. There were family parties, and above all there were funerals, which were a time when the wider family got together. And if necessary there was mediation to fall back on. Closer to our own day, people who were young in the 1920s talk about having enjoyed a degree of freedom, but it led to the same consequences as regards choice of spouse. It was at weddings, people said, that one met one's husband or wife-to-be; it was there that the first encounters took place between the youngsters of two families that were engaged in forming an alliance through their children.

Are there wedding rituals specific to Bigouden?

It is essential for anyone studying the way in which unions were contracted to know how marriages presented themselves. What was the symbolism of their ritual language? A more difficult task is looking for a concordance between rites and social usages. The cultural sphere often has its own peculiar and relatively independent manifestations. Also we saw in the previous chapter how a change occurred in the 1880s in the way in which marriages were made. Did the ritual change as well? And if so, can such changes be narrowly attributed to the changes in matrimonial practice? Ought we not also to allow for the effect of wider socio-economic changes?

The way in which wedding rites have changed is not an easy subject to deal with. The reseacher is often at a loss. Questioning old people in the 1970s, one can only expect to be told about weddings that took place in the 1920s at the earliest. For earlier periods one has to fall back on documents written by local observers. Of these there are a great many, but their shortcomings are notorious. They are sometimes poorly located, for example, and lack precision as regards the social group observed. One cannot be sure whether the person giving the description has been told about a wedding or was actually present, observing it at first hand. Nevertheless, such descriptions constitute irreplaceable documents as far as pinning down changes is concerned.

Breton marriage fits into a pattern of peasant rites common to the whole of France and to much of Europe. They may be described in terms of the system of classification of Arnold Van Gennep's rites of passage.[5] Whether we take the various stages of the nuptial scenario or the nature of the customary acts relating to marriage, we can be sure of one thing: from the young couple's first meetings to the final moments of the wedding ceremony, Breton customs and Bigouden customs are among the cultural expressions of behaviour patterns most often attested to in peasant societies in which there is a culture with a powerful oral ingredient based on the more or less faithful repetition of the previous generation's models – in other words, tradition.[6]

The illusion of a culture specific to Brittany (and more particularly Lower Brittany) may, it seems, derive from a kind of archaicism that caused the region to cling for longer than other parts of France – longer, even, than Upper Brittany – to its behaviour patterns with regard to fertility as well as to its costumes and its traditions of music and dance. Observers at the end of the nineteenth century and in the early years of the twentieth could still witness those interminable processions in which the newly-weds, accompanied by their families and their guests, wended their way to a wedding banquet set out in the open. What a godsend such sights represented for all the picture-postcard publishers in search of the exotic and the photographers who kept them supplied with material!

However, we ought not to deny the region a degree of cultural speciality linked mainly to the Breton language which exudes, transmits, and continually renews an original cultural expression. The accounts of folklorists and travellers provide present generations with their only authority for the existence of, for example, those endless ceremonial speeches, veritable contests in eloquence, that served to slow down the progress of the wedding ceremony. Quoted at length in Bouët and Perrin's 1918 study of Breton life, *Breiz-Izel*,[7] and referred to as late as 1840 and 1841,[8] such orators had disappeared from the memories of the people born around 1870 whom Jean-Michel Guicher interviewed in the 1950s.[9] Nevertheless, Brittany's cultural specificity continued to assert itself. It came out, for example, in the repertoire of oral, musical, and choreographical works. As well as these cultural features there were certain social particularisms that helped to lend a specific quality to the Lower Breton nuptial scenario: the size of the wedding party, the role of young people, the existence of the go-between – all features that have survived down to our own day.

It was the custom in Lower Brittany to invite a large number of relatives and neighbours. The eighty or 100 guests usually recorded elsewhere in France here became 200, 500, even 1,000 in exceptional cases. This phenomenon may be attributed to demographic circumstances, notably to the high level of fertility in Brittany, where people had more relatives than in regions that practised contraception. It also surely reflected the scattered pattern of habitation. Guests were drawn not only from the hamlet itself (the *ker* containing the farm where the reception was to be held), the *quartier* (embracing several hamlets), and the commune of residence but also from the other communes throughout which kin were dispersed. Another reason for inviting large numbers of wedding guests was that the size of the group assembled possessed a strategic value, redounding to the family's honour and constituting a sign of social recognition. That accounts for arrangements peculiar to Lower Brittany that were necessary for the organisation of wedding banquets. It would have been difficult to hold such occasions inside; they had to be out of doors. The large number of guests also accounts for a further peculiar feature encountered only in certain regions, of which Bigouden is one, namely the *écot* arrangement under which each guest 'paid his score'.

As in every peasant society, a wedding marked a moment of rupture with everyday life. The importance of the wedding ritual reflected the importance of the economic and social functions of marriage. Moreover, the wedding itself provided a privileged occasion for cultural activity and invention.

Part of the function of wedding days was to effect social integration. The guests included close and distant relatives, neighbours or workmates on the agricultural labour team, and the local young people, who played a leading part in the wedding ceremonial. The helpers (cooks, those who waited at table, musicians) were associated with the ceremonial in a different fashion and at a different point. There was even a place for those not invited, as there was for beggars.

In another feature peculiar to Lower Brittany and to Bigouden in particular *la jeunesse* was not organised in the same way as we hear of in many other French regions.[10] There was no formal grouping of youth associated with a defined territory. There were no shared functions performed in common, notably those of organising festivals or exercising social supervision over young people themselves or with respect to a moral code. For instance, the region had no rites comparable to that of the *charivari*, giving public expression to the fact that a couple had infringed society's rules, either because the wife beat her husband, or because husband and wife were ill-matched, and effecting a kind of vindication of popular morality. There was no counterweight to parental decisions in the form of an organisation of young people capable of demanding freedom of action and defending members who refused an arranged marriage. The absence of any such organisation had a further consequence in terms of choice of spouse at the local level; there were no spinsters reserved for a village's stock of bachelors.

Moreover bachelors and spinsters were kept separate and were closely supervised. They met only at *pardons*, wedding dances, and processions. The clergy gave parents their support in suppressing all seemingly improper social intercourse. Every informant confirmed this; young people had to wait for 1920 and the advent of the bicycle to gain a degree of autonomy. We cannot talk in terms of cause and effect; we can only note the co-existence of two social phenomena: no geographical endogamy, and an absence of the kind of organisation of young people found in the rural societies of western central France.[11] It is tempting to draw a further connection between that absence of any organisation of young people and the absence of the barrier, barrage, or even welcome rites described at length by Van Gennep (but said by him not to have existed in Lower Brittany) as having been observed when the fiancé came from outside the village.[12] It is a well-known fact that Van Gennep was not really interested in Brittany and that his sources on the region are greatly inferior to his sources on others that he studied either at first hand or through the medium of local informants (Savoy and Burgundy, for example). However, postcards exist depicting barrier or welcome rites in which the boys of the village, glasses in hand, greet arrivals from another commune (see plate 7).

The most we can say is that these welcome rites, standard ingredients of the nuptial scenario, did not in Bigouden take the violent form they sometimes took elsewhere, when feelings among young people ran very high and the bachelors of a village looked with so proprietory an eye on 'their' girls that they exacted compensation from the incoming bridegroom in the form of money and drink.

Was there a connection between the existence of the go-between, the dispersed pattern of habitation, and the absence of any organisation among young people? Go-betweens are mentioned in other peasant societies that do not combine these three features. What was striking about the Bigouden go-between was how long such people lasted as an institution; their presence was still attested to in 1950. Le Doaré, writing in 1896, said that the tailor or beggar 'formerly' fulfilled this function.[13] Professionally obliged to tramp the countryside, putting up at farms, they knew all the inhabitants of each one and were in a better position than anybody to suggest marriages between well-matched fortunes and political opinions. Later, relations served as matchmakers, or the job was done by specialists, such as the innkeeper. In the early years of the twentieth century the woman who ran the café in Saint-Jean-Trolimon organised many marriages in order that the wedding parties might, by way of saying 'thank you', come and eat in her new restaurant. Known in many parts of Lower Brittany, notably in the Quimper region, as *baz-valan* ('broom-stick', from the distinctive attribute he or she carried around) the go-between subsequently had no need of a distinctive characteristic; his or her intentions were quickly divined. In Bigouden, as we have seen, the go-between was called *koritour*. Virtually all old people who were married before the Second World War had had their union mediated by a *koritour*.

The *koritour* combined two tasks that were sometimes distinct but at other times merged. The first was to suggest marriage partners to families when those families were unable to find them for themselves (for example, when widowed persons remarried). The second was to organise the preliminary matrimonial arrangements in order to avoid putting families in touch that had no chance of getting on with each other.

Large wedding parties, a lack of any structured organisation of young people, and the presence of go-betweens were all features that tied in comfortably with Bigouden marriage customs. Large numbers of guests were invited because there were so many kin. Society did not shepherd its youngsters because the mobility of domestic groups was too great and the structuration of kindreds occurred at the level of a territory that was clearly too large for any kind of grouping to work (such a grouping presupposed a relatively restricted number of persons able to meet frequently). Finally go-betweens were necessary because the area of spouse selection was so large and because it was important to bring together similar levels of wealth.

Wedding rites and social change

The problem of how wedding rites evolve is a complex one. Such evolution may be purely formal, the meaning of the rite remaining unaffected. Conversely, a rite may keep its formal content and change its social significance.

It would be pointless to try to rehash the accounts that Alexandre Bouët wrote to accompany Olivier Perrin's drawings, describing wedding rites in the Quimper region.[14] No author has ever gone more closely into the question of Bigouden customs, and readers familiar with French are referred to his descriptions for the earliest period.

In the nineteenth century, Paul du Chatellier and E. Le Doaré drew attention to several details of wedding ceremonies in Plonéour-Lanvern.[15]

The role of the *koritour* has been mentioned but how was an agreement sealed?

When the parents seemed disposed to accept the candidate announced, he came round one night after dark with his father, his brothers, and some friends, bringing a bottle of brandy [*eau de-vie*]. The girl's parents set another bottle on the table, together with bread and butter. People talked about nothing in particular, then a few days later the young man sent word to the girl that he would be in the village [*au bourg*] on such and such a day. If she came there with her parents, the couple became engaged and they all sat down to a meal together.

Once both parties had accepted the deal but before it had been finally agreed visits were either exchanged between the two families (if the young people had no farm of their own) or made to the farm that the couple would be occupying.

Nicolas Le R. declared: 'When my betrothed came to see my parents with her parents she was taken round the fields, shown all the lots. I never made such a visit myself. I knew full well I'd never be moving into my father-in-law's place. There was another son who'd be taking that farm.' Nicolas Le R. married in 1919, so what the folklorists used to call *les accords au cabaret* ('pub deals') and *la visite des lieux* ('the site inspection') went on for a very long time.

Is it true that the parties tried to get the better of each other? Quite possibly, in so far as the well-to-do peasants at the centre of our study, being scattered over a fairly large territory, were not necessarily familiar with the precise value of the suitor's family's assets. Armand du Chatellier, writing in the early 1860s, tells how 'every day the notaries, as the natural depositaries of their [the peasants'] savings, are confidentially enjoined by their clients to mention twice or three times the amounts they may have on deposit in the event of the other side coming, at their suggestion, to verify what they say they are worth'. He mentions a 'wealthy heiress swindled in this fashion'.[16] Was not relinking between lines that already knew each other aimed precisely at mitigating this very problem? The first marriage, by enabling the two families to evaluate each other's assets, paved the way for a second marriage, negotiations for which could be conducted in full knowledge of the situation.

Discussion bore mainly, as we have seen, on the size of dowries. Hence the many precautions taken to protect the respective contributions. Hence too, the forfeits provided for by notarised deeds in the event of the projected union not taking place.

Did rites evolve, or did observers get things wrong? The *Fête de l'armoire* ('wardrobe party') described by Bouët and Perrin was also noted by Le Doaré and Paul du Chatellier, but in more ambiguous terms:

> When the landowner marries off his daughter, all the tenant farmers have to bring the bride's trousseau and her wardrobe or *arbel* on their carts. To thank them, the said landowner throws a big party or *fest an arbel* to which he invites men, women, children, and the groom who drove the car. The landowner's lady receives him in person, gives the bride a glass of brandy, and places a piece of bread in her basket for those left at home. At the meal, which lasts all day, it is the members of the family of the lord-landowner [*seigneur–propriétaire*] who serve the guests.[17]

Who were these 'landowners'? Was du Chatellier referring to the aristocratic landlords of which he was himself one, residing in the *château* of Kernuz and owning dozens of farms in the region? One can hardly see a daughter of du Chatellier marrying a tenant farmer and taking part in a *fest an arbel.* So was he talking about the wealthy *domanier* farmers, the very ones whose marriages we are interested in? But then why does he use the term *seigneur-propriétaire*? The date of his account of this custom also seems very late. Did people, as it were, still 'stage' the actual furniture removal in this almost theatrical way at the end of the nineteenth century? Pierre Hélias, who is careful to date the customs he records, makes no mention of such a party. According to him, it was the groom's parents who ordered the bed and the bridal wardrobe.[18]

The obligatory wedding at the *mairie* was for a long time a formality; it was the religious ceremony alone with its solemnity that formed an integral part of the festive patterns of behaviour designed to underline the importance of the occasion. The call on the mayor was something that had to be done but was not going to interrupt the daily round of work on the farm. Only the bride and groom, their parents, and the witnesses went to the *mairie.* Sometimes they did not even do that, and stories are told of how the mayor, an ordinary farmer to whom the formalities were just as inconvenient as they were to those getting married, occasionally pronounced the ritual words in the corner of a field where he had hastily assembled the bride and groom together with their parents and witnesses. The deed was drawn up in all its artificial solemnity the following Sunday: *Devant nous ont comparu en la maison commune . . .*

At the church there is nothing in particular to record, except that 'the bride and groom bring a basket containing bread and wine and place it on the altar steps, on the gospel side. After the wedding these are distributed in the vestry to all their friends, starting with the priest who has blessed their union.'[19] It was a classic aggregative rite using food.

Until 1945 it was rare for the mother to be present at the wedding mass and informants confirmed that godparents escorted the engaged couple to the altar. Mother was at home, busy with the final preparations for the wedding meals and the party. In the years leading up to the Second World War, Bigouden weddings were still substantial events. The first thing they were noted for was their length. This moment of celebration and rupture opened with the inevitable preparations for the meals and went on for as long as the food lasted. This was *le fricot*. The wealthier the families, the more numerous the guests, the more time the preparations took, and the longer the feast lasted.

Generally celebrated on a Tuesday, weddings went on for at least two days and usually three, the idea being to give every member of each household a chance to participate in turn.

In the 1850s it was customary in the wealthiest families to have a final series of meals a week after the wedding. During these the newly weds waited at table. On the Thursday the leftover food was distributed among the poor. These would then recite prayers (*Pater* and *Ave Maria*) for the prosperity of the newly-weds and a *De Profundis* for deceased relatives.[20] In fact there was a permanent association between the cult of the dead and nuptial rites, particularly through a mass for the dead being celebrated on the day after the wedding.

Weddings were occasions that marked a break with the daily round by providing a large number of copious meals featuring suitably exceptional food. When café proprietors took over in the late nineteenth century, they continued to serve the traditional wedding menus: *pot-au-feu*, tripe, and *cochonnailles* (various pork preparations) followed by puddings made of eggs and flour. If there was anything left over, people came back next day with the youngsters for a *fricot bihan*. The move from farm to restaurant did not affect the length of Bigouden weddings, which still extended over three days in the 1930s.

These wedding banquets with their many guests would have represented a considerable financial burden on the host families. So this was a region where guests paid their *écot*, their own score, generally in the form of a coin handed to the master of the house at the end of the evening meal. The custom was retained after the move to restaurants, with each guest paying for his own meal (as is still done today). People called it *pea diouz taol*, literally 'paying at table'.

The high points of the nuptial scenario were marked with music and dance. There was dancing in the square outside the church after the party had done the rounds of the cafés of the *bourg*. Such dances put the families concerned on show and helped to publicise the marriage. They also gave expression to a sense of the sacredness and seriousness of the knot that was in process of being tied, as in the *dans an dud nevez*, the dance of the newly-weds. Music, too, accompanied all the key moments of the wedding. There was the *air des rôtis*,[21] for example, a special tune to greet the entry of the meat course.

Up until the end of the nineteenth century horse races were organised for big weddings, as they were for *pardons*.[22] The custom died out, however, when the horse-drawn charabanc replaced the horse as people's means of getting about.

Between the meals, everyone was free to dance. Non-guests, beggars, and young people from other communes all joined in. In the nineteenth century these *bals de noces* were forbidden as much by the religious authorities, who saw in them the influence of the devil, as by the municipal authorities, which feared them as a source of public-order disturbances. The priests excommunicated tavern-keepers without mercy. In 1875 the Pont-l'Abbé police chief wrote to the prefect to complain about the crowds that overran the wedding dances held in the town's cafés and taverns. 'As soon as there is a marriage in prospect, the café proprietors organise a dance floor. It is then a question of which one will catch the engaged couple and the young people close to them and stand them cups of coffee in an attempt to get the dance held at his place rather than at someone else's.'[23] According to him, at one wedding the dancing went on for four days!

Youth as a body may have had no structured organisation, as it did in other parts of France, but we must not forget the pre-eminent role played at weddings by the groomsmen and maids of honour who were paired off in the bridal procession for two, three, or four days. Such couples were matched with the greatest care by relatives and by the future bride and groom as if in preparation for fresh weddings. Often a relative of the groom would be paired with a relative of the bride with a view to a future relinking. Attention was paid to compatibility of age and character. Many people now aged fifty and over met their spouses at weddings: 'We were invited as the younger generation.' In the 1920s, when the role of the go-between was becoming less important, bridal processions offered young people the opportunity, as they escorted – and imitated – the newly married couple, of making their own matches.

It was the role of youth to receive the guests, escort the newly-weds, and bring them the bread and milk that formed the symbolic prelude to the consummation of their marriage. Some observers claim that married couples kept three nights of continence following their wedding, known as *Nuits de Tobie* or 'Tobias Nights'. Cambry mentions them for the late eighteenth century,[24] and according to Le Doaré and du Chatellier they were still observed at the end of the nineteenth: 'On the evening of the wedding the bride and groom returned to their families. It was only the next day or several days later that they began to cohabit.'[25] The bread-and-milk rite was thus pushed back to the third evening, which presupposed that the *jeunes*, the 'young people' of the wedding, returned to the farm.

Bigouden marriage rites cannot be described outside time. They were constantly under pressure from religious and civil constraints and from external cultural forces. In other words, they were malleable; they changed. For instance, like so many other peasant societies, Bigouden society absorbed the rite of the wedding photograph. The rule in most parts of rural France was for the photographer to take a picture of all the guests together and a picture of the newly married couple. In Bigouden, because of the size of weddings more photographs were taken: group shots, a grandfather with his descendants, shots of the young people.

This is one rite that has not changed. A Bigouden wedding party will still pose for the camera as a group, even though today's photographic techniques make it possible to take pictures in a more modern, soft-focus style.

Happy processions wending their way through the village also continue to publicise weddings today, the only difference being that now they are processions of cars. As in the past, the guests still assemble at the groom's house. The procession forms up and moves off to fetch the bride. Then, after the brief ceremony at the *mairie*, it makes its way to the church. Emerging after the mass, the party stops for a drink at the bar in the square. It is customary for the wedding banquet to take place in a different commune than the wedding. Every Saturday in summer is booked up months ahead at such restaurants as 'Au refuge' in Saint-Jean-Trolimon, 'Au Gai Papillon' in Kerbascol, and 'A la Caravelle' in Pendreff. And as in the past each guest settles his own score for a meal that is both lavish and long-drawn-out – and so costly that often only one half of a couple invited will attend, the other coming along in the evening for the dancing, which is still public. Before the bride and bridegroom retire there are the ritual 'promenades' by car; champagne and special cakes are taken at the groom's house and then at the bride's. Finally, as in the past, the young people see the newly-weds off to bed.

These rites are still traditional in structure. They are rejected by the young, however, who just want a reception, like people have in towns, or who quite simply shrink from the whole ceremony of marriage. Like the rest of France, South Bigouden is seeing more and more young people cohabiting with the consent of both families.

Wedding costume: tradition and change

The development of the Bigouden bridal costume is a particularly visible sign of social change going hand in hand with ritual change. Traditional costume offers a way of publicly affirming social identity and, on the occasion of a wedding, of displaying wealth, René-Yves Creston stressed the fact that Breton styles of dress became diversified at the time of the Revolution[26] and have evolved steadily ever since. As far as Bigouden is concerned, he drew attention particularly to the *coiffe*, the size of which poses a number of problems as to the origin of so monumental a head-dress. Creston remarked that there was nothing special about the composition of the *coiffe* to set the Bigouden region apart in any way. He said this in reply to those who claimed to see some Asiatic influences that had supposedly helped to marginalise the Bigouden population. Nowadays the debate has run its course, and the subject is closed. It is enough for us here to note the changes that occurred in traditional costume as seen in the photographic documents collected in the course of this investigation. Tradition, change, and fashion are not incompatible.

The most striking aspect of the photographs, which were taken between about 1880 and 1940, is the way in which they reveal how rapidly the different parts of the costume evolved while still retaining the overall proportions that make it possible for us to talk in terms of a 'traditional' costume. Nowadays, when someone says of a woman, 'C'était une belle Bigoudenne', it is to stress that she used to wear the costume with elegance, unlike those who around 1920 went *kiz ker*, following 'town fashion' (in other words they cut their hair short and abandoned both costume and coiffe). The women who remained in the district made the departure from tradition in the 1950s.

Let us take the Bigouden coiffe first of all. In the early years of the century it was described as a 'diadem' or 'tiara'; it was very wide, covering the forehead from temple to temple. Subsequently it grew steadily taller and narrower until René-Yves Creston referred to it as a 'mitre'. Its decoration changed, too, because every year the embroideresses of Pont-l'Abbé launched a new fashion.[27] Yet the constituent elements remained the same. So did the functions of the coiffe, which were decorative (the beauty of the lacework) and social (for example, a particular colour indicated that the wearer was a widow) as well as practical. In practical terms it served chiefly to conceal the hair, being attached to the traditional support of the cap, *coef bleo*.

As with the version of the garment worn by countrywomen in general, the underlying purpose of the coiffe was to tame the woman's hair (a sexual symbol). Piled up in a chignon, it was carefully tucked away beneath the cap, to which the coiffe was then attached. Gradually, however, we find the hair starting to appear in wedding photographs. Drawn back over the temples in the early years of the twentieth century, by the 1930s it is waved and a lot of it is left showing. Nowadays, old women who take care over their appearance give themselves two round loops, one on either side of the cap, to soften the lines of the face. One wonders what condition women's hair was in after a lifetime spent tucked up under a cap! Many old women wear hairpieces behind, so imagine my surprise when one day, visiting the home of a very old fisherman over towards Penmarc'h, I found the man's wife had a splendid head of curly white hair spilling out abundantly from beneath her little black cap. Because the caps worn today are black; you no longer see the embroidered, ornamental side-pieces that used to constitute the principal form of decoration in the days when coiffes were low.

René-Yves Creston pointed out that the women of the Pont-l'Abbé region, like those of Quimper, gradually abandoned gilets and layered dresses, though cuffs were retained.[28] Up until the late nineteenth century embroidered gilets were covered with braided *manchou* of brightly coloured ribbon. In the photograph dating from 1880 (plate 5), the only person wearing one is the bride, and according to the (female) informant who supplied the photograph the same clothes had already served the bride's mother. So this was part of the very old-fashioned clothing that fell into disuse with the arrival of the twentieth century. The embroidered fronts of these gilets, the mark of Bigouden fashion, featured highly distinctive motifs (the peacock feather or *plum paon*, for example). In the early years of the twentieth century brides used gilets with old embroidery that they proceeded to enhance in response to what appears to have been an exceedingly fickle fashion. For several years gilets were dotted with balls of strass on the front and on the outsides of the sleeves (plate 9). Just after the First World War they were very dark; when they started to be trimmed with fur or to have floral motifs added to the background embroidery (plates 10, 11).

The shape of the gilet changed too, with the neck becoming wider (a tendency that has become increasingly accentuated down to our own day). More flexible materials began to come into use; cloth that was quilted and stiffened with close embroidery gave way to softer velvets that tended to reveal the figure rather than bundling it up in fabrics so unyielding as to make the bride look like a doll with rigid arms.

Skirts went through a similar development. As the neckline widened and the coiffe rose higher, the skirt grew shorter. However, the shape remained fundamentally the same, with the pointed slash and the gathering over the stomach helping to give it a rounded look. Here too fabrics became more flexible, and the ephemeral fashion of a period that might last no more than a year or two found expression in the ornamental apron, a very full affair wrapped round the entire skirt.

Towards the end of the nineteenth century the apron was braided and girdled with ribbons; before the First World War it was made of white or black moire and girdled with a long ribbon to which was attached a bouquet and a garland of orange blossom (plates 3, 9). The machine embroidery around the bottom of the apron was done in a uniform style. After the war this changed, becoming more flexible, more festooned; or it might be embellished with pearls.

The way in which costume developed makes it possible to trace the synthesis between tradition and modernity. While preserving its old structure in terms of elements and forms, costume altered its proportions, fabrics became softer, liberating the figure in the same way as was happening in urban dress. Women living in towns rejected the corset and the bustle, raising their hemlines and adopting materials that were pleasant to wear and did not inhibit movement. The women of Bigouden did likewise.

A distinction should be drawn, however, between coastal and inland communes. As far as costume is concerned, one tiny detail can betray the difference to the tutored eye – how much hair is left showing beneath the coiffe, perhaps, the style of decoration on the apron, or the way in which the *lacennou* of the coiffe are tied. The influence of the Quimper region particularly affected the wedding costume of North Bigouden, In the 1940s the North Bigouden bride often wore white (though the shape of her dress as well as her coiffe still made her a *Bigoudène*). Moreover, the garment was made of a silk that would have been perfectly suitable for a *kiz ker* dress. We ought also to differentiate styles of dress related to degrees of wealth, though that is more difficult in view of the fact that girls were in the habit of borrowing dresses from friends who were better-off than themselves. When Alour Le G. married in 1919, for instance, his wife Augustine D. borrowed a dress belonging to the mother of Marie-Thérèse N., a neighbour of Le Haffond.

Male costume shows the same sort of evolution, but it was more rapid and occurred earlier than that of female costume. In the earliest photographs (1880; e.g. plate 3) the men are wearing trimmed hats, embroidered waistcoats, and jackets with double facings that give the impression of several waistcoats worn one on top of another. Wealthy bridegrooms stuck a *pardon* pin in their collars and proudly sported what was evidently the chain of a pocket watch. The waistcoat – embroidered or of black velvet, with a white collar facing or a celluloid collar – continued to be worn until well into the 1930s, giving the groom a very splendid air. After that date young men abandoned the waistcoat completely in favour of a stiff-collared shirt and bow tie.

Studying the way in which wedding costume evolved reveals the changes of ritual that echoed the social transformations of marriage. Those changes show evidence of the influence of customs adopted from the town such as could be traced by analysing the musical repertoire played at wedding dances between the wars and after 1945. However, unlike traditional music, which is experiencing a revival in very precise social and cultural circumstances, traditional costume has died out. Hardly a woman below the age of sixty ever wears it nowadays. Brides at today's weddings wear the same wide-brimmed *capeline* hats as town girls wear. In fact, they very often are town girls, returning to the village to celebrate their nuptials in rural surroundings.

6

The sea in abeyance

'The study of matrimonial exchanges merges with the economic and social history of the families concerned, of which the genealogical diagram recreates only the skeleton', wrote Pierre Bourdieu in 1972.[1]

Matrimonial exchanges are unintelligible outside their social and economic context. This is so even in primitive societies, where the marriage system is accounted for in terms of kinship: a man marries such and such a relative because she is his relative, the relationship being laid down by the marriage rule. It applies even more in peasant societies that have no prescriptive marriage rules and where social organisation is closely linked to the economic sphere. In South Bigouden, if a man marries his first cousin or his father's first wife's niece it is not because she is a consanguine or an affine. Either she is in line to succeed to a farm that is about to become free or her dowry, added to the capital already accumulated, will suffice to set the couple up on their own. Here, too, matrimonial regularities are impossible to explain in isolation from the social relationships of production and outside the economic context within which they occur.

A rather unusual economy

To study the development of the Bigouden economy in the eighteenth, nineteenth, and twentieth centuries is to watch social relationships slowly succumbing to a new kind of economic logic – and to wonder what kind of resistance kinship systems put up to the changes involved. However, it is difficult to conduct such a study in the absence of the necessary analytical concepts. We must beware of judging the people of Bigouden as the travellers of the late eighteenth and early nineteenth centuries were so ready to do. Jacques Cambry, Arthur Young, Villermé and Benoiston de Châteauneuf, and Armand du Chatellier, writing over a period of seventy years, reiterated the same old diagnosis and the same old vision of the stick-in-the-mud peasant, incapable of any kind of innovation and more interested in dancing and drinking than in working. The theme of the peasant as blinkered, whether by idleness and 'Celtic day-dreaming' (the pejorative judgement) or by ignorance and poverty (the positive judgement), is one that recurs over and over again.

But surely what we are dealing with here is a different kind of economic logic, one wholly alien to observers who were imprisoned in pre-capitalist, capitalist, productivist, and modernist ideological strait-jackets? May we not gain a clearer view of the relative stagnation of the Breton economy if we examine it in the light of some of the analytical concepts that have been evolved for dealing with primitive economies? Primitive economies, as such anthropologists as Marshall Sahlins have shown, are characterised by abundance. In them a few persons suffice to feed the group and need devote only a limited amount of time to the work of agricultural production or gathering. Granted, Sahlins himself reserves the word 'primitive' for 'cultures without states, without a constituted body politic, and only where historical penetration by states has not modified the economic process and social relations'.[2] Bigouden no more escaped the state with its bureaucracy and its social hierarchy than did other peasant societies. However, that state and those non-peasant social classes were so remote as to exert only a feeble influence on this relatively self-contained world. Karl Polanyi has shown how, prior to the Industrial Revolution, the traditional economic system was absorbed into the social system and even tied in with the presence of markets. Where these were most developed 'they prospered under the direction of a centralised administration that encouraged self-sufficiency in peasant households'.[3] A market economy is not incompatible with the domestic production system. The characteristics of this mode of production were also found in the peasant population studied here. Can we not see this as a society pursuing non-productive goals, and would that not enable us to account for people's 'stick-in-the-mud' behaviour with regard to agricultural production and to reassess their 'idleness'? The traditional judgement is shown to be even more wide of the mark when we realise that the Bigouden peasantry had to adapt to a slowly but surely evolving economic situation. Though partly open to maritime activities, for two centuries the region turned its back on fishing; the sea remained 'in parenthesis', as it were. The economy had to withstand considerable demographic pressure and find ways of feeding ever-increasing numbers of mouths. Furthermore these developments were accompanied by the political upheavals of the revolution, repercussions of which were still being felt in this region in the 1830s.

Cultivated and uncultivated land

Fishing and maritime trading activity having declined through the eighteenth and nineteenth centuries and come to a virtual halt by about 1880, the only resource available to the people of South Bigouden was their land.

To the east of the sandy coastal strip lay the vast expanses of the *traon*, land suitable for growing barley and wheat. The *gorré* with its lighter soil was mainly covered with heath, though that did not mean it was wasteland (it provided grazing for livestock). Even in the eighteenth century cultivated land exceeded uncultivated land in area, which was

Fig. 24. Kinship and marriage ties between the people photographed at a wedding at Kerhervé in Loctudy, 1904 (Plates 1 and 2).

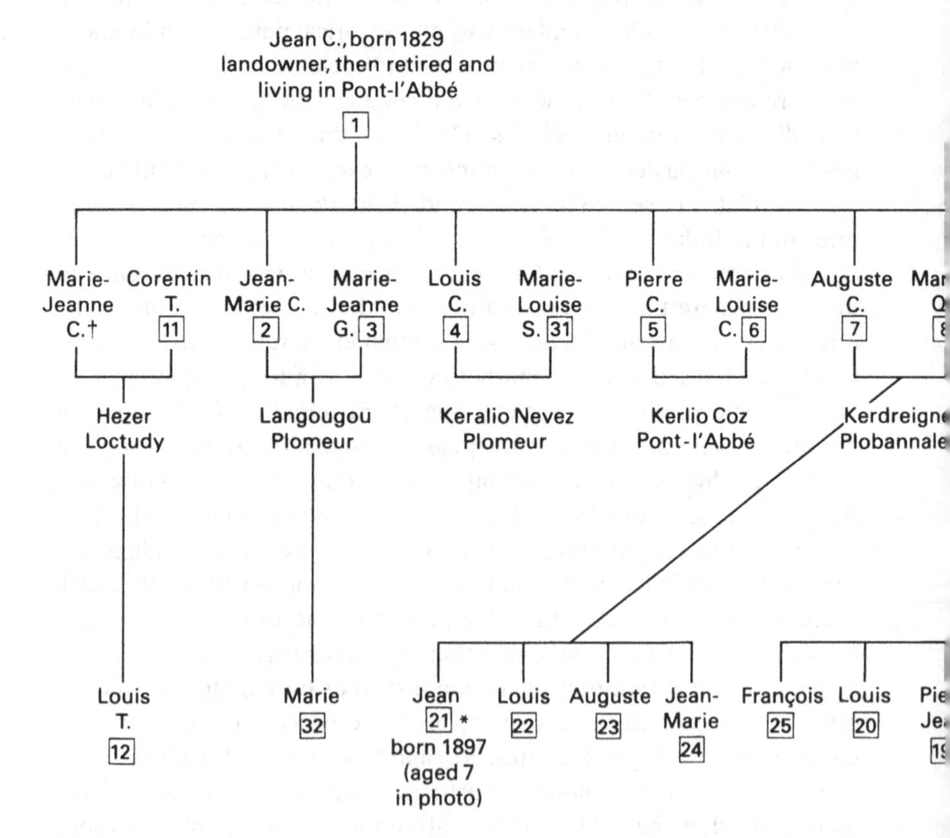

the reverse of the situation obtaining in Brittany generally and in the *département* of Finistère in particular.

M. de la Tour, the provincial administrator of Brittany, calculated the proportions of cultivated and uncultivated land in the subdelegation of Pont-l'Abbé as follows:

Area in *arpents* (roughly = acres)	7,047
Cultivated land	3,747
Uncultivated land	3,300[4]

At the beginning of the nineteenth century, 28 per cent of the total surface area of the five Breton *départements* consisted of uncultivated

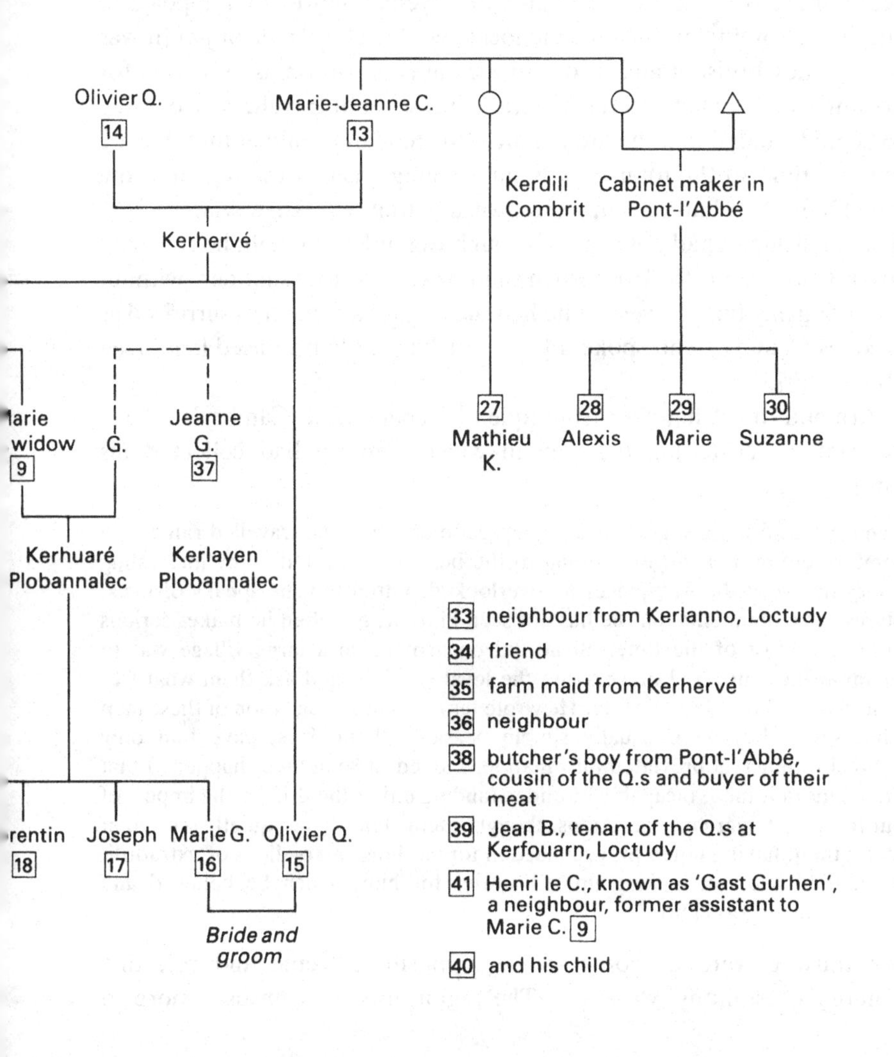

land. That was nearly twice the national average (14.5 per cent). Finistère had a particularly high proportion with 42.5 per cent. In Quimper *arrondissement* the distribution was as follows:

Arable land	42.5 per cent
Permanent grassland	8.3 per cent
Woodland	5.0 per cent
Heath	39.5 per cent
Other	4.7 per cent[5]

Travellers had different, even contradictory perceptions of the Breton countryside. In his 1906 study of 'The rural classes in Brittany from the sixteenth century to the Revolution', Henri Sée quoted from a letter written in 1768: 'I saw much uncultivated land in Brittany. All that grows there is broom and a plant with a yellow flower that appears to consist of nothing but thorns: the local people call it *lande* or *jan* [it was gorse]: they bruise it and feed it to the animals. Broom is used only for heating ovens.' That was the general impression of travellers, according to Sée. He added that the area of uncultivated land could at that time be put at a third of the total. But Arthur Young, visiting the region in the late 1780s, had had an impression of a 'surely excessive wilderness'.[6] The Englishman had not passed through Bigouden, though; he travelled from Châteaulin to Rosporden and then on to Quimper without venturing any farther west. If he had, he might have been as surprised as Jacques Cambry, who spoke a few years later of 'a promised land' (see above, p.6).

Armand du Chatellier subsequently challenged Cambry's idyllic description, criticising the way in which Cambry had collected his information:

> Cambry was often a superficial and imprecise observer; he travelled rather too comfortably in a carriage, keeping to the beaten track, and many interesting things in out-of-the-way places he overlooked. Often too, he speaks of facts, places, or monuments that he has not seen himself, and then he makes serious mistakes. Most of the time, all he did on arriving in a large village was to summon the municipal officers and the local notables and ask them what was remarkable about their locality. He wrote his notes at the dictation of these men who, since they were usually simple peasants themselves, gave him only imprecise, much abbreviated information. Indeed, it sometimes happened that these ignorant men, incapable of understanding either the drift or the import of our traveller's searching questions, thought them childish and quietly scoffed at them; then, having noted his predilection for all things marvellous or extraordinary, they had fun making up tall stories for him, which he believed and repeated in all good faith.[7]

We must beware of going to the opposite extreme and rejecting Cambry's testimony wholesale. The region *was* prosperous – more so than most parts of Brittany. However, contrary to what was related by this 'ethnologist before his time', it was cereals that accounted for the wealth of Bigouden, not its fruit and vegetables, which merely served to improve the local diet.

Bigouden was in fact a region that produced cereal surpluses and exported them. This was not incompatible with its retaining certain traits characteristic of the organisation of production in societies that pursue non-productivist goals – in other words, where the economic and social spheres merge. The Bigouden economy had certain features in common with the pre-capitalist economy of the eighteenth century. At the same time it remained geared to the satisfaction of the needs of the group while producing excesses that were ritually squandered. It effected a synthesis of the two types of assigning different purposes to its products, depending on their economic nature. The first distinction was at the level of cereal production. Two sorts of cereal were grown: the so-called 'noble' cereals of wheat and barley and the 'poor' cereals of rye and buckwheat. Jean Meyer noted: 'In a normal year the poor cereals are used to feed the populace and the noble cereals to pay taxes, farm rents, manorial rights, and tithes. There is a tendency towards a relative dissociation of the cereal market.'[8]

So the surpluses were a direct product of the peasant milieu. They were commercialised by a class of merchants who were often not Bigoudens and who during the eighteenth century amassed substantial wealth. The subdelegation of Pont-l'Abbé figures alongside those of Quimper, Redon, La Roche-Bernard, Vannes, Auray, and Hennebont as one of the regions of Brittany producing cereal surpluses that were exported or used to balance the local internal deficit.[9]

As well as making this distinction between types of cereal, the region was characterised by what agronomists considered a poor use of livestock. Animal production was wholly geared to subsistence farming. If a farmer sold his calves, it was in order to be able to go on using the cow's milk to make the pancakes and gruels that were his staple diet. Jean Meyer noted that in Brittany butter and milk were 'as much and more the poor man's food as the rich man's'.[10]

'Peasant ignorance' was blamed throughout the nineteenth century for the vicious circle that prevented any kind of cross-fertilisation between agriculture and animal husbandry:

> And notice how everything is interconnected here and how effects spring directly from their causes. Lack of manure leads to less produce; lack of produce means that only a few animals can be kept, for many could not be fed without plenty of forage. Few animals mean little manure, and little manure means little produce . . .[11]

The Breton peasant was quite simply impervious to such logic because he was held to his traditional behaviour by the structure of the mode of production as well as by the constraints of the method of land tenure imposed on him.

The problem of the kind of economic lift-off and increased production that were to become so necessary in the context of the nineteenth-century population expansion was bound up with the clearance of heathland. Superficial observers recommended that more heath be taken under cultivation, not understanding that it complemented agricultural activity and helped to balance the animal/crop cycle. The 'warm' land was regularly tilled; the heath was burn-beaten about every seven years and put down to pasture in between. It was sown with gorse and broom which were cut and crushed for use as fodder, the stems providing firewood. After the herds had grazed it, the scrub mixed with dung could be used to manure the fields.

So the Bigouden heath was not synonymous with 'wilderness'. It performed several essential functions in the economic cycle, notably as a source of fuel. Cambry, indeed, drew attention to the fact that 'the scarcity of trees, nearly all of which have been used up by the sabot makers ruined by the saltpetre works, means that the people are forced to burn peat, broom, cow-dung, gorse, and clumps of seaweed from the beach'.[12]

It was not only the economic advantages that peasants derived from it that prevented individuals from clearing heath. There was also the legal aspect; the nobility collected dues (*droit de champart*) on 'cold' land and had no wish to lose them.[13] In any case, the *domaine congéable* system did not favour heath cultivation because domanial tenants, at risk of eviction, had no incentive to work for their landlords. For his part, the landowner was unlikely to encourage tenants to make improvements to the land for which he was going to have to reimburse them when they left.

The 'warm' lands, on the other hand, were manured regularly with a mixture of dung and seaweed known as *fumier mellé*. Barley grew in Loctudy and Tréogat; in Lanvern it was mainly rye. Crops were rotated in the order wheat, barley, oats or rye, and buckwheat.[14] Contrary to what might have been supposed – and in spite of the efforts of local notables and particularly of the Breton clergy – potatoes were not yet widely grown at the end of the eighteenth century. Armand du Chatellier reckoned that it was only under the Second Empire (after 1852) that the potato was really absorbed into farming practice,[15] though that does seem a little late, judging from the evidence of post-mortem inventories. Besides encouraging the clearing of heath agronomists sought tirelessly to introduce the growing of new crops, invariably meeting with fierce resistance from peasants. We can only salute the latter's wisdom when we learn that attempts were even made to acclimatise rice![16] Sensible projects suffered from being mixed up with experiments that were doomed to failure, so that one begins to appreciate peasant suspicions.

Other agricultural products included the textile plants flax and hemp (spun at home and given to the local weaver to be made into cloth), fruit, such leguminous plants as peas and vetches, and the vegetables that so impressed Cambry and that were intended for home consumption. They were produced in the *liorzou* or cottage garden that lay alongside the farmhouse.

In the eighteenth century sheep were numerous – small, black-woolled animals that Jacques Cambry observed grazing the sandy coastal strip in Penmarc'h, Plomeur, and Beuzec.[17] Their meat was tender, and they supplied tallow for high-quality candles. However, the colour of their wool explains why skirts and jackets could only ever be of dark cloth; not until white sheep's wool came into use did it become possible to dye cloth in bright colours.

One other resource of Bigouden folk came from the sea – not in the form of fish, Penmarc'h having stopped fishing during the seventeenth century, but in the form of objects washed up from wrecks. On the principle that everything on the beach is common property, the local people got hold of by no means negligible quantities of cash, oil, wine, salt, and tallow from ships that foundered on the rocks fringing this extreme south-western tip of Brittany. Not until 1794 was the first small lighthouse erected at Penmarc'h,[18] and shipwrecks were plentiful. Twenty-six such disasters occurred between 1716 and 1738.[19] But of course they were not disasters for everyone. Not only the inhabitants of the coastal strip but people from inland, too, rushed to lay claim to what was washed ashore. Following the wreck of the *Prince Royal*, eighty-four tons, out of Copenhagen, Jean Helias of Castellou in Gorré-Beuzec (a hamlet in the enclave of Saint-Jean-Trolimon) was fined forty *livres* for carrying home a cask of wine.[20] In 1785 the parish priest of Beuzec, one Loëdon by name, acknowledged receipt of a reward of sixty-eight *livres* for the effects of persons wrecked in the *Marie-Hélène* of Amsterdam, 180 tons, on her way from Seville to Amsterdam.[21] Jacques Cambry himself benefited indirectly from this additional resource when the parish priest of Penmarc'h invited him to dinner and served him 'the most delicious Segura wine found on the beach and swapped by the locals for a few bottles of inferior cider'.[22]

Violent struggles often took place over ships' cargoes between shipwreck survivors (where there were any) and the local people as well as among the locals themselves – proof that these gifts from the sea played a not unimportant part on their lives.

The *domaine congéable*: a special kind of land tenure

No study of the resources of the Bigouden peasant in the eighteenth century can stop at an inventory of what he gained from arable farming, animal husbandry, and shipwrecks. This is because his mode of production was powerfully constrained by the way in which land was held. Without a cadastral survey (drawn up for the region in 1806 and 1835) it is difficult to give an overall view of land tenure. As in many other parts of Brittany, land was owned by the nobility, the church, and the urban middle class, and as we shall see in the next chapter the sale of *Biens nationaux*, the lands and property confiscated by the state during the Revolution, did not greatly alter the situation. In relatively few instances was land actually farmed by its owner, although in Plogastel-Saint-Germain, in 1771, owner–farmers did hold 32 per cent of all landed property.[23]

In his study of land ownership in eighteenth-century Plonivel, Vincent Le Floc'h showed that 85 per cent of the land in the parish belonged to the nobility, with the church and the bourgeoisie sharing the rest more or less equally between them.[24] The ecclesiastical landowners were the Carmelite church in Pont-l'Abbé and the parochial church council of Plonivel, which had received gifts of land in exchange for the saying of masses for departed souls. With some degree of variation from parish to parish, peasant ownership of land can be shown to have been generally on a limited scale.

However, this statement requires some qualification in that, like much of Lower Brittany, Bigouden had a peculiar system of land tenure. Known as *le domaine congéable*, this in fact made for two owners of any one farm: one who owned the land and another who owned the buildings.

The *fonds* (which in addition to the sub-soil comprised all trees above a certain size) would belong either to a lord, to the church, or to a bourgeois landowner. The 'edifices and surfaces' were the property of the domanial tenant under what were known as 'reparative rights'. These covered the farm buildings and the arable layer of the land together with the hedges, bushes, and any improvements that the domanial tenant managed to make to the soil. Landowner and domanial tenant were parties to a lease that, in the eighteenth century as well as in the nineteenth century, ran for a period of nine years, measured from Michaelmas to Michaelmas and was known as a *bail à convenant* or *baillée d'assurance*. This stipulated the amount of domanial rent that the tenant would pay. He had further to pay a commission, a one-off money payment for each lease that constituted a kind of right to the lease. His proprietorial status did not render the position of the domanial tenant secure, because on expiry of the lease the landowner was entitled, after refunding him to the value of his reparative rights, to give him notice to quit (the *congéable* part). On the other hand, if the domanial tenant quit his holding without the landlord's consent, he lost his entitlement to such a refund. When he moved to a new holding he had the further financial burden of evicting the former tenants by purchasing their reparative rights from them. That purchase even had to take place when a child inherited his father's lease, so to speak – in other words, when a domanial tenant had his son or son-in-law take his place. The value of the reparative rights would be assessed in order that the successor might buy them from his father or compensate his brothers and sisters. Finally, every lease included an *aveu*, which gave a full description of the premises about to be occupied. Valuable documents for the researcher, these 'avowals' lent a flavour of subjection to the relationship between landowner and domanial tenant. As Henri Sée noted, 'it was in their capacity as subjects of the landed nobles that they [domanial tenants] were required to make this declaration'.[25]

In addition to the conditions imposed by the landowner, the peasant was subject to the evolution of general market conditions – including the level of supply and demand as expressed by that of prices. The eighteenth century was marked by rising prices in the 'noble' cereals market; they virtually doubled, in fact, between 1735 and 1789.[26] Vincent Le Floc'h found the same trend on Plonivel, where the price of a *boisseau* (about 13 litres) of wheat stood at 7 *livres* in 1757, 9 *livres* 2 *sols* in 1765, and 11 *livres* 5 *sols* in 1782.[27] Trade was growing. The 2,000 tons of grain exported through Pont-l'Abbé in 1732–1733 had become more than 5,000 tons by the eve of the Revolution. Cereals were exported to Holland, Spain and Portugal.[28] The notables of Brittany campaigned through the 'Society of Agriculture' for free trade in cereals, and this was eventually granted in 1774.

At the end of the eighteenth century, concurrently with these rising cereal prices, there was an increase in ground rents.[29] In the seventeenth century these had been paid in cash. They were later converted to payment in kind (capons or hake, measures of wheat or barley). Fixing rent in kind made it possible to peg it systematically to the cereal price. There is nothing surprising, then, about the stability of domanial rents expressed in measures of wheat, barley, or oats. The amount of the rent was in proportion to the size of the farm, and since that did not change, the amount did not change. Its yield, however, being pegged to the cereal price, increased as the latter rose throughout the eighteenth century.

As an example, let us look at the rent for Kerbonnevez farm, held by Thomas Mourrain and Renée Lancoudrez in June 1755 from the lord of the manor of Châteaugiron, the marquis d'Epinay. The farm consisted of the manor and noble seat of Kerbonnevez together with lands in a single tract with a *méjou* or open field, warrens, and meadows (approximately forty-five acres). The rent was sixteen heaped measures of white wheat, sixteen heaped measures of barley, and 6 *livres* in cash, plus the usual corvées, which had now become money dues and were estimated at 9 *livres* 12 *sols*.[30] This former noble manor was already occupied by the tenant. Today all that is left of it is a splendid well and a stone with heraldic bearings that has been reinserted in a wall.

Or take the 'commoner holding' of Kermatheano Creis, of which it was possible to examine the leasehold and subsequently freehold deeds from 1638 up to the 1904 deed of sale.[31] This twenty-seven-acre farm was split in the eighteenth century between a number of owners. In 1779 Pierre Le Perennou and associates held Kermatheano from three owners, paying a total rent of sixteen *boisseaux* of wheat, eight *boisseaux* of barley, and 9 *livres* 12 *sols* worth of corvées. Fixed at that level in 1638, the rent remained unchanged until 1847, when after various transactions among the owners the amount was paid in cash. What we see increasing on this farm in the eighteenth century are the corvées. We also see the establishment of a 'gracious non-returnable' commission (100 *livres* at Kermatheano), paid each time the lease was renewed.

Rents remained in the hands of the same families. Where, as in the case of Kermatheano, owners and occupiers can be traced over a long period, we find the generations succeeding one another in direct or, if demographic circumstances did not permit the existence of a direct heir, in collateral line of descent. Vincent Le Floc'h made the same observation for Plonivel. As the years went by, however, landowners moved away from the farms whose reparative rights they held, delegating the renewal of leases to notaries, who were on the spot and consequently wielded considerable power. At the end of the eighteenth century some of the Kermatheano landowners lived at Guengat and the others at Penhars. The latter, rather than rely on a notary, appointed a proxy who lived in Pont-l'Abbé.[32] The distance – physical and social – between owners and tenants increased steadily throughout the nineteenth century.

In the eighteenth century there was intense speculative activity in connection with reparative rights. Those who held them did not always exercise them, sub-leasing them instead either to relatives or to other peasants. According to a deed of 1785, for example, the reparative rights to Kermatheano Creis had been in the hands of the same peasant family for perhaps as long as the land had belonged to a noble line: 'The avowers [those making the *aveu* of the domaine, i.e. Alain Le Perennou and others] solemnly declare that they hold the said [reparative] rights from their common ancestors but that, having suffered a fire, they do not have the relevant papers to support their ancient ownership, which has been handed down from father to son for more than a century.'[33] However, up until about 1780 the domanial tenant was non-resident. In 1775 Pierre Le Perennou was living at Kerguillec Glas in the parish of Tréméoc.[34] His daughter Anne Le Perennou, widow of Hervé Goascoz, took possession of the premises in 1785. For a long time, then, these reparative rights were sub-leased by the titular domanial tenant. A peasant who gained possession of several such *convenants* could add considerably to his wealth by sub-leasing them.

In Plonivel parish a domanial tenant could recover his outlay within the term of a lease, namely nine years. He paid the landowner ground rent and commission amounting to 57 *livres* 15 *sols* per annum. He sub-leased his reparative rights to another tenant, who paid him 436 *livres* 12 *sols* for the same period. This left the domanial tenant with 3,400 *livres* in nine years, which was virtually the value of his reparative rights.[35]

If the *rente convenancière* went up, peasants speculating on the backs of sub-tenants grew wealthy. The late eighteenth century was marked by intense speculation in reparative rights, which did not let up in the nineteenth century as long as peasants were able to invest their ready cash in rural property. This led in the second half of the eighteenth century to the emergence of a class of well-to-do peasant farmers who in fact constituted the majority of the ancestors of the genealogies studied. In contrast to the comfortable position enjoyed by this class of peasants, life for their sub-tenants was hard. They had not only to pay their rent but also to perform manorial services in the domanial tenant's stead and carry out repairs.[36]

Not all farms were held *en domaine congéable*; some were straight tenancies, some were freehold. But these two types of holding were relatively rare in comparison with the infinitely complex *domaine congéable* system that distinguished between the land and the 'edifices and surfaces'. On top of the legal complexity, which often involved three individuals or rather three family groups pursuing different strategies with regard to the same piece of land – the landowner, the domanial tenant (who held the reparative rights), and the sub-tenant – there was, secondly, the complexity that arose out of the nature of the land, the distinction between 'warm' land (under cultivation) and 'cold' land (to a greater or lesser extent grazed), and thirdly the distinction associated with the two types of agrarian landscape: *méjou* (open field) and *bocage* or *parcou* (enclosed field). The relative value of farms thus depended essentially on a combination of these three factors, the reparative rights being sometimes worth more than the land itself, sometimes about the same, and sometimes very much less. The quality of the land in open fields was superior to that found in enclosed fields. At any rate, more than the distinction betwen 'warm' and 'cold' land it appears to have been that between *méjou* and *parcou* that made the difference. The former (comparable to the *openfield* of eastern France) involved abutters, if not in an obligation to work in common, at least in the adoption of common working practices with crop rotation, fallow year, and collective grazing rights. Enclosed fields or *parcou* had a far higher value because of the length of the *édifices* (the hedges, embankments, and ditches) that bordered them. Any trees growing on the embankments were also the property of the domanial tenant (unlike big trees, which belonged to the landowner) and increased the value even further. Lastly the *liorzou*, the plots around the farmhouses where flowers and vegetables were grown, had a very high value due to the fact that they were proper gardens.[37]

There were ecological, psychological, and economic reasons for the large number of embankments. They shielded crops from the wind, improved drainage, facilitated herding, and marked out and enhanced the value of someone's property.

The legal framework of Bigouden farming did not change much in the nineteenth century, nor was there a great deal of technological change in the way the farms of this granulitic region were worked. On the other hand, the rural landscape did undergo some modification along the coast, where vast sandy tracts form a strip lining Audierne Bay. Beginning in the late eighteenth century and continuing into the 1830s, these *palues* became objects of bitter rivalry between parishes and between farmers with conflicting goals.

The *palues* in dispute

These windswept stretches of wild and magnificent country belonged to Baron du Pont, and 'for his hunting pleasure the baron had huge preserves in the *paluds* [the Breton spelling] of Plomeur and Tréguennec, notably that of Tronoan'.[38]

Abutters took advantage of these wastelands, grazing their animals on them and using them as a source of natural manure in the form of washed-up seaweed. Here is how the subdelegate of Pont-l'Abbé replied in 1768 to the inquiry into common land:

> We have in the department of this subdelegation lands bordering the coast from the parish of Tréoultré-Penmarc'h to that of Plovan. These lands are called Pallues, and the inhabitants of all the adjoining villages use them to extract peat, which they mix with dung to manure their fields . . . No use could be made of these tracts other than for the pasturing of horses, cattle, and sheep, enormous herds of which may be seen in these Pallues, representing the wealth and even the fortune of the farmers of this same Pallue. Apart from these advantages of the land, the parishioners of Plomeur, Tréoultré-Penmarc'h, Beuzec-Cap Caval, and Tréguennec derive further considerable benefits therefore, using their freedom to avail themselves of and frequent these Pallues to reach the beach, where they extract seaweed from the sea . . . And without the aid of this seaweed, which is a wonderful manure, the abutters of these tracts would be unable to sow their fields.[39]

The abutters were not unaware of the benefits to be derived from these tracts of land, and they applied themselves to defining what the different parishes owned or at least to establishing rights of use. An old baptismal register for Plomeur provides an interesting insight here. In listing those present at a baptism in February 1729, the priest revealed details of a legal dispute then in progress between the various parishes of the coastal strip regarding possession of and exclusive rights to this land. His account specifically mentions pasturing livestock, cutting peat, and drying seaweed.[40]

Not only were the *palues* jealously guarded by those who used them; they were also protected against reclaimers who sought to enclose and take possession of lands to which they had only a shaky title. In 1792 there was a major court case between the parishioners of Beuzec-Cap-Caval and a number of farmers who had begun enclosing several pieces of the coastal strip that the authorities had probably granted by subinfeudation for a moderate annual rent.[41] The beneficiaries of the grant were non-parishioners, and a struggle began between the habitual users, who wanted to go on pasturing their animals on the land without restraint, and the settlers, who considered themselves entitled to establish their individual rights of ownership. The complaints of the habitual users mentioned 'lumps of flesh sliced from their animals' backs with sickles, other animals with broken legs, and threatening shots'.

The laws of 28 August 1792 and 10 June 1793 allocated to local councils the common lands that had formerly belonged to the king, and the commune became the landlord of its domanial tenants in the coastal strip. These property rights did not go unchallenged, however, and up until 1870 the local council of Saint-Jean-Trolimon was continually involved in lawsuits with private individuals regarding the ownership of this or that plot of land.

The lawsuits are a measure of the importance of these stretches of land that, having formerly been used collectively, came to be let individually to settlers. Such struggles for living space also bear witness to the population pressure experienced from the 1770s onwards as well as to the constant opposition between the inhabitants of the coastal strip and the farmers who worked the more or less fertile lands of the *traon* or the *gorré*. The term *tud ar palue*, the 'people of the coastal strip', is one that local farmers often still use today to denote the large families existing on the edge of beggary, breeding-grounds of agricultural labourers who, living a long way from anywhere, certainly from the *bourgs* of Plomeur or Saint-Jean-Trolimon, often without their own transport and even their own tools, were reduced to hiring out their muscle. The opposition between the *tud ar palue* and everyone else continued until the 1950s, when depopulation brutally affected the area. Today the ruins of humble hamlets in the coastal strip have been converted into second homes set amid scenery of outstanding beauty, particularly when bathed in the rays of the summer sun as it sets behind the Pointe du Raz.

The territory is thus seen to have been subject to a series of agronomical oppositions: *traon/palues*, cultivated land/heath, open fields/enclosed fields.

The demographic pressure of the nineteeth and early twentieth centuries was tolerated in so far as the coastal strip was gradually occupied, heath taken under cultivation, and fields enclosed. However, a number of obstacles associated notably with the way in which land was held stood in the way of this tendency, and the 1789 *cahiers de doléances* or 'books of grievances'* provide an excellent picture of the various difficulties encountered by Bigouden peasants towards the end of the eighteenth century.

Peasant grievances at the time of the Revolution

The researcher is best advised to go straight to the *cahiers de doléances*. Many complaints were not retained in the *cahier général* of the senechalsy of Quimper when they ran counter to the interests of the middle-class authors of that digest. An example was the peasants' wish to have the use of some of the timber they had planted for fuel, for making farm implements, and for maintaining buildings.[42]

* The popular petitions that members of the States General took up to Paris whenever that body was summoned, as it was in 1789 for the first time since 1614 [Translator].

I

An example of a relinking marriage

1 A wedding at Kerhervé in Loctudy, 1904. The bride and groom are in the back row, flanking grandfather.

2 Another family group, taken the day after the wedding. The bride and groom are seated at bottom right.

3 and 4 A double wedding at Plobannalec in 1894, with two brothers marrying two sisters. The men are wearing hats trimmed with lace and have *pardon* pins attached to their collars. The brides are wearing crosses round their necks, ribbons at their waists, and embroidered gilets. The good pupils are wearing their *croix d'honneur*.

III

5 Wedding at Kersoc'h, Plomeur, in 1880. The bride is wearing her mother's wedding gilet.

6 Wedding at Saint-Jean-Trolimon in 1909. The party is grouped in front of one of the cafés of the *bourg*. The second woman from the left is wearing a mourning *coiffe*.

IV

7 and 8 Postcards, *c*. 1900. Complaisant nuptials or stage wedding?
The welcoming ceremony; the gavotte.

V

9 Wedding, *c*. 1912. The groom's waistcoat and the bride's gilet are decorated with strass or paste jewellery.

VI

10 Wedding in 1918. The groom is wearing a fob-chain, the bride a belt of orange blossom.

VII

11 Wedding at Plonéour in 1929. The elegant daughter of a well-to-do Plonéour farmer marries the Saint-Jean-Trolimon blacksmith.

12 and 13 Wedding at Plonéour in 1929. The girls grouped around the bride and the young men grouped around the groom.

IX

14 All that remains of the manor of Kerfilin in Saint-Jean-Trolimon – an arch . . .

X

15 . . . and a well carved with heraldic bearings and the date 1463 (?).

16 Stone trough used for bruising gorse in the courtyard of Rugaoudal. The inscription dates from the time of Sébastien Le Berre and Jeanne Le Donge, who lived there between 1846 and 1876.

XI

17 Old barn at Le Steud, seventeenth century.

18 Cowshed at Kerstrat farm with double-ribbed ogee doorway, seventeenth century.

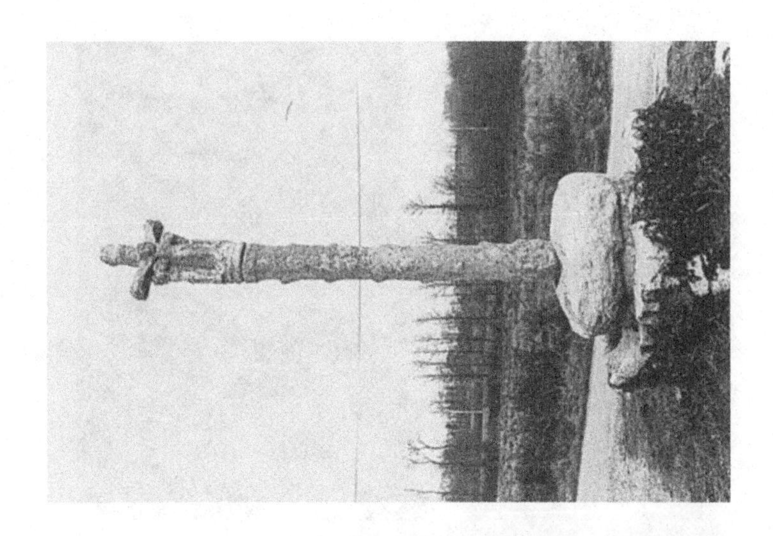

XII

Crosses in Saint-Jean-Trolimon:

19 The Saint-Evy cross in Kerbascol;

20 The *bourg* cross near the church;

21 The Kerbleust cross;

22 The 'abcess cross' or *Croas an Esquidi* in Kerfilin.

XIII

23 In the early years of the twentieth century farmers adopted the habit of posing in front of the photographer's camera in his Pont-l'Abbé studio.

XIV

24 First communicant with his ceremonial candle, *c.* 1918.

XV

25 The Tronoën *pardon*, held on the third Sunday in September. The procession from the chapel to the fountain. Postcard, *c*. 1900.

XVI

26 The parish church of Saint-Jean-Trolimon at the beginning of this century, showing the thatched houses of the *bourg*. Postcard.

27 The parish church today. A factory making lined slippers has replaced the thatched houses of the beginning of the century.

Apart from Penmarc'h, which because its population was emigrating requested some lightening of its tax burden, its taxes having always stayed the same 'although most of the inhabitants are today either day labourers or fishermen by trade',[43] demands were directed more towards a fairer distribution of taxation among the three orders and some reorganisation of the mode of production. It was the system of the *domaine congéable* that was challenged:

That noble landlords, in operating *domaines congéables*, shall no longer be able to evict their domanial tenants by giving them notice of dismissal or to exact lease commission and that all timber at present or in future standing on what have up to now been *domaines congéables* shall belong to the domanial tenants (article 11, Plomeur).[44]

Or again:

That the excessive rents and leases of *domaines congéables*, that the shackles they place upon their [i.e. the peasants'] labours, and that the hatreds and divisions they produce among them prompt them to ask that such leases be abolished and that they may no longer be evicted, not even by the noble landlords, pointing out, moreover, that it should be greatly to the advantage of the State that they should be allowed to plant on the domains for their own profit (article 6, Beuzec).[45]

Road construction and the state of tracks and paths are often mentioned as obstacles to the movement of cereals. Draught horses and carts abound in the post-mortem inventories of the seventeenth and eighteenth centuries,[46] and it was by cart that the farmers drove their wheat to Pont-l'Abbé, whose trade of course underwent an enormous expansion in the eighteenth century. But the state of the roads was appalling. Article 2 of the Tréguennec *cahier* requests:

that negotiable roads be made between the villages and towns, paid for by the three orders. These roads are particularly necessary in the parish for transporting cereals both to Pont-l'Abbé and to Quimper and Pont-Croix, all quite substantial trading towns. Our narrow little tracks are so uneven that for five or six months in the year it is quite impossible to drive carts along them. In view of the exceedingly onerous corvées that we have been performing since 1763 to repair the road from Quimper to Douarnenez, four leagues [sixteen kilometres] from here and a road that we never use, we hope and trust that this request will be granted.[47]

Statutory labour duties for the improvement of main roads were indeed tolerated with difficulty. Plomeur's article 6 called for their abolition and requested that they be 'replaced by joint taxation of the three orders'.[48]

Other complaints related to conditions of existence for the rural population, in particular to infant mortality: 'That doctors, surgeons, and midwives be established in rural areas, where they are becoming necessary in view of the increasing numbers of fatal confinements' (article 8, Tréguennec).[49] One complaint of the Bigouden parishioners concerned brandy, which they wished to have available at the same price for everyone. Plomeur's article 12 requested that 'all those who make up the Third Estate, farmers, artisans, and other tradesmen, shall be entitled to buy brandy [eau-de-vie] at the distribution points operated by the provincial tax farmers in pots or pints at the same price as that fixed for gentlemen, clergymen, and other privileged persons'.[50] The request was granted.

What emerges from the *cahiers de doléances* is the voice of progress demanding reform of the structure of land ownership, improved communications, and the beginnings of a right to health. At the same time these documents speak repeatedly of the deep-seated desires of a population that sought to cushion the rigours of its existence wth the consolations of strong drink.

Extensive agriculture, then, areas of land in dispute, and demands that provide a glimpse of a new economic logic were three characteristics of the Bigouden economy at the end of the eighteenth century. However, the period of affluence during which a class of well-to-do farmers was able to grow wealthy did not last long. The Revolutionary period and the early years of the nineteenth century were marked by profound economic and social crises that shook society to its foundations. The hesitant indigenous boom was brought to a halt, particularly since the population now began to take off and the inheritance system dispersed the leases that had been laboriously concentrated in the hands of individual lines during the eighteenth century. Furthermore, the poverty of the mass of the people and the impoverishment of the better-off were not – as happened in a number of regions of France – offset by the sale of *Biens nationaux*.

The difficult years of the early nineteenth century

If, unlike other parts of Brittany, Bigouden experienced no serious disturbances during the Revolution, neither did the peasants of the region benefit from the sale of *Biens nationaux*. The descendants of the well-to-do lines living at the time of the Revolution would have had the means to come forward as purchasers of the property that the new government confiscated from the nobility and the church but were able to do so only to a very limited extent. According to Alain Signor's 1969 study of 'The Revolution in Pont-l'Abbé',[51] the urban middle class helped itself to the lion's share. However, Signor does seem rather to have over-estimated the proportion of bourgeois purchases. Numerous 'purchasers' were merely lending their names to transactions on behalf of more modest tenant farmers. In fact several ancestors in our geneaologies were able to buy property at the time of the Revolution. The very wealthy R. family, for example, acquired three holdings from the marquis de Baude.

In Saint-Jean-Trolimon twenty-four estates were declared *Biens nationaux*. Eight farms belonging to the *émigré* Alleno de Saint-Alouarn were bought back by his widow for the sum of 217,800 *livres* (they included Kermatheano, Kersine, and Kerstrat, among others). Kerfilin, the *métairie* where Madame Audouyn de Pompéry went dancing,[52] was repurchased for 13,000 *livres* by Marie Penfeuntenyo of Kervereguin, whose family hailed from Loctudy.[53] These two noble families were still living in Plomeur at the time.[54]

A dozen estates belonging to the Du Pont barony or to the Carmelites and various church possessions in Beuzec-Cap-Caval, Saint-Jean-Trolimon, and Tronoën went to bourgeois purchasers – notaries or merchants of Pont-l'Abbé or Quimper. A piece of land in the hamlet of Kerbascol, for example, which had belonged to Beuzec parochial church council was purchased by Le Bastard de Kerguiffinec-Maubras. His daughter married a du Chatellier, a noble line that was to play an important part in the region in the nineteenth century (the owner of the *château* of Kernuz, Armand, was prefect of Finistère; his son Paul became an archaeologist). One of the Kervouec farms, formerly the property of the Saint-Jean-Trolimon parochial church council, was purchased by Kerillis-Calloc'h, a lawyer who already held the leases of Kergonan, Kerbascol, and other Saint-Jean-Trolimon farms, for which he received 1,818 *livres*.[55] Only two Saint-Jean farmers acquired property: Louis Durand, a domanial tenant who purchased the freehold at Pen An Ilis from the chapel of Tronoën for 1,219 *livres*, and Sébastien Le Garrec and Corentine Le Gloannec, who became the owners of various lots for 386 *livres*.

In Saint-Jean-Trolimon, as in a good many communes, not only did the urban middle class buy up confiscated land; former nobles, too, under cover of a recently acquired 'citizenship', recovered possession of their own property. The cadastral plan of the commune makes plain how in the nineteenth century property remained essentally in noble hands; only rarely were well-to-do peasants able to acquire their land, which prolonged the precariousnesss of their situation. Because of this an already fragile economy became even more fragile in the early part of the nineteenth century as the economic stuation became destabilised and disturbed the social fabric.

The population had increased towards the end of the eighteenth century, and apart from those peasants who had grown wealthy an ever larger section of the population was growing poorer. The Saint-Jean-Trolimon 'books of grievances' echo this state of affairs. They point out that the local authorities

are living amid a mass of poor and indigent whose condition is pitiable, and that despite all the assistance that their charity moves them to provide their situation still merits the attention of the government; that in consequence they petition the King and the States General to advise in their wisdom and beneficence as to the means of relieving them, asking for themselves and for their poor the special protection of the government.[56]

At the very beginning of the nineteenth century an endemic poverty was exacerbated by falling cereal prices. The evidence of government representatives and prefects with regard to the early nineteenth century was in the same vein:

There has always been a large number of beggars in a department [Finistère] that has no industry and virtually no manufacturing. When a poor farmer meets with some misfortune, when he finds himself supporting a large family, when old age or some other infirmity renders day labourers unable to keep up with the rigours of agricultural work, come a year of high prices, which happens every three or four years, such unfortunates have no alternative but to beg or to send their children out begging, who once they have embraced this idle, vagabond existence are not easily reclaimed for farm work; there is no more scandalous but nevertheless common sight in the countryside than men or women who are fit for work asking for alms. They were to be found in droves during the famine that afflicted the department in Year IX [1801], when in most communes between a third and a quarter of the population were short of food. The measures adopted by the Prefect brought the department out of this truly alarming crisis without upheaval: he forbade all beggars to leave their communes, and he encouraged the better-off to assist their less fortunate brethren. The local councils met and divided the communes into three classes of citizen; the first class comprising those individuals who were in a position to assist others, the second those who were capable of self-sufficiency, being unable to provide assistance for others but having no need of it themselves. They then determined the number of poor that each member of the first class would feed, depending on his resources. Everyone was brought under this arrangement, and it is impossible to praise too highly this act of beneficence and hospitality of which the inhabitants of the countryside have provided the most affecting example.[57]

The law of 16 ventôse, Year II, had made it obligatory for local councils to fill in a form supplying information about 'needy patriots of both sexes', giving their age, civil status, domicile, trade (if any), and number of children.[58] Saint-Jean-Trolimon, which numbered some 630 inhabitants at that time, recorded twenty-two needy persons. They included eight women, mostly elderly childless widows, which meant that they could not count on any family support, and a number of ten-year-old orphans who with no father or mother were still too young to provide for their own needs (though they had not many years to wait). The rest were either old men with no means of support or relatively young men, aged about forty, who had dependent children and were no longer managing to make a living by their day-labouring or by practising their trade of tailor, weaver, or cooper. It happened occasionally, of course, that requests for assistance came from people who might enjoy some family support. One such was Michel Le Donge, a fifteen-year-old cripple from birth living at Kerbascol; he should have been supported by his direct ascendants, who were well-to-do peasants. In general, however, these were people who did not figure in our geneaologies.

As indicated in the report quoted above, Year IX was particularly disastrous. The minute book of the Plonéour council notes that 'the majority of farmers who in other years have enjoyed easy circumstances find themselves compelled to get rid of stock and prone to other effects of the very direst necessity in order to spare their families the horrors of starvation'.[59]

Following the famine of 1801 there was a fresh crisis in 1806. It arose out of a cereal glut that caused prices to slump.

The number of needy has increased alarmingly; it is out of proportion to the well-to-do class. The countryside teems with beggars, and on certain days in the week these spread into the towns, which furnish a certain number themselves.

To these hosts of beggars must be added an even greater number of families who have no other means of livelihood than the fruit of the labour of their days and who in the course of an entire year can only just put by the modest sum that goes to pay for a poorly built hut roofed with thatch and barely capable of accommodating six people. That will give some idea of the situation in this department.

I am not talking about tenant farmers here. Most of them are glad to be able to pay their taxes and find the price of their leases. A large number of tenant farmers are currently several years in arrears with their landlords.[60]

There was a further crisis in 1818–1819, caused this time by a wheat shortage followed by a famine. In addition to the climatic factors that might contribute towards upsetting cereal markets, there was a structural cause having to do with religion. This, as Yves Le Gallo has observed, was the absolute nature of Sunday rest, which could be disastrous when farmers were late getting in the harvest. Le Gallo develops this point on the basis of the example of Tréguennec, quoting a letter from Chaulieu, the prefect of Finistère, to the bishop, Monsignor Dombidau de Crouseilles:

The mayor of Tréguennec tells me that the white wheat harvest is only just beginning in his commune and that last year it was finished by the end of

August; that because of the unsettled weather people fear they may not be able to save the whole harvest this year, and that in order to avoid this misfortune there is occasion, Monsignor, to ask you graciously to permit that, if need be, farmers may, having attended mass, perform on Sundays the duties required by the harvest. He fears a repeat this year of the losses experienced in some recent years. Permission to work to save the harvest on Sundays after mass did not prevent part of it from being lost, because that permission was requested and obtained too late. The continual variations in the weather appear to me to justify the mayor of Tréguennec's fears, and I feel I must entreat you, Monsignor, to be so kind as to grant people throughout your diocese permission to work on Sundays, after attending mass, but only in case of necessity, at gathering in the harvest.

Le Gallo found confirmation of the strictness of the Sunday ban at a time when it had generally become more relaxed in a letter that this very mayor of Tréguennec, Le Bastard de Kerguiffinec-Maubras, wrote to Monsignor de Poulpiquet on 1 August 1838: 'This is indeed gloomy and most distressing weather for the beginning of harvest. It occurs to me that if in bad years we had had, for distribution to the poor, the grain that was lost in this commune and that we could have saved on Sundays, poverty would not have been so great.'[61]

The early part of the nineteenth century appears as a period of stagnation, not to say recession, in the economic and cultural spheres. It corresponded, if you recall, to the period of very pronounced family endogamy, notably as a result of marriages between spouses who were already related by an older matrimonial link that their own renewed and tightened. It was also the period of extremely young marriages. Contracted with a view to securing exemption from military service, these contributed towards an increase in the birth rate.

Matching this economic and social recession, there was also a cultural decline. Yves Le Gallo again: 'Revolution and Empire made the department a wilder place. The political and military authorities promoted an atmosphere of police mistrust with regard to populations suspected of treating with the counter-revolution.' The first targets were the priests and with them the education that they dispensed: 'The Lower Breton clergy, whose magisterium was exercised with exceptional authority in the field of teaching, suffered dramatic tribulations that led to closures of seminaries and presbytery schools and brought about the ruin of secondary and primary schools.' Le Gallo adds: 'The old want of intellectual harmony between Lower Brittany and the kingdom was accentuated by the Revolution and the Empire, which even gave rise to a sharp decline in this respect.' At Cleden-Cap-Sizun, for example, the parish schools of the eighteenth century, where boys had received their education, disappeared between 1798 and 1830.[62] This cultural ebb is strikingly apparent from the marriage certificates of the lines investigated. These show that the number of spouses capable of signing their names was greater in the eighteenth century than in the nineteenth. The grandfather could write; the grandson was illiterate.

The early decades of the nineteenth century were marked by a continued low level of agricultural productivity. Increased productivity calls for increased division of labour and specialisation of functions, and those conditions simply did not exist. Far from it, the production process followed traditional lines either within the domestic group or in the context of a collective organisation of labour. Moreover, Chayanov's law applied perfectly: an increase in production does not follow from an increase in population; on the contrary, within each production unit production drops as the number of members of the household goes up.[63]

The principal innovation of the early nineteenth century was the potato, cultivation of which spread throughout Bigouden. Towards the end of the previous century great efforts had been made by Monsignor de La Marche, who earned himself the Breton nickname of *Eskop ar patatez* ('the potato bishop'),[64] to popularise this plant which was capable of meeting the steadily growing food requirements of an expanding population. Gradually those efforts were rewarded. At first it was grown in the *liorzou*, the vegetable garden around the house. Then during the 1820s it spread rapidly until it was common throughout Finistère.[65] The heavy soils of Penmarc'h, Plobannalec, and Loctudy greatly favoured 'Parmentier's manna';* the lighter soils of the northern part of the region were relatively less well suited to growing the new crop. Nevertheless, it was planted everywhere, notably on the *palues* of the sandy coastal strip.

For a long time country people distrusted the potato, which fitted ill into a diet based hitherto on gruels and pancakes combining cereals (generally poor ones), butter, and milk. In the initial phase of its expanding cultivation the potato was sold to the local potato-starch works that grew up rapidly around 1830. There was one at Pont-Guern in Pont-l'Abbé, owned by the Le Guay brothers, another in Loctudy, started by the owner of the château of Kerazan, Le Normant de Varannes, and another established by Le Bastard de Kerguiffinec-Maubras alongside the manor house[66] (it was known locally as *ar fecal*). Like most innovations, the commercial exploitation of the potato was undertaken by notables, the peasants confining themselves to production. Pont-l'Abbé enjoyed a period of intense activity during the middle

* So-called, like the soup, after the man who did so much to popularise the potato in France, Antoine Parmentier (1737–1813) [Translator].

part of the nineteenth century, notably shipping potatoes to England. It exported 400 tons in 1829, 1,200 tons in 1859, and 4,000 tons in 1870.[67]

Gradually the potato was incorporated into the local diet. Once accepted, it became a basic foodstuff for both man and beast. Armand du Chatellier wrote in 1849, 'This tuber is used among the more well-to-do in soup, mixed with bread; it is hardly eaten by the poorest people except cooked in water: they are lucky if they can get hold of salt to season it. Some I have seen who were obliged to cook it in seawater.'[68] The plant required hoeing and as such was more labour-intensive than cereal-growing. That is why there is a kind of functional symbiosis between potato-growing, the consumption of potatoes, and population growth. The potato takes more labour to cultivate but feeds more people than poor cereals. Potato-growing was added to cereal cultivation; it did not replace it. It was connected, in fact, with the land-grubbing that went on during the nineteenth century.

Until the 1850s, however, the Bigouden economy remained fragile. The year 1846 saw the last serious agricultural crisis in the region's history (subsequent crises affected the fishermen around the turn of the century and the labouring population of Pont-l'Abbé in the 1930s).

'In consequence of dry springtimes and very wet summers, the harvest of the three main bread cereals, which in an average year had stood [for Brittany] at 120 million hectolitres [1 hl = 2¾ UK bushels], declined to 113 million in 1845 and 91 million in 1846.'[69] In 1844 mildew ravaged the potato crop; in 1845 there was a very bad winter and the harvest was negligible. Lower Brittany experienced the kind of food shortage it had known in previous centuries, which accounted for the uprising in Pont-l'Abbé on 22 January 1847, when potatoes were being loaded on a ship bound for Plymouth. As is often the case in popular uprisings provoked by famine the female insurgents were the fiercest. The mayor of Pont-l'Abbé was 'jostled, battered, scratched, and bitten by women'.[70]

The nineteenth century was marked by large numbers of beggars in every commune, day labourers in a precarious situation, and tenant farmers afflicted by a series of misfortunes. Poverty and destitution were widespread.

The crises of the early decades of the nineteenth century were followed by a period of rather greater affluence. This lasted until about 1880, when the region once again turned towards the sea and its economic equilibrium shifted onto fresh foundations.

The period 1850–1880: a generally favourable economic climate

Eventually agricultural prices improved. After one last slump around 1852, the price of wheat began an upward trend that continued into the

1870s. Cereal prices as a whole rose by 35 per cent, while meat prices increased at an even faster rate: 81 per cent for beef, 66 per cent for veal, 56 per cent for pork.[71] The more southerly communes continued to export their cereal surpluses. Plobannalec, after satisfying its own requirements, sold 2,000 hl of wheat, 1,200 hl of barley, about 1,300 hl of oats, and 400 hl of buckwheat; Loctudy sold 6,000 hl, and Treffiagat sold three-quarters of its harvest.[72] However, since land prices doubled in thirty years,[73] the boom benefited landowners very much more than it benefited tenants. It also restricted any possibility of the latter acceding to the status of the former. Moreover, leases also went up, which ate into the margin of surplus accumulation. So despite a generally favourable economic climate, because of the legal conditions of the tenancy system and because of the expanding population the situation of those who actually worked the land did not improve.

The relative prosperity of the region stemmed from the development of potato-growing, which took its place alongside cereals in the traditional production cycle. The change was made possible by peasant land-grubbing on a large scale. Saint-Jean-Trolimon constitutes an excellent example of this.

There was a substantial increase in land clearances throughout Brittany, where between 1840 and 1880 the area of heath was reduced by half, from 1 million to 500,000 hectares [1 ha = 2.471 acres].[74] Granted, the coastal regions of Brittany were already more cultivated than inland areas, and it is difficult to assess precisely how much land was newly taken under cultivation. Apart from the heath belonging to each holding, the extension of cultivation primarily affected the sandy coastal strip.

According to the cadastral plan drawn up in 1833, the land in Saint-Jean-Trolimon was utilised as follows:

Arable	507.53.60 arpents*
Meadows and reedbeds	63.27.79 arpents
Open pasture and marshland	476.33.08 arpents
Heath	229.65.09 arpents
Gardens and vegetable plots	22.58.94 arpents
Coppice	12.50.40 arpents
Orchards	16.65.70 arpents
Woodland	2.60.70 arpents
Houses and outbuildings (floor area)	7.61.57 arpents

Arable land, pasture, and heath occupied respectively 34 per cent, 32 per cent, and 15 per cent of the total area. The proportion of land under cultivation in the commune was lower than the average for the

* 1 *arpent* = approximately 1 acre.

arrondissement (Quimper) or even the *département* (Finistère). The reason was that about a sixth of the surface area of Saint-Jean-Trolimon is accounted for by the sandy coastal strip, which was about to be progressively exploited.

The laws of 28 August 1792 and 10 June 1793 awarded the common land that had formerly belonged to the Crown to local councils. The *palues* of Saint-Jean-Trolimon were auctioned in the 1830s to farmers who then held these lands under the *domaine congéable* system (the difference here being that the landlord was the commune).

The letting of the sandy coastal strip transformed the way in which the land was used. Sheep-rearing declined and was gradually abandoned. Fields were enclosed and put under cultivation. At first they were used to produce a combination of poor cereals and potatoes, but we shall see how, towards the end of the nineteenth century, as the canning industry developed, peas became the main crop.

Land-grubbing was doubly useful. It benefited the commune, for which it was a source of wealth, and it benefited the poorer sections of the population who had been hit by the economic difficulties of the early nineteenth century exacerbated by demographic pressure. The fact that these infertile lands came to be occupied is an indication of the level of poverty that obtained, and the peasants who worked the *palues* were regarded as poor wretches indeed. Remember the major lawsuit that Beuzec instigated against 'outsider' settlers in 1792 and that revealed the need for fresh living space for what was already a growing population at the end of the eighteenth century.

The minutes of the local council recorded in 1847: 'For more than ten years now the commune has been leasing to homeless poor people portions of these lands [the *palues*] that their industry has turned into arable fields.' The minutes also mention 'various individuals who have built huts and cleared a bit of land'.[75]

The poverty of the *tud ar palud* (as they were called by those who were not of their number) was further underlined by outside observers. The archaeologist Paul du Chatellier, excavating at Kerveltré in 1878, found 'huts, most of them built of turves, the dwellings of poverty-stricken people'.[76] One of the occupants' main problems was having no wood. They used anything they could lay hands on for fuel, including potato and cabbage roots, but they were desperately short of kindling, which grew on the embankments and as such belonged to the domanial tenants.

Cultivation of the coastal strip was relatively intense. An observer towards the end of the nineteenth century, referring to the 1835 cadastral plan of the commune 'marking off a flat area of more than a kilometre square bounded by the *étang* of Saint-Vio to the north, by the hamlet of Le Stang at the north-east corner, by Tronoën chapel at the southern corner, by the sea to the west' (and, we might add, open to the south, where it is continuous with the *palue* of Plomeur), said: 'Now this huge area is cultivated and divided up, with a number of cottages dotted about.'[77] However, occupation of this land fluctuated. People abandoned it in time of crisis as being too infertile. And its exploitation was further held up by a series of legal disputes over ownership.

The farms lying farther to the east, being situated on fertile land, ought to have benefited from the generally favourable economic climate. They did not, however, and a study of a few cases will reveal the delicate balance of their individual economies. To grasp this it is not enough simply to compare cereal prices and the costs of leases. The latter varied substantially, even within the same small area, depending on the quality of the land. For example, eight ha of warm land and eight ha of cold land were necessary to generate an income of 300 francs in the commune of Pont-l'Abbé, while at Plobannalec, a few kilometres to the south, five ha of warm land and one ha of meadowland brought in the same figure.[78]

What we need to do is to relate both the price of land and the income derived from it to a single farm. That sort of comparison is not easy to achieve. The cadastral plan shows the surface area but is vague about the type of holding – whether a simple lease or a *domaine congéable* arrangement. Furthermore, the production and yield of that farm are not known. One possibility is to use post-mortem registry documents, which worked out incomes using the rule: capital times twenty. This information is inadequate, however, in that it does not show the capacity of the farm. Comparing cadastral plans and registry documents is an endless headache because of the mobility of domestic groups and possessions. One never knows precisely which holding one is dealing with.

Finally, post-mortem inventories sometimes give the amount of current rent still due. But to what area does it relate? Documents bearing on the population as a whole, though plentiful, are of no use here. There is no alternative to studying the cases that allow surface areas to be compared with prices over a long period. These files of documents bring part of the economic life of a farm alive in terms of its relations with the landowner.

For two farms it was possible to trace the movement of rents and *domaine congéable* leases through the nineteenth century. Studying these fascinating documents is a complicated business. At Kerbascol no fewer than four different leases were required to make up a farm of some fifteen hectares held by a single domestic group. Whereas the farmers succeeded one another down the years from father to son or son-in-law, ownership of the four holdings changed hands and eventually became concentrated.[79] At Kermatheano ownership of the reparative rights (as distinct from that of the land) became concentrated in the same hands in the middle of the nineteenth century; at the beginning of the observation period the farm was leased as a *domaine congéable*, at the end as a simple tenancy.[80]

In 1836 René Lagadic, fifty-eight, and his wife Elizabeth Trebern, sixty-four, were working one of the thirteen farms that made up the sizeable hamlet of Kerbascol. They were assisted in this task by René junior, thirty-three, and his wife Marie-Catherine Andro, twenty-six, who had three children aged six, four and one, and by another couple, Jean Riou, thirty-seven, and his wife Marie-Louise Lagadic, thirty-two, together with their five children aged thirteen, eleven, seven, four, and two years. This complex household also included another relative, Jean-Louis Lagadic; aged fifty-two, he was sometimes given as René's brother, sometimes as a servant, and he was probably either physically or mentally handicapped.

Fig. 25. Composition of the domestic group working Kerbascol farm in 1836.

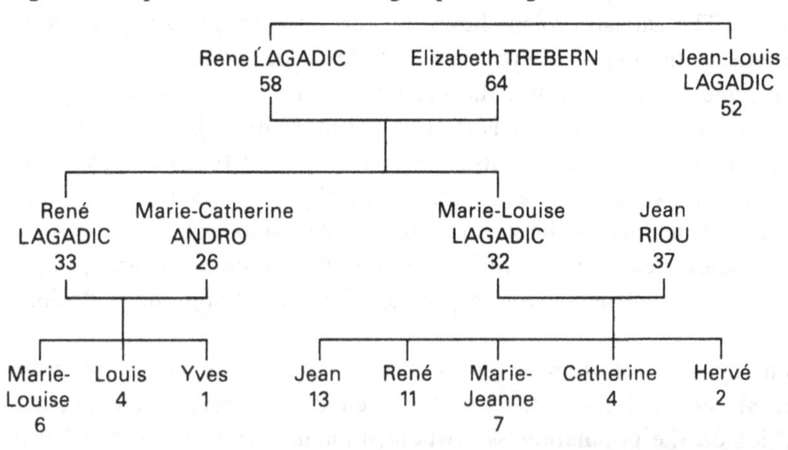

These six adults and eight children worked and lived off three distinct holdings of which they possessed the reparative rights, as well as an additional plot of land held on a simple lease. The total area was seventeen hectares forty-three ares twenty centiares, made up of the three holdings – Jaffelou, 10 ha, Lagadic, sixty-six a twenty ca, and Jolivet, five ha seventy-eight a eighty ca, all held under *domaine congéable* agreements – and the additional plot of one ha seventy-seven a.

In 1836, then, our household had dealings with four separate landlords whose rents ran for different periods. The price of farm leases rose steadily throughout the nineteenth century. In 1832 Jaffelou was leased for ten *boisseaux* of wheat and ten of barley plus twelve francs commission – as in 1678, when the lord of the manor of Kerambiguette leased to one Etienne Jaffelou (did he perhaps give his name to the farm?). In 1834 the new owner raised the rent to twelve *boisseaux*. From 1856 the three holdings of Jaffelou, Lagadic, and Jolivet went for a combined lease price of 700 francs rent per annum plus a non-returnable commission of 1,050 francs, which in 1873 was raised to 1,500 francs. In 1876 the rent went up to 900 francs and the commission to 1,700 francs.

Land prices were also rising all the time, which worked to the disadvantage of the peasant. In 1832 Jaffelou was sold for 5,000 francs. In 1841 Lagadic and Jolivet were sold for 8,350 francs. The three farms together were subsequently sold by auction to an urban middle-class family from Audierne. They acquired the fifteen ha sixty-six a twenty ca for 27,100 francs, having outbid the domanial tenant, who according to oral information received had offered 23,000 francs. The additional plot of one ha seventy-seven a went for 400 francs in 1840. At that date the rent increased to one *boisseau* of wheat and one of barley; in 1873 and 1882 it was fixed at 100 francs per annum. Kerbascol was characterised by great stability of succession, both on the part of the landowners and on the part of the domanial tenants who occupied it down the generations. But a steady increase in the cost of the lease made it impossible for the person working it to put together enough capital when the land eventually came up for sale.

A similar development can be traced at Kermatheano Creis, a farm of eleven ha sixty-six ca situated on the Plomeur road not far from the *bourg* of Saint-Jean-Trolimon. It was held as a simple tenancy from 1847, before that as a *domaine congéable*. Documents exist enabling us to trace the development of the rent, sale prices, and the nature of the tenancy from the beginning of the seventeenth century.

Up until 1847 the land belonged to several owners, one of whom held his inheritance from an ancestor mentioned in a lease of 1636. In 1822, for example, the farmer paid his two owners a total rent of sixteen *boisseaux* of wheat, eight of barley, and nine francs eighty in cash, on top of which there were 'gracious and non-returnable commissions' of 250 francs. The rent as thus expressed in cereals remained stable until 1844.

Pierre Le Berre, a domanial tenant descended by blood and marriage from the 'honourable man' Pierre Perennou,* who had acquired the reparative rights to Kermatheano Creis in 1742, got into serious financial difficulties. He had already contracted loans in 1836, 1839, and 1840, which he had managed to repay. But he was unable to meet debts accumulated from 1841, debts that were secured by mortgages on his real estate, which consisted of the reparative rights to various farms (in addition to Kermatheano Creis, which he had occupied only since 1841, he held the rights to Meout Creis, Keryoret, and Kerbleust in Saint-Jean-Trolimon and Kerjezequel in Plomeur). Those debts amounted to 4,000 francs. On 25 March 1847, therefore, Pierre Le Berre was obliged to sell his reparative rights to the owner of the land, Madame Léocadie de Roquefeuille, who lived in Quimper, for 5,400 francs. Two years earlier she had acquired three-eighths of the land for 4,000 francs. Applying the rule of three, we can estimate the total value of the Kermatheano land to have been around 11,000 francs.

To recapitulate, a farm of some ten hectares of good quality land – five ha sixty-nine a of warm land, four ha forty a of cold land, and fifty-four a of meadowland and pasture, the rest of the surface area being taken up by buildings, farmyard, and dunghill – had a value of 17,000 francs, land and rights combined, which put the average price per hectare around 1,500 francs. The reparative rights represented only a third of the total value of the farm, whereas often in the late eighteenth century their value was the same as or even greater than that of the land. This drop in the value of reparative rights was a factor in the relative impoverishment of peasants.

Those ten hectares were rented out to the former domanial tenants and their heirs at a price that went on rising throughout the second half of the nineteenth century – from 800 francs in 1847 to 900 francs in 1856, 950 francs in 1865, 1,000 francs from 1873 to 1891, and 1,050 francs in 1900.

In 1895 the farm was sold for a total of 24,000 francs. It was sold again in 1904 for 28,200 francs to the parents of the people farming it at present, who came from Plobannalec (see table 21). The figure was comparable to the value of the Kerbascol lands (see p.173).

It is possible, on the basis of figures in post-mortem inventories for the amount of rent still outstanding on the death of the head of the household or his wife, to put forward an estimated average level of farm rents for the second half of the nineteenth century. Depending on the quality of the land, which varied considerably from farm to farm, an

* *Honorable homme* was a kind of title given to a peasant who had bought formerly 'noble' land.

annual rent of 1,000 francs would have secured about twelve hectares.[81] If the farmer owned the reparative rights, his lease under the *domaine congéable* arrangements would have amounted to only 700 francs per annum, to which must be added commission of around 1,000 francs payable once every nine years.[82]

Rents went on being expressed in cereals for much of the nineteenth century and remained surprisingly stable, sometimes even in relation to the eighteenth century. As we have seen, this was as a result of rents being tied to the cereal price. On the other hand, the 'gracious commissions' went up. In return for this extra payment the lessor undertook to renew his tenant's lease. As far as the latter was concerned, it was the price he had to pay for a secure future. For instance, when a lease on Jaffelou farm was renewed for nine years from Michaelmas 1838 to Michaelmas 1847, the deed was actually signed in 1834.

This general trend of rising rents and rising land prices runs right through the documents relating to inheritance. Those documents also confirm another important phenomenon, which is the relatively low value of reparative rights in relation to land values. For the farm of Kerlaouedec Bras, for example, as for Kermatheano Creis, the ratio was one to nearly four and a half (land 13,300 francs, reparative rights 3,000 francs) for around eighteen hectares.[83]

Remember that the position of the domanial tenant was very restricting. His freedom of action was limited by measures relating to buildings – 'lessees may not erect any new buildings where none exist at present nor make any alterations to the size or to the quality of materials of the old ones' – and to the use of timber – 'all timber regarded as pertaining to the land belongs to the landlord . . . without lessees being permitted ever to cut or pollard any tree on pain of [liability for] full damages';[84] it was further specified that oaks suitable for planking as cask wood together with walnut and chestnut trees were the property of the landowner.

On the other hand, the domanial tenant sometimes enjoyed a particular advantage because of his geographical situation. The lease for Jolivet, for example, which lay on the edge of the sandy coastal strip, granted the lessee the right 'to take from the *palue* of Saint-Vio, on the east side of the old road leading from this property to Tréguennec parish church, turves for the sole purpose of manuring the fields of the said farm'.[85]

A further heavy obligation on the part of the domanial tenant was that of making the *déclaration à domaine* at his own expense. In other words, he had to pay a notary to come and describe the movables and

Table 21. *Owners and occupiers of Kermatheano Creis*

Owned by	Date	*Lease*	Occupiers
Messire François Geslin	1694		
Dame Suzanne-Louise de Glemarec and dowager ladies de Kerulut			LE PERENNOU
Le Perennou buys ⅛ of woods belonging to Dame Françoise Avril, widow of Mathurin-René Le Pape of Kermorvan	1742		Pierre LE PERENNOU
Three owners Mme de Lagadeck de Roquefeuil Demoiselles de Kermorvan-Le Pape The mayor	1779	1775 to 1784	Alain LE PERENNOU = Anne LE GUILLOU
		1785 to 1813	Anne LE PERENNOU, widow of Vincent CARROT then of Hervé LE GOASCOZ
Two owners ⅝ belonging to Jérôme de Roquefeuil and Charlotte du Rocheret, ⅜ belonging to Gabriel Malherbe and Dame Marie Pitot	1822	1813 to 1840	Clémence LE GOASCOZ = Pierre LE BERRE
Pierre Le Berre has ⅛ of the rights to the woodland (subtracted from the ⅝)			Pierre LE BERRE = Corentine LE BLEIS
Pierre Le Berre sells the ⅛ of his rights to the woodland to Mme de Roquefeuil, widow	1839	1840 to 1849	
Mme de Roquefeuil, widow, purchases the ⅝ from Mlle Pitot for FR 4,000	1845		
Pierre Le Berre sells his reparative rights to Mme de Roquefeuil, widow, for FR 5,400	1847	1849 to 1882	Corentine = Mathieu DURAND; Jean-Marie LE BERRE = Marie COSSEC; Marie-Perine

Mme Le Borgne de La Tour, née Léocadie de Roquefeuil, sells the whole farm to Henri Ponthier de Chamaillard for FR 24,000	1895	1891 to 1926	Jean-Marie, Corentine, Marie-Perine, Marie-Jeanne, Marguerite, Pierre LE BERRE = Anne-Marie TANNEAU, Henri
The latter sells it to Auguste C. and Marie Q. for FR 28,200	1904		Pierre-Jean, Anne, Marie-Jeanne, Noël, Henri, Marie-Louise
		1926 to 1977	Jean C. = Catherine C. son of Auguste C.

real estate of which he had taken possession. Finally, the instability of the *domaine congéable* was essentially bound up with its legal character. The fact that 'edifices and reparative rights shall always be considered personal estate with regard to landowners' meant that the latter could sell them in the event of the rent not being paid. Owners then had the right to evict domanial tenants on expiry of their leases, reimbursing them in the amount of the professional valuation of 'edifices, surfaces, straw, and manure'. The landlord was in fact all-powerful as far as the domanial tenant was concerned. It would be wrong to assume, however, that this one-sided relationship dictated the quality of dealings between the two parties. Some lines of landowners were concerned to keep a line of domanial tenants, too; others were always on the lookout for an occupier who would make a better job of developing their capital.

The economic balance of the farm was bounded on one side by the level of rent that the respective owners imposed on their tenant and on the other by the produce prices fixed on the regional market over which peasants had little control. Between the two the peasant had to release a surplus each year to pay his rent and to effect transactions connected with inheritance and succession, giving dowries, paying *soultes* or compensatory balances to equalise shares, and buying up his siblings' shares. An analysis of the prices of produce sold by the farmer can be made on the basis of post-mortem inventories, which mention quantities harvested or crop reserves not yet harvested, together with prices. However, this was not a proper market price but an estimate that would be included, along with the movables and farm implements, in the total assets of the joint estate of the marriage that had just been dissolved by the death of one of the spouses.

It followed that prices could be falsified upwards (to ensure a maximum share for the children) or downwards (with the object of paying less tax). Each family might pursue a different strategy, and in any case this is something we can never know today. Moreover, harvests varied in quality, and in fact the occasional inventory will specify 'good' or 'poor' corn and distinguish between one pile of potatoes and another. Finally, the price might also vary according to the time of year when the inventory was drawn up. For all these reasons we must make do with noting trends relative to price movement between the beginning and the end of the period under observation (1836–1900), hypothesising that in spite of the reservations outlined above they nevertheless do reflect market prices. Of all the inventories examined, about thirty contained information on the subject at issue:

	Average price		
	barley	*wheat*	*potatoes*
1830–1850	FR 8	FR 16	FR 2
1850–1860	FR 12	FR 22	FR 4
1860–1890	FR 15	FR 20	FR 5

Average prices of barley, wheat, and potatoes often fluctuated widely from one year to another, particularly in the case of potatoes. One thing that emerges from these estimates is that cereal prices did not increase as much as has been estimated for France as a whole. In a word, the generally favourable economic climate was very much more so for the landowner than for the farmer who worked the land, since the prices that the latter obtained for his produce rose more slowly than his rent. Landowners were made even more powerful by an expanding population that placed them in the position of lessors in a market where demand was high and the supply of farms was limited. They did not get out of land until the very end of the nineteenth century and particularly the early years of the twentieth century, when other sectors of the economy looked like showing higher profits. In other ways, too, the nineteenth-century domanial tenant was in a very much less favourable position than his late eighteenth-century forbears. The value of reparative rights was crumbling, and it was difficult for him to acquire numbers of them and speculate. Squeezed by rising rents and land prices that were well beyond his capacity to save or to contract debts, the farmer could scarcely replenish what he had been able to inherit. A line's capital would dwindle from generation to generation.

Despite what was on the whole a positive economic trend, the explanation of the gradual disappearance of peasant hierarchies begins to emerge here. The conditions of production were unfavourable to the farmer, nor did the collective conditions of life and of labour evolve in any fundamental way.

To sum up, then, the economic development of Bigouden followed an unusual pattern as far as the timing of events was concerned. If it is difficult to set a starting-date (in this case the existence of solid investigations led me to begin in the final decades of the eighteenth century), on the other hand an overall study of Bigouden must not be extended beyond the 1880s, when fishing and all the activities associated with it profoundly altered the economic, social, and demographic balance of the region. In the course of this century of gradual change, the economic climate did not always operate in the same way as far as the peasant population was concerned. It was favourable to farmers in the eighteenth century, particularly towards the end. But on the whole the Revolutionary period and the early years of the nineteenth century served them ill, the economic climate adding its disastrous effects to those produced by the disorganisation of the administration and of the clergy, which at the time was responsible for recording civil status as well as for education. When at last that climate showed some improvement, the peasants derived little advantage. Instead a process of social levelling set in, while the framework of production changed hardly at all, as will be seen from the next chapter.

7

Frozen hierarchies and social relations

The economic changes of the period 1750–1880 were accompanied by social changes, but these were gradual. Despite the impoverishment of individual lines, hierarchies froze. Opportunities for social mobility were small, and peasant society set in patterns that linked those at top and bottom of the ladder in a symbiotic relationship.

In these conditions the mode of agricultural production evolved a little. It resembled that found in economies of abundance, as described by Marshall Sahlins. Analysis of the yearly round shows that work did not take up the whole of people's time. Far from it: activities of pure social ostentation, whether in connection with the religious sphere (*pardons* and the festivals associated with these processional pilgrimages), the economic sphere (fairs and markets), or the matrimonial sphere (weddings), occupied a substantial place in the calendar.

Work in South Bigouden was based mainly on human physical strength (it continued to be so until the 1950s). The post-mortem inventories examined later in this chapter will show large numbers of hoes, shovels, spades, and other simple tools wielded by hand and often manufactured and maintained by the user. Physical strength and skill were what counted above all. Production and the techniques of production were learned from first-hand observation, as fathers passed on advice and tips to their sons and sons-in-law. The milieu proved resistant to any abrupt intervention, accepting only the kinds of innovation that came up slowly from the base rather than being imposed from above. The consequences were a state of relative underproduction and an economic instability that manifested itself in the crises of the early part of the nineteenth century (the last of which, as we have seen, occured in 1848). Surpluses were regularly consumed in processes of ostentation that were more important to the peasant population than any pre-capitalist or capitalist strategy of accumulation.

Further evidence of this different economic logic lay in the interest accorded to food, with its ritual imbalances. Several times a year wedding banquets or meals associated with *pardons* marked a break with daily routine by their abundance and most notably by the presence of meat. Wealth was not accumulated for its own sake or in order to generate further wealth. Wealth was for squandering, for showing generosity, particularly towards the most deprived (i.e. beggars); it was for asserting one's status.

Looked at in this light, the economic behaviour of the peasant population is easier to understand. So too is the astonishment or indignation of travellers or agronomists faced with the habit-bound, drunken Breton. Their verdict is not unrelated to that of the early anthropologists who used to declare that 'the native is congenitally idle'.[1]

Manual labour and extensive production went hand in hand with domestic organisation of work. As in most rural societies, domestic group, unit of production, and unit of consumption merged into one. Production was based on a division of labour according to sex that explains the fundamental necessity for the creation of this mixed cell. Certain occasions called for a large work force, which was often recruited on the basis of kinship. As a result the *ad hoc* assembly of the group lost its strictly economic purpose and became a force for reactivating temporarily overstretched social and family ties.

On the other hand, at the core of the peasant domestic group Chayanov's law applied; the forces of production were employed with an intensity that diminished with the number of persons present in the unit of production.[2] This is certainly the picture we get of farms in years gone by; the farmhand worked while the farmer himself did little more than secure his social place in the sun – something of prime importance in rural logic.

Is it pushing the comparison between primitive and peasant societies too far to ask whether, in Bigouden, the economy existed structurally? And whether, rather than a separate, specialised organisation, 'the economy [was] not something that social relations and groups constitute[d] in their generality, particularly kinship groups and relations'?[3] Our post-mortem inventories will show that there was no great difference between the way of life of the wealthiest and that of the less wealthy. Moreover, social mobility, which usually decreases as population increases, tended to level out such differences as did exist.

Wealthy farmers who speculated

The social structure of eighteenth-century South Bigouden was hierarchical. In addition to nobles, bourgeois, and propertied clergy there were wealthy farmers and tenants. The latter were exploited by the former, who by the end of the century had managed to amass considerable wealth. Most of the ancestors in the genealogies studied belonged to the category of farmers who speculated in reparative rights. It was they who exercised political control at the local level. And they were all related by marriage.

Who were these wealthy *laboureurs*? Bouët and Perrin, in their study of the Bretons of Armorica,[4] spoke of a 'rustic aristocracy' and of the 'nobility of our countryside [*noblesse de nos campagnes*]', referring to the old nobility that had fallen into decay. Parts of Brittany still had a high percentage of impoverished aristocrats in the early eighteenth century, but in the Bigouden region they seem to have disappeared by the time this study begins. Jean Meyer advances the hypothesis that 'the absence of any reference to this nobiliary plebs [in Lower Brittany] is possibly explained by the fact that its elimination from the region was already complete'.[5] When the lord of the manor ceased to live there, he leased out his manor as a *domaine congéable*, together with its dependent farms. And it was these wealthy farmers who moved in. In 1732, for example, in the avowal of the Du Pont barony, the lord leased out the manor of Kerdegasse (Kersegas in Saint-Jean-Trolimon) for 180 *livres*.[6] Kerbonnevez was similarly leased out by the lord of the manor of Châteaugiron.

The territory once teemed with manors. They were not always large, imposing houses; some were in fact quite modest buildings, even single-storey or with a simple attic. But they were all built of dressed stone. Once a manor was occupied by a farmer, it soon fell into disrepair. It would be poorly maintained, and any rebuilding tended to spoil the effect of the original design. Where the manor was pulled down, either the stones were used to build a humbler dwelling or they were sold, the new house being constructed of 'simple masonry' with dressed stone used only for door and window surrounds. Sometimes all that is left of past glories is a well or a fine doorway surmounted by an ogee arch – now looking rather out of place in a barn or a cowshed (see plate 18).

In 1789 there were a few noble families resident in Bigouden. Jean-Maurice de Penfeuntenyo inhabited his château of Kervereguin, near Loctudy. Ange Le Gentil de Rosmorduc occupied the manor of Kerazan. And there were Guillaume du Haffond de Lestriagat, living in Treffiagat and Plomeur, Alleno de Saint-Alouarn, living in La Ville-Neuve, and the Marchioness Duplessis de Grenedan et Penfeuntenyo, living at Kerfilin (in the Saint-Jean-Trolimon territory).[7] These nobles do not figure in our marriage networks.

However, the eighteenth-century marriage certificates of the genealogies studied do often mention an 'honourable man'. The title is given to Pierre Le Perennou, for instance, domanial tenant of Kermatheano. These 'honourable folk and farmers and householders' were the lessees of large farms, enjoying tenancies determined several years in advance.[8] The terms *honorable homme* or *noble homme*, which in the Middle Ages indicated noble birth, meant no more in the eighteenth century than a member of the middle class. 'According to the decree issued by the Parliament of Brittany in 1772 [writes Jean Meyer], this title was not a mark of personal nobility and carried no exemption from common taxes.'[9]

These farmers were sometimes given the title *laboureur du lieu noble* (literally 'worker of the land of the noble seat'), which makes it quite clear that they were peasants who had been able to purchase a nobleman's estate or (much more commonly) acquire the *convenant* relating to such an estate (giving them the reparative rights). Apart from this mention, which is not systematic, there is no special term for them. In certain parts of Lower Brittany such wealthy peasants did have collective appellations: in Haut-Léon they were called *Julots*,[10] in Plouzané *Pinvidig ha Goz*, or in Goulien *Pochou Gwiniz*.[11] In Bigouden there was no such name for them, suggesting that they were too numerous or disappeared too quickly to constitute a true class.

In order to assess the number of wealthy farmers and relate it to the genealogies studied, a search was made of the catalogue of lay leases for the period 1733–1791. The same document was also useful for assessing their patrimony.[12] Alain Signor pointed out in his study of the Revolution in Pont-l'Abbé that prior to 1789 large numbers of Bigouden farmers owned their own land. He put the figure between 500 and 700. There were, he wrote, 'comfortably off, even wealthy farmers who, allowing for the different types of tenancy, held 476 farms that were the object of leases'.[13]

The conclusion drawn from a nominal inventory of the same batch of records was slightly different.[14] These wealthy farmers did not necessarily own their holdings; they appear far more often to have owned a number of *convenants* relating to farms scattered throughout South Bigouden, acquired by the speculative process analysed in the previous chapter. Accounting for individual holdings leads to error because there were far fewer peasant lessees than farms held under lease. Each peasant's account comprised three, five, ten, sometimes many more holdings. Alain Signor appears to have gone astray when he reckoned the number of peasants to have been greater than the number of leases.

On the contrary, the latter seem to have far outnumbered the former, since each lessee held several.

And remember that Alain Signor was talking about the whole of Bigouden, whereas the present study is centred on the southern part of the region. The number of peasant owners seems in fact to have been greater in the south than in more northerly areas around Plozévet or Pouldreuzic. The numbers of South Bigouden peasants leasing out land in the eighteenth century were as follows (though the figures can give no more than an indication because of the high incidence of homonymy and the resultant problems of identifying the holders of accounts):

1730–1750: 177
1751–1770: 182
1771–1790: 149

If we put the population of South Bigouden in the late eighteenth century at approximately 6,500 inhabitants and if we assume that the proportions of tenant farmers and day labourers were more or less the same as those observed at the beginning of the nineteenth century – that is to say, 60 per cent and 40 per cent respectively – we reach a figure of 3,900 tenant farmers (men, women, and children included). Within the category of tenant farmers, still according to the proportions observed at the beginning of the nineteenth century, only a quarter – that is to say, 975 persons – were wealthy. The figure needs to be further divided by four (the average size of a household) to find the number of heads of household. In all some 240 men could be described as *paysans aisés*.

This figure is close to that for peasants who granted leases. It is further comparable to that for the individuals identified in our genealogies as founder ancestors. Between 1610 and 1770, 1,190 founders were identified (see table 16). Between 1730 and 1790, 517 founder ancestors were found, but these generally went in couples so that the figure must be halved: 250 men.

Comparison of the list of founders in generations 6, 7, 8, and 9 with that of the lessors of land reveals a degree of coincidence. Three factors prevent that coincidence from being complete. First, there is some uncertainty about the name of farm lessors. Secondly, the definition of founder ancestor means that the names of those persons who already have ancestors in generations prior to the sixth do not appear; in other words, a number of peasants who leased out land were lost for the purposes of comparison from this chronological section because their ancestors had been identified during the period prior to 1730. Thirdly, tracing back genealogies did not necessarily lead to the wealthiest individuals when spouses appear who, though acceptable after 1850, were nevertheless descended from peasants who, being poor, were by definition absent from the index of lessors of farms. For all these reasons the coincidence of something like 30 per cent between founders and lessors of farms is relatively satisfactory. It also enables us to qualify our genealogies in social terms. Tracing back the ancestors of the fifty farms selected in 1836, we reach the wealthiest stratum of peasants united in one vast kindred practising relinking between lines.

At the same time there is reason to think that we have grasped this social group of the population virtually in its entirety because of a consistency between demographic calculations, the list of peasants who leased out farms, and the list of founders. So the body of genealogies studied is more than just a sample; what we have apprehended is the whole or virtually the whole of a certain stratum of the population.

How many holdings did each peasant own? The number varied, though it was always more than one. And any assessment is clouded with uncertainties regarding the type of ownership involved. Some mentions are in fact highly explicit, others much less detailed. There was a striking degree of dispersion; in Bigouden, the more holdings that were owned, the more widely they were scattered. This is a good indication of the presence of speculation, with peasants buying and selling reparative rights. Further proof is provided by comparison of the accounts of fathers and sons, which reveals that they do not contain the same holdings, some having been sold and others bought in.

Altogether the average number of holdings per account can be put at five, which means that some peasants had ten or more and others only two or three.

Here, for example, with all its uncertainties and difficulties of interpretation, is the Quittot account from Combrit. The account runs from 1733 to 1777 (it is hard to tell where the father's ends and the son's begins).

Between 1733 and 1750 Louis Quittot leased out fifteen holdings spread between Tréméoc, Pluguffan, Plonéour, and Combrit. They brought in between 42 and 140 *livres* apiece. There is no mention of the type of ownership, only of the place. Moreover, these were different holdings – all except three, which were renewals of leases. Kervasiou or Kersiou in Pluguffan, for instance, was leased out for 140 *livres*, Kergouaré and outbuildings in Tréméoc for 45 *livres*, and Fao Glas in Plonéour for 42 *livres*. Between 1751 and 1770 Louis Quittot Jnr. leased out eleven holdings that were similarly scattered between Plonéour, Combrit, Pont-l'Abbé, Tréguennec, Lanvern, and Beuzec. The location of each holding is even less precise in this account. There is mention only of 'inheritances' – for quite large sums, in fact: 177 *livres* rent, 210 *livres*, and so on. Of the holdings of the previous account only four recur. Moreover, their rent has gone up; Kervasiou in Pluguffan from 140 *livres* to 168 *livres* in wheat and cash, Kergouaré in Tréméoc from 45 to 60 *livres*, and so on.

Another son's account runs over the same period. This was François Quittot, whose property shrank and became concentrated around a single holding, the lease of which was renewed regularly. Between 1751 and 1770 two leases are mentioned: 'pieces of land in Tréguennec' for 24 *livres* and the manor of Kervillic in Tréguennec for 165 *livres*. Between 1771 and 1783 there were leases for an 'inheritance in Lanvern' for 12 *livres* and five leases all pertaining to Kervillic, the manor and parts called *huella* (located in the upper part), renewed in 1772, 1780, and again in 1783, each time for a larger sum. The lessee secured himself against eviction by renewing his lease in advance, several years before it expired. The large sums involved suggest that François Quittot actually owned the land.

Or take the account of Michel Penven, mentioned six times between 1759 and 1770 and four times between 1771 and 1782. His property was substantially less. There were inheritances for 12 and 42 *livres* in Plomeur and Plonéour. In 1770 there is mention of a lease of 132 *livres* and rents at Kerlouazec in Plonéour, which still figured in his account in 1775. Between 1771 and 1782 the holdings mentioned were situated at Le Haffond in Beuzec and at Keryoret. In 1782 Michel Penven was described as a 'farmer [*laboureur*] at Beuzec' and leased out to his son Guillaume the reparative rights to Kergreac'h in Saint-Jean-Trolimon for 204 *livres* 14 *sols*.

Oddly, this process of peasant speculation seems to have gone unnoticed at the time. That fine nineteenth-century observer of local customs, Armand du Chatellier, made clear reference to it in relation to a late period, which in fact marked the end of this behaviour:

> If the farmer, instead of looking for a farm and putting both his capital and his agricultural equipment to use, is looking purely for an investment, the reparative rights he has acquired, rather than giving him occasion to go and live on and work the holding of which he has taken the lease, constitute only an object of simple speculation that he will sub-let after exercising his right to evict at his own risk, making a profit or a loss according to whether the price of the sub-lease covers or fails to cover his outlay in compensating the evicted party. And it has to be said that the Breton farmer never makes a mistake at this game and rarely fails to obtain a very large return on his money. I have learned this to my cost on numerous occasions, letting pieces of land for very small rents that clever peasants then tripled by sub-letting as soon as they were in possession of the leases I had granted them. On the other hand I have several times had my revenge by acquiring lands under lease [*à domaine*] and doubling or tripling the

income from them with fresh leases, myself evicting the former superficiaries [domanial tenants] on expert advice.[15]

The extent to which these peasant speculators enriched themselves varied. In the case of one line, that of the R., they did so to a spectacular degree, giving rise to a legend that is still very much alive today. The family's wealth in the eighteenth century was the result of favourable demographic factors (a single surviving heir in each generation between 1650 and the end of the eighteenth century), good marriages, and successful speculation enabling them to make 100 per cent profit and more in each generation.

Between 1736 and 1739 Pierre R. and Marie H., domiciled in Tréméoc, declared four holdings. Between 1741 and 1774 their son Pierre and his wife, Marguerite L., declared twenty-six, nearly all different from year to year. Their son Pierre (1749–1808), who married in 1766, declared the same number between 1780 and 1791. These holdings, many of which comprised land *and* reparative rights, were leased out for substantial sums. The hamlet of Kerlanno in Loctudy, for example, went for 586 *livres*, a holding at Le Ezer for 473 *livres*, and the manor of Kergambae for 432 *livres*. Alain Signor reckoned that all these holdings together brought in 4,216 *livres* per annum.[16] In fact the income from them will have amounted to something like 5,675 *livres*. These various holdings were handed down from father to son, for we find the same ones occurring in the father's account and in the son's account. But fresh ones were also acquired as opportunity offered.

This accumulation of capital probably reached its zenith at the time of the Revolution, after which the demographic fortunes of the line went into reverse; the descendants of Pierre, who died in 1808, differed from their ascendants in being highly prolific. In 1793 Pierre had a house built at Kervergit, a veritable manor of a place, and had inscribed on the lintel 'built by Pierre R. 1793'. The memory of this inscription was preserved only by word of mouth, because in later years the stone was turned around and whitewashed by another occupant of the farm, a distant descendant of that splendid forbear.

When Pierre's brother Corentin shared out his property among his three children on 16 April 1790, there were seven landholdings, six reparative rights, and three houses to be distributed. Each child received the equivalent of 5,900 *livres*,[17] The legend according to which Pierre R. derived substantial wealth from the sale of *Biens nationaux* is without foundation. He did purchase a number of holdings, but the greater part of his fortune derived from earlier speculations. An example was his acquisition, on 10 May 1774, of the *rente foncière et dominiale* of 6 *livres* on a farm called Jolivet in the Saint-Jean-Trolimon hamlet of Kerbascol. He acquired it from the lord of the manor of Lestrediagat, a dragoon officer in Quimper, and got rid of it two years later by making an exchange with Yves Calloch of Kerillis. The bulk of the patrimony was thus acquired in the eighteenth century.

The pre-Revolutionary fortune of the R. family started to become dissipated with the advent of the nineteenth century. The matrimonial strategies of this exceptionally wealthy line failed to offset the effects of the partible inheritance system. Those strategies can be traced – not in this instance through the formal marriage figures but by making a close study of the accounts of the father, sons-in-law, and daughters-in-law of Pierre R. Unable to find affines as wealthy as himself, he had to look among the wealthiest available. The Peron, Calvez, Guichaoua, Lay, and Cleac'h families, for example, owned four, five, or six *convenants*, never more. They were all recruited among the lines of founder ancestors figuring in the genealogies studied, which confirms that the investigation identified a social group virtually in its entirety in the eighteenth century.

The matrimonial behaviour of the R. line well illustrates a certain peasant mentality. According to Cambry, around 1770 an R. had married a daughter of one Gauder, a 'famous surgeon' of Pont-l'Abbé.[18] Why did they not look for marriage partners among the urban middle class at the zenith of their wealth? The reason is that we are dealing with peasants, people whose economic logic was not really geared towards acquiring wealth for its own sake. It was governed, like that of the well-to-do farmers of the same period, by a strategy of keeping up appearances, of maintaining rank at the level at which one belonged – that is to say, among the peasantry – rather than by any vying with such representatives of the urban middle class as shop-keepers, notaries, or public officials. Nor, as far as one can tell, was this material affluence distinguished by a particular life-style or by innovative farming. Pierre R. simply built himself a very fine house, signing it with his name, as others had a patronymic plate made.

The *laboureurs aisés* of the eighteenth century slowly lost their patrimonial superiority during the course of the nineteenth century. But they left a deep impression on popular awareness; such names as Guirriec, Coïc, Daniel, and of course R. live on in local memory even today. The semi-legendary origin of the latter, whose patronymic goes back to a single ancestor, is still an object of debate. Some attribute the reputedly dark complexion of members of the R. family to a Spanish extraction. Another source holds that, once upon a time, the sole survivor of a shipwreck at Saint-Guénolé 'could not speak but could only point to the open sea', repeating the same word over and over again; that word formed the basis of the new patronymic (which does not sound like a Breton name). Others claim that three brothers came from the Auvergne region of central France or possibly from Scotland. The researches of the family's genealogist show that the patronymic goes back a very long way; the cartulary of Quimper church mentions a priest of that name in 1360.[19]

This attribution of what was probably fanciful origins underlines the specificity of the ancestor. Yet the descendants in no way form a line or clan with any feeling of belonging together. It is simply that, though they no longer have any recollection of their common origin, all who bear this name know that, however remotely, they are related. There is gossip about the family, too: 'All that intermarrying . . . used to hang themselves . . . finally woke up to the snags of that kind of marriage, what with the number of cases of insanity and nervous breakdown.'

Apart from this high incidence of consanguine marriage, which is peculiar to them, the R. are described in much the same terms as the other well-to-do ancestors of the eighteenth century: 'They were proud people . . . arrogant . . . They were great families once upon a time, people who married amongst themselves because no one else was worthy of marrying into such families. They were folk who had plenty of property, folk who had themselves served separately at table . . . and sent their children to religious schools when others did not go there.' The speaker was Louise L., who came of a branch of the R. family so poor that a gulf now separated her own parents from their ancestors.

There is a special, untranslatable Bigouden term for them: *tud cheuc'h*. It suggests simultaneously wealth, a certain stand-offishness, and a degree of condescension towards others. By extension the term also means 'smart', 'distinguished', 'elegant'.

The interesting thing about the R. is that they constitute an extreme case as far as our founder ancestors are concerned. They were wealthy speculators, certainly, but of a particular kind since they failed to maintain their wealth in the nineteenth century. And of course it was they who held such political power as could be exercised at the local level. The correlation between wealth and local political power was in evidence all through the nineteenth century and into the twentieth, as we shall see when we come to examine the particular case of Saint-Jean-Trolimon in chapter 9. Using a different geographical base, it can also be demonstrated for the eighteenth century by comparing the list of founder ancestors and their descendants with that of property-owners who sub-leased *convenants* and that of the members of those assemblies that more or less filled the role of the municipal councils of the nineteenth century. It was not possible to make these comparisons for the early part of the eighteenth century; that would have meant embarking on fresh searches aimed at identifying the councillors of the parishes under investigation, which the high incidence of homonymy would have rendered problematical. Having found a person called Nicolas Le Lay, how can one be sure that he is the same man as appears in the genealogical index? For an affirmative answer it is necessary for several clues to tally, notably the place of residence of the individual or his parents at the time of his marriage.

As regards the late eighteenth century, on the other hand, a study of the names of the signatories of the *cahiers de doléances* enables us to identify the local notables. Not surprisingly, they also appear among our founder ancestors and are at the same time usually described as *laboureurs*. Actually, the process of identification is still tricky. How, for example, is one to identify the twenty-six members of the electoral assembly of Tréoultré-Penmarc'h, summoned in 1789? There is less difficulty about identifying the two *deputés*: Claude Keraudren and Hervé Le Cloarec. We know nothing about the latter, but the former is one of our founder ancestors, and we have his account: he sub-leased land or *convenants* for more than 1,600 *livres*, and he could sign his name. One of the three deputies of the Loctudy electoral assembly was of course Pierre R. At Tréguennec the two deputies Grégoire Le Goascoz and Louis Quittot were both well endowed; we have just been looking at the latter's account, and Grégoire was the son of Alain, who had a titular income of more than 700 *livres*. Of the two Beuzec deputies Pierre Tanneau, referred to as 'de Resnaout' (actually Rosnaon), had an income of nearly 400 *livres* in rents. He signed the declaration issued by the Tréoultré-Penmarc'h assembly.

The list of members of the Beuzec electoral assembly differs slightly from the one that was summoned at the time (1792) of the court case regarding the enclosure of common land in the sandy coastal strip (see chapter 6, pp. 158–60). On the latter, three-quarters of the members are identifiable. The five council officers and the mayor appear on our lists of propertied peasants. The mayor of Beuzec, incidentally, appeared at the trial in two roles: as an officer of the council and as a plaintiff whose animals had been attacked on the commonland by the settlers attempting to enclose parts of it. For it was not only the poorest farmers who put their cattle out to pasture on the former royal lands. Henri Carrot, the mayor, leased out three holdings in Beuzec and Plobannalec. Henry Durant of Lanvenaël was in the same position, being both a council officer and a plaintiff. A farmer in Beuzec, he declared six holdings in Loctudy, inheritances in Beuzec, and the manor and outbuildings at Pen An Ilis (Beuzec) for 344 *livres*. The other officers of the council,

François Le Lay, Yves Gloaguen, Nicolas Le Failler, and Pierre Lelgouarch, appear to have been equally well endowed, as do the other leading citizens mentioned. Most of them declared between three and five holdings bringing in incomes of between 300 and 500 *livres*.

These *laboureurs aisés*, often educated men who could sign their names, were all related by marriage; the marriage and power networks overlapped. We shall find this observation confirmed all down the nineteenth century through the detailed example of the transfer of power in Saint-Jean-Trolimon.

A hierarchical social structure

Studying the social structure of the nineteenth century is easier because the documentary sources are plentiful and can be cross-referenced. So plentiful are they, in fact, that it proved appropriate to narrow the field of investigation to the commune of Saint-Jean-Trolimon, which provides a good example of the organisation and operation of social hierarchies in Bigouden.

Examining the distribution of landed property offers an initial approach to analysing social structure. As an approach, it is both simple and complicated: simple because the bulk of the land belonged to noble families whose names have already been mentioned (Duplessis de Grenedan, for example, or Alleno de Saint-Alouarn), complicated because of the presence of the domanial tenant, interpolating his claim between those of landowner and lessee. Sometimes domanial tenant and lessee were one and the same; sometimes the domanial tenant did not farm his holding himself but sub-leased it to someone else. This makes it difficult to compare the list of names of owners with that of residents; sometimes they overlap, but more usually they do not, the domanial tenants living outside the commune, in Plonéour or elsewhere in South Bigouden.

An illustration of the phenomenon of multiple ownership is provided by the following case. In 1853 the commune purchased a piece of land in the centre of the *bourg* on which to build a school. It paid 240 francs to Madame de Cargouët for a plot of three acres sixty centiares. She was the owner of the land and resided at Lamballe. To Marie-Louise Gloaguen, non-resident holder of the reparative rights to the plot, it paid fifty francs; lastly, ten francs were paid to Louis Trebern, the tenant of the locality and owner of the two parcels named *Liors ar Hastel* and *Liors an dreon an ty* ('garden of the château' and 'garden behind the house').[20]

According to the 1836 cadastral plan, the bulk of the landed property was in the hands of eight noble families. They were non-resident, and their farms were scattered throughout the territory.

Duplessis-Grenedan	85 *arpents* (= acres)
Dupont-Desloges	41 *arpents*
Urvoi de Portzamparc	34 *arpents*
De Roquefeuille	22 *arpents*
Saint-Alouarn	73 *arpents*
De Cargouët	31 *arpents*
Bastard du Mesmeur	9 *arpents*
Du Marc'hallac'h	27 *arpents*
	322 *arpents*

Much of the land was in middle-class hands:

Jean-Marie Desban (mayor of Saint-Jean-Trolimon until 1841, residing at Le Steud Bihan)	28 *arpents*
Jacob Coïc (residing in Pont-l'Abbé)	32 *arpents*
Kérillis (notary in Pont-l'Abbé)	20 *arpents*
Astor (at the château of Kerazan in Loctudy)	11 *arpents*
Laënnec (barrister in Nantes)	15 *arpents*
François Moizan	12 *arpents*
	118 *arpents*

In addition to these private landholdings there was the land owned by the commune. This amounted to 252 acres (more or less), mostly in the sandy coastal strip at Tronoën and Kervellec. There were also communal lands belonging collectively to the hamlets. These were sometimes quite small, but occasionally they accounted for substantial areas such as the twenty-nine acres belonging to Gorré-Beuzec or the forty-one acres belonging to Kerbascol and associated hamlets.

No analysis of the cadastral plan, however minute, can remove all doubts about the extent of landownership by farmers.

Twenty-nine heads of household are designated *propriétaire*, with a measurement indicating the size of their landholdings. They may also appear as domanial tenants, in which case the area indicated probably relates to the farm, while the actual land belonged to someone else.

Of those twenty-nine 'proprietors', seven heads of household held in excess of two acres. In alphabetical order, they were Joseph Daniel of Rupape with seven acres, Michel Donge of Rugaoudal with two, Jean

Loch of Kerbascol with sixteen, the widow of Jean Marechal of Kergreac'h with twenty-eight, Yves Poullelaouën of Le Steud with eight, Louis Quittot of Kerstrat with six, and Isidore Tanneau of Quernel with fifteen. These heads of household and the lines to which they belonged are at the core of the sample of genealogies, as they constituted the backbone of the political life of the commune.

It is possible, however, to follow the structure of occupations in detail right through the nineteenth century and down to the present day by analysing censuses in detail. Not that the information they provide is always strictly comparable. In some years, status is mentioned (owner, tenant); in others the census taker merely noted the branch of activity, entering every farmer as *cultivateur*.

With these qualifications, and confining ourselves here to the censuses of 1836, 1851, and 1876 (leaving analysis of the censuses of the latter part of the century until the next chapter), there is much that such comparisons can tell us.

In 1836, of 155 households in Saint-Jean-Trolimon, ninety-six were headed by *cultivateurs* (owners, domanial tenants, and ordinary tenants were all lumped together); that was 62 per cent. There were seventeen day labourers (11 per cent), ten beggars (7 per cent), ten weavers (7 per cent), and seven tailors (4 per cent).

The remaining heads of household followed various trades essential to the religious, social, and economic functioning of the village; there were a priest and a verger, two publicans, two builders, two millers, two sabot-makers, a cooper, a cartwright, and a farrier.

The poverty of a section of the population at that date is indicated by the large number of beggars as well as by the small number of day labourers (a figure that could increase if work was more plentiful). We also find a relatively large number of weavers working from home and making a living from their proto-industry – an activity that was to disappear rapidly as a result of competition from manufactured cloth. Tailors were also numerous. Until the adoption of urban dress they would continue to have work. Their semi-nomadic existence, travelling from farm to farm, made them ideal marriage go-betweens. Lastly a solid body of tradesmen performed the necessary tasks of building houses, shoeing horses, making barrels, binding wheels, and dispensing strong drink and tobacco. Usually they lived in – were in fact the sole inhabitants of – the *bourg*.

A similar analysis conducted in 1851 shows a substantial drop in the number of weavers and above all a reversal of the proportions of tenant farmers and day labourers; this time tenant farmers account for 34 per cent, day labourers for 44 per cent. The figures reflect the endemic crisis that shook the region at the beginning of the nineteenth century, with the wealthiest farmers becoming poorer and a destabilised rent situation hitting the worst-off.

The increase in the number of day labourers in the 1851 census is a sign of the relative pauperisation of the village at this time, the more so since more than 5 per cent of heads of household are still declared beggars. Many small tenant farmers had arrears of rent and were in a precarious position. Either they were sons of small tenant farmers already known from previous censuses who had not been able to find a tenancy, or they were former weavers, builders, or cartwrights recorded in censuses five or ten years previously who now had no customers for their particular skill. Many of them came from communes lying to the south, Penmarc'h or Loctudy (thirty-one out of eighty-six or 36 per cent), in an instance of the way in which economic crises accentuated mobility, as we have seen. Several of these households had left by the time of the next census, which accounts for the fact that in 1856 the proportion of day labourers was already down to 20 per cent. In fact it continued to decline until the end of the century. The number of tenant farmers stabilised at that time around 60–70 per cent, corresponding to an economic revival and a reorganisation of the legal framework of land ownership. The beggars, by an analogous phenomenon of loss of social position, were day labourers who had fallen victim to the economic crisis and adverse family circumstances. In 1851, out of eleven such households, three were wandering widows or spinsters who stayed only for a single census period; the others were day labourers who were too old, too weary, or simply superfluous to requirements. These beggars lived all over the commune, not just in the *bourg* but wherever a charitable farmer made it his duty to offer them hospitality – at Guin-Gorré, at Kervouec, wherever it might be.

After 1860 beggars disappeared from the census figures – though not yet from social life. On the other hand the same artisanal infrastructure remained in place until the end of the century.

Poverty remained endemic throughout the nineteenth century. The post-mortem records reveal large numbers of these unfortunate creatures passing from census to census, wandering from one commune to another before death suddenly struck them down. When Françoise Gouien died in Saint-Jean-Trolimon on 10 December 1858, for instance, her husband declared on her behalf forty francs' worth of furniture: kitchen and hearth implements, a bed, a bench, a few clothes, and a property of twenty-five acres forty centiares in Plomeur, leased out on the basis of a verbal agreement for an income of twelve francs per annum. This declaration shows, incidentally, that poverty could go hand in hand with ownership of a small amount of property.[21] Or take the case of the order of the prefecture council authorising the Saint-Jean-Trolimon church council to sue one Duchatellier for outstanding burial fees in respect of one of his tenants, 'the latter's widow having been unable to pay them, Mr. Duchatellier having seized everything she possessed on grounds of arrears of rent'.[22]

Alongside the farming community there was a shifting infrastructure of tradesmen. From the middle of the nineteenth century the particular class of artisan represented by the weavers tended to disappear as their domestic production came under strong competition from industrial production. They abandoned their trade and in subsequent censuses were listed as day labourers. The number of tailors, on the other hand, represented as a percentage of the population as a whole, remained relatively stable until 1880.

The seven tailors recorded in 1851 were partly spread over the territory of the commune, each one at the centre of his network of clients. Most of them, however, were concentrated in the *bourg*, closer to the wealthiest families and also closer to Pont-l'Abbé, where they had other clients or obtained their supplies. It can be calculated that each tailor had about 160 clients. Two millers produced flour for local needs, but there was no baker as yet, indicating that bread was still manufactured domestically and also that the traditional pancakes still formed the basis of people's daily food intake.

Only one smith was needed to shoe the horses, which seems surprising. There are several possible explanations: farmers also used the services of those of Plonéour, Plomeur, Le Stang, and Beuzec; or else the state of the roads was such that horses were very often able to go unshod. Horses, being a sign of wealth, were not in fact all that common, as the post-mortem inventories confirm. An 1878 declaration of horses and mares lists thirty-one stallions, forty-eight geldings, and ninety-eight mares – scarcely one animal per farm.[23] Ploughs were usually drawn by oxen. Carts were still uncommon in the mid nineteenth century, and people travelled on foot more often than on horseback. There was only one cartwright in the village, and he was sometimes declared as such and sometimes as a day labourer. The post-mortem inventories reveal details of tools and products of the cartwright's trade (pipe boxes, wheel rims) used and made on the farm. The same artisan occasionally combined the trades of cartwright and blacksmith. Even when two people were involved they were often close relatives as well as living in the same vicinity, cartwright and smith being obliged to collaborate closely over such tricky jobs as binding wheels, for example.

Some sixty people resided in the *bourg*, grouped in fifteen households whose status was somewhat peculiar. The *bourg* of Saint-Jean-Trolimon, which lies on the eastern edge rather than at the geographical centre of the commune, housed the tailors, the verger (who combined his ecclesiastical functions with such civic tasks as 'calling out and repeating the bids' during public auctions),[24] small farmers, day labourers, and two beggars. All these inhabitants lived in small houses of cob and thatch that appear to have gone up in smoke at frequent intervals. Alongside these half day-labourer, half beggar households lived the artisans of iron, fire, and strong drink, all united in one solid kindred. In addition to the cartwright and the smith, the censuses of the 1850s listed two female publicans, one of whom kept a sort of hotel, providing accommodation for a weaver's lad and a lady who seems to have had no trade or calling.

A study of these households reveals a degree of specialisation by sex. If iron and fire were male trades, strong drink was predominantly female. The complementarity of the two activities was in keeping with the sexual distribution of functions. Close consanguine and affinal ties were woven among all these households living in the *bourg*, which seems to have monopolised these trades. While we find a publican established at Kergonan between 1880 and 1900, together with his blacksmith son, it is not until the early years of the twentieth century that we see small traders spreading out to the hamlets. In other words, for more than fifty years the *bourg* concentrated the same kindreds as are still there today.

The post-mortem inventories of two innkeepers drawn up in the mid nineteenth century afford a glimpse inside their premises. In the case of Jean-Baptiste Blayau, who died in 1851 at the age of thirty-two, the total assets of the inventory were small,[25] the only indication of his trade being the presence of four candlesticks (whereas lighting was often confined to a single candle), twenty-three plates, a pitcher, nine forks, ten iron spoons and twelve wooden ones, thirty glasses or tumblers, and two sets of measures. His clientele was hardly numerous, judging from the equipment he had available. Like all tradesmen, he worked a rented plot of land, raising two cows and a calf. He had no horse, and one imagines that he did the journey to Pont-l'Abbé (only four kilometres, after all) on foot to fetch his supplies. The household had deposited 180 francs in cash with the notary, but it was also burdened with various debts – owing the wine merchant eighty-one francs, for example.

Jean-Marie Le Berre, the village's other publican, died in 1848, aged forty-seven.[26] He had slightly more in the way of furniture and household utensils. In particular there was a stock of brandy (*eau-de-vie*) worth thirty francs, and the man had also owned two looms, worth

Table 22. *Saint-Jean Trolimon: households broken down by number of servants (as a percentage)*

No. of servants	1836	1841	1846	1851	1856	1861	1866	1872	1876	1881	1886	1891	1896	1901	1906	1911	1921	1926	1931	1936	1946
0	58.8	59.5	59	60	54	61	64	72.1	66.7	68.8	66.3	50	67.75	76.65	75.8	81.6	78.3	81.7	86.7	89.6	91.3
1	16.3	23	18	14.5	15.5	14.5	13	12	15.8	17.3	16.9	21.8	16.7	11.2	14.5	9	17.9	14.3	10.5	8.8	8.2
2	13.7	7	9	11	7	12	9.5	6.8	10.8	7.5	11.4	6	9.15	9.1	7.3	7.6	2.1	3.5	2.6	1.6	0.5
3	0.6	3.5	8	9.5	12	5	9.5	6.2	5	4.3	3.3	7.2	5.4	2.55	0.5	1.8	1.7	0.5			
4	0.2	5	4	4	9.5	5	2	2.3	1.7	1.6	1.6	3.3	0.5	0.5	0.5						
5	0.1	1	1	0.5	1.5	0.5	1			0.5	0.5	1	1								
6	0.06	0.5	0.5	0.5	0.5	1	0.5														
7		0.5				1		0.6													
8			0.5																		

thirty francs apiece. His livestock comprised a cow and calf and beehives valued at forty francs. He too had no horse. The joint property of the household was also burdened with numerous debts, notably to Messrs. Alavoine and Desban of Pont-l'Abbé for goods supplied and to the notary for a loan.

Simply listing the trades of the inhabitants of Saint-Jean-Trolimon suggests the social structure of the village. There are several other ways of uncovering it. We can find out, first of all, which householders were most heavily taxed. In 1834, when the local roads maintained by the commune needed to be repaired, sixty farmers were required to supply a number of days' work in terms of men, horses, oxen, and carts; seventeen were required to give six days' manual labour, twenty-three gave four days, and twenty gave two days.[27] Also the original of the register of direct taxation gives a list of farmers with their income from land.[28] However, this is more difficult to use because domanial tenant and freeholder are often lumped together in the same account. Forty-four farmers are listed with an income from land of around 150 francs. Some of the farmers who figure in the sample of genealogies are mentioned: Jean Le Loch of Kerbascol with more than 800 francs, Louis Durand of Kervouec with 345 francs, Isidore Tanneau of Quernel with 223 francs.

There is another way of assessing the social structure of the village, which is to distinguish between those farmers who had servants and those who did not. The use of wage labour was linked to the evolution of the family life cycle in that number of servants diminished as the head of the household advanced in age and was able to replace them with his children. Nevertheless, the number of servants figuring in each census provides a good snapshot of the situation at various moments during the nineteenth and twentieth centuries, enabling us to confirm the distinction between tenant farmers and day labourers. The former, even on a small farm, were in a position to employ someone; the latter were not. Between 1836 and 1866 around 60 per cent of households had no servant, while 40 per cent employed between one and six wage workers (see table 22). In 1836 it was of course the owner–farmers who employed the largest numbers of them: one had six, three had four each, seven had three, seven had two, and eleven had one. Most farms had only one wageworker, though a few were able to employ up to three until 1891. The days of the large estate with four or more servants were over after the 1870s. The number of farms employing wageworkers slowly diminished from that time on, and the trend was confirmed in the years following the First World War. Well-known throughout the farming world, the phenomenon was perhaps felt less cruelly in this region because of the presence of numerous day labourers spread throughout the territory and the sandy coastal strip. Right up until 1960 it was still possible to find someone to lend a hand. Nowadays there is a serious crisis in this regard, and the farmer and his wife do all the work themselves.

Masters and servants

The living conditions of the farm servant in the nineteenth century depended on the living conditions of the employer household. It was probably better to serve on a wealthy farm than on a poor one. Young servants and the children of the family shared the same work, the same food, and sometimes the same bed. It often happened that, from one census to another, a younger sister or brother of the head of the household or a half-sister or half-brother or perhaps even a nephew was listed on one occasion as a relative and on another as a servant. Usually farm servants were the offspring of poor tenant farmers unable to feed their own children beyond the age of ten or twelve. Working on the farm, they underwent an apprenticeship that enabled them to progress from *petit valet* to *grand valet* (*mevel bras* in Breton). A feature of the system of farm service was the rapid turnover of personnel. The census, of course, only recorded the population once every five years – too large a gap to capture this phenomenon satisfactorily. Even so, it is clear that servants never spent longer than that period in the service of the same household. The observations of du Chatellier confirmed their mobility, which was in fact linked to the family life cycle: 'it is regarded as axiomatic that at the end of the year the two parties know each other too well, the one to give orders and the other to obey willingly'.[29] Du Chatellier saw this as 'one of the most striking signs of the degeneration of local manners', whereas we know how deeply such mobility was inherent in the economic and social system of the group. It is unusual, for example, to find elderly servants; they quit the servile conditions as soon as they had saved enough to be able to set up on their own as day labourers.

The wages of farm servants increased throughout the nineteenth century. In 1835 a farm foreman received between 105 and 120 francs per annum, a farmhand 75 francs, and a 'woman for the cows' 39 francs; more generally, a male servant received 75 francs per annum and a female 60 francs with board.[30] In 1913 the annual wage of a farmhand with board and lodging stood at between 200 and 400 francs, while a woman earned between 200 and 250 francs.[31]

The post-mortem inventory of Corentin Le Donge in 1866 gives precise indications regarding the wages drawn by farm servants on large farms. Under liabilities, it lists what was due to the servants:

To Yves Le Reun [aged 22 in 1866], FR 21.50 as a proportion of wages based on the sum of FR 100 from 25.12.1865 to 12.03.1866; to the same, FR 4.45 as a proportion, during that period, of the value of three metres sixty centimetres of cloth, two pairs of sabots, and four days' tailoring, reckoned together at FR 21.70; to Ambroise Guéguen, servant, as a proportion of wages, FR 5.05; to the same as a proportion of various goods supplied, FR 3.40; to Marie-Anne Stephan [also aged 22], as a proportion of wages, FR 10.50; to the same as a proportion of various goods supplied. FR 5.10; to Perrine Le Tirilly, servant [aged 25], FR 8.85 as a proportion of wages; to the same as a proportion of the value of various goods supplied, FR 6.20.[32]

What emerges from this list is that the annual wage of a *grand valet*, also known as a *valet-cultivateur*, could be in excess of 100 francs. Wages of *petits valets* diminished with age, and women's wages, themselves graded according to jobs, responsibilities, and age, were lower.

The farm servant received wages in money and in kind. He could also reckon to share in all collective celebrations such as fairs, *aires neuves* (when the threshing-floor was renewed), and other occasions of collaborative labour and general rejoicing.

The symbiosis between the life styles of masters and servants, the richest and the poorest, was noticed by all nineteenth-century observers. Armand du Chatellier revealed how he was 'one of these old farmers who despite his laboriously acquired 12,000 franc income went off to work with his servants every morning, like them carrying only a plain piece of black bread in his pocket to still his first hunger'.[33]

This identity of behaviours masked the rigidity of hierarchies. Demographic pressure robbed the social system of its relative mobility. What du Chatellier had to say about a kind of hypothetical life cycle supposedly leading from the condition of farm servant to that of day labourer and then to that of tenant farmer would seem to have been closer to a pious wish than to social reality. A degree of social progress may have been possible at the end of the eighteenth century; in the nineteenth century it was no longer so. We find here the same sort of anachronism as characterised his analysis of farmers who speculated in reparative rights, which seemed to apply more to the end of the previous century than to the time when he was writing. So the following passage should be regarded rather as a Utopian model envisaged by an enlightened agronomist than as a true account of social mobility in the nineteenth century:

Boys and girls are placed as wage-earners from the ages of twelve to fourteen with freehold and tenant farmers who want them for their board and a few garments such as linen shirts, trousers, and skirts together with three or four days' tailoring, one or two pairs of sabots, and 6, 7, or 8 crowns [*écus*; the *écu* was worth three francs], depending on the strength of the individual concerned. From fifteen to eighteen they receive a higher wage. Gradually they acquire young rearing animals, which they tend with the other animals of the herd; sometimes they are given a few furrows, which they sow themselves.[34]

At the age of twenty-four the servant would be getting 100, 150, or 200 *écus*. He then married and attempted to set up on his own in a *Penn-ty* (literally a 'house end') consisting of a small dwelling, a vegetable plot, and a piece of land on which he would raise one or two cows and grow hemp. Such land was farmed by the principal tenant, who sub-leased it to him. The young married servant thus moved into the category of agricultural day labourer. The latter was in a very precarious position, being dependent on the farmer to whom he hired out his labour. His income was governed by the amount of work available, so that it was ultimately a function of the whole cycle of agricultural production.

According to the ever-optimistic du Chatellier it was possible for the agricultural day labourer, after a few year's work, to acquire his own farm implements and draught animals:

Later, with four or five hundred crowns saved up, he will buy a cart, a plough, a draught horse, and a pair of oxen. He is then a farmer and may aspire to any land leases available that would be appropriate to the manpower at his disposal and the young family that he has managed to establish.[35]

Not only was access to land ownership impossible, contrary to what du Chatellier maintained; the condition of the day labourer tended rather to be subject to a double decline, both social and geographic, with the poorest leaving to find work at the coast.

We are dealing with an agricultural society, and the study of social hierarchies cannot be dissociated from that of modes of production. The post-mortem inventory offers a reading of the domestic organisation of production at a particular moment in time, providing a further opportunity of comparing rich and poor and also, by taking account of the synchronic dimension, of assessing such technical changes as occurred.

Post-mortem inventories: a reading of the household

Post-mortem inventories constitute a rich source[36] for studying the domestic and agricultural equipment of a particular farm. What makes them even more interesting is the fact that the information they contain can be compared with that on the household index cards drawn up on the basis of the census lists. For example, we can compare the number of occupants of a farm with the equipment used. Because of the difficulties of tracking documents down, only seventy-five post-mortem

Table 23. *Inventories broken down by age of deceased*

	under 30	30–40	41–50	51–60	61–70	over 71
1820–1830			1			
1831–1840		2	1	1	1	1
1841–1850	3	3	3	2	3	
1851–1860	1	5	7	3	2	
1861–1870	1	1	6	3	2	
1871–1880		1		2		1
1881–1890		1	1	1		
1891–1900		4		3		
1901–1910		3	3	2		

inventories were found between 1821 and 1920. Given the average figure of 180 households per census, that is not a lot. Why so few? Deficient data gathering is to blame, as are the circumstances in which these documents were drawn up (one spouse of a couple had to die leaving children, which was always the case). The irregularity of these documents thwarted a plan to observe the development of one household's standard of living through the nineteenth century and to compare it at given moments with all the households in the village. The fact is that the post-mortem inventory did not catch every household at the same stage in its cycle – and a household did not enjoy the same patrimony when death came to it at the age of thirty as when it struck at sixty.

Moreover, the inventories record only movables – and probably not all of those, as we have seen. The separate property of both spouses was excluded, so we find no mention of jewellery, pipes, penknives, or devotional objects. One also wonders how much was concealed – cash kept in a stocking, harvests, livestock, and so on.

The distribution of our inventories by period looks like this (for distribution by age of head of household, see table 23):

1821–1830	1 inventory
1831–1840	6 inventories
1841–1850	14 inventories
1851–1860	18 inventories
1861–1870	13 inventories
1871–1880	4 inventories
1881–1890	3 inventories
1891–1900	7 inventories
1901–1910	8 inventories (including 4 'estimated schedules of assets' doing duty for post-mortem inventories).

The typical Bigouden farm was as small in the nineteenth century as it had been in the eighteenth century. There was the same house consisting of two rooms with mud floors, situated in a hollow near a vegetable plot. Each end of the house (the *bout levant* and the *bout couchant*, according to the inventories) was fitted with a fireplace. The main room did duty as kitchen, bedroom, living-room, and dining-room. The other room might be used for cooking the animals' food if it did not accommodate a second household. The animals occupied an adjacent building; in Bigouden men and beasts were separated by floor-to-ceiling walls, in contrast to other regions where the wall rose only to a height of one and a half metres.

Habitat advances, judged in terms of salubriousness and technical efficiency, were very slow. In 1849 du Chatellier noted:

> A few well-off proprietors have enlarged their dwellings, but for the most part without discernment – that is to say, without enlarging the apertures or increasing the height of the rooms, and above all without isolating the dung and rotting vegetable matter that in the wettest months of the year surround their houses with a stagnant pool into which all manner of refuse is thrown every day. As for the dwelling of the small tenant farmer and the poor man, its value cannot be put above 800 to 1,200 francs, furniture included, for a farm of between 300 and 400 francs' rent.[37]

A small house, then, with outbuildings arranged around a yard containing the dunghill. It is a sight that can still be seen today in certain hamlets of Saint-Jean-Trolimon (just such a farm was chosen as the location for Claude Chabrol's film *Le Cheval d'Orgueil*). Most houses, in fact, were rebuilt before the First World War, when another storey was added and the rooms were better separated. But it was not until the early years of this century that the change occurred.

The post-mortem inventories will reveal the reserves stored in the upper storey of the house, the sheds being used mainly to house tools and vehicles. The animals were counted in their stalls, which were either adjacent to the house or in a separate building.

Substantial disparities between farmers

The function of the post-mortem inventory was to value all the movable belongings of the household; furniture, livestock, cereals, linen, farm implements, and harvests stored or still in the field. After pages and pages of description, we find a figure representing the assets of the household. The first striking fact to emerge from an examination of the inventories collected is the degree of variation in that figure. It ranges from

FR 617.80 to FR 6,134.10 for the period 1841–1850
FR 303.05 to FR 14,823.20 for the period 1851–1860
FR 1,020.00 to FR 45,184.35 for the period 1861–1870
FR 364.25 to FR 66,813.55 for the period 1871–1880
FR 1,539.00 to FR 27,306.75 for the period 1881–1890
FR 1,025.30 to FR 4,004.22 for the period 1891–1900
FR 2,088.00 to FR 18,777.35 for the period 1901–1910

The assets of the various successions cover a range of which we can be sure of having apprehended the two extremes because the inventories with the highest asset balances invariably relate to couples whose marriages fell within our field of observation as belonging to the most well-to-do section of the peasantry. The inventories showing the lowest asset balances relate to the most deprived category of the population, identified as such in the censuses. The latter figures closely match those recorded by du Chatellier. The range itself extends from 1–10 to 1–200, according to period. These figures reflect the gap that might exist between two households living at the same period. However, the difference is not due to amounts of livestock or cereal reserves; it stems essentially from sums of money that were either kept in cash, placed with the notary, or lent to third parties.

For example, in one post-mortem inventory, which shows assets of 66,813.55 francs, credits accounted for 55,685 francs and cash for nearly 10,000 francs. This particular inventory seems exceptional from every point of view. Yet hoarding of money appears to have been the rule in all the inventories recording assets in excess of two or three thousand francs. The joint estate of husband and wife held a part of its variable asset balance (between 5 and 50 per cent) in credits. Occasionally a large sum of cash is recorded, as in the inventory drawn up at Treganné Creis on 3 May 1864, in which the asset balance of 11,714.75 francs included 4,976.25 francs in ready money. Possibly death came to the man unexpectedly just as he was preparing to make a deal of some kind.

An important corrective qualifies any conclusions that may be drawn from the value of the assets recorded, namely the amount on the debit side. Once he had established the assets of the joint estate, the notary proceeded to calculate its liabilities in the shape of debts of all kinds owed to third parties, including loans from notaries. Where this figure exceeded the value of the assets, it may have signified the end of a particular farm, the departure of the household from the house it had occupied, financial ruin, and possible loss of social status. This was not always the case, however, for the difference may have been made up out of the separate property of the surviving spouse through a combination of transactions not recorded in the inventory.

An unbalanced inventory does not necessarily imply a move. For example, following the death of Corentin Le Maréchal in 1855, his inventory showed a total valuation of 3,608.20 francs. However, there were debts of 13,948 francs consisting of sums owed to the landowner who had made over the farm at Kergreac'h as well as various loans contracted with notaries in Pont-L'Abbé and Quimper. When his wife, Marie Courtès, died ten years later, the valuation was about the same but the amount of debt recorded in the post-mortem inventory was less – about 3,500 francs. In the ten years that had elapsed between the two deaths the farm had managed to pay back nearly three-quarters of the debt and was now being worked freehold.[38]

Table 24 summarises the balance between assets and liabilities and gives a figure expressing the ratio of debts to total assets. These estimates cannot provide a complete picture of the economic balance of all the farms of the commune. But we do find a majority of farms with assets greater than their liabilities, albeit not by much. On the other hand indebtedness was generally not catastrophic, though there were exceptions. It seems then, that throughout the nineteenth century the farms of Saint-Jean-Trolimon hovered around a fairly precarious point of equilibrium.

These figures are evident of the stability, not to say the stagnation of the economic situation during the nineteenth century. It was certainly possible to accumulate capital in the form of credits, but not to any great extent because of the value of land, as we saw in chapter 6. When a farmer was able to purchase the land he farmed, he had to go into debt and was often obliged to resell immediately some of the land he had bought. If we compare the price of land with the total amount of assets inherited, the practice would seem to have been unavoidable.

Finally, analysis of the ratio of assets to liabilities is disappointing in so far as, even if it covers a broad spread of peasant society, it cannot serve to identify clearly the various social categories. On the other hand, leaving debts on one side, analysis of the composition of assets, breaking them down into furniture, livestock, and crops, is extremely interesting when it comes to studying the operation of the internal hierarchies of such a society.

It is in studying the composition of the assets recorded that we come face to face with the idea of the life cycle of the household. Two asset balances in the same amount may in fact capture their respective households at different points in their existence: at the beginning, before the full potential of the farm had been developed, or at the end, when transmission has already been effected. So to be sure that we are comparing like with like we have to exclude the inventories of deceased

Table 24. *Ratio of assets to liabilities of farms taken from post-mortem inventories*

1831–1840	1 indeterminate instance (where the debts are mentioned as 'reminders') 5 instances where assets exceed liabilities, including 3 that come out at the following levels: 0.07, 0.009, 0.34
1841–1850	10 farms where assets exceed liabilities, including 2 where the ratios are 0.56 and 0.9 4 farms where liabilities exceed assets at 1.15, 1.53, 2.06 and 4.80
1851–1860	4 instances impossible to determine ('reminders') 11 farms where assets exceed liabilities (in 6 instances the figures are: 0.55, 0.45, 0.76, 0.36, 0.49, 0.19) 3 farms in debit at 1.77, 3.86 and as much as 22.98 in one instance where assets of FR 1,071 face liabilities of FR 24,627!
1861–1870	10 farms where assets exceed liabilities (in 7 instances the figures are 0.07, 0.19, 0.23, 0.36, 0.12, 0.06, 0.58) 2 farms in debit (1.15 and 1.52)
1871–1880	impossible to determine
1881–1890	1 instance of a farm in debit 1 instance of a farm in credit
1891–1900	5 farms where assets exceed liabilities (0.36; 1.56; 0.11; 0.15; 0.4 and 0.006)

who were either too young or too old and work only on those belonging to the 41–70 age group.

Table 25 represents a comparison of forty-six post-mortem inventories, showing the age of the head of household, the total amount of assets, including credits and cash, and the balance of the valuation. Within this estimate three items are analysed: livestock, cereals and potatoes (stored or in the ground), and domestic furniture – beds, tables, wardrobes, *banc-coffre*.* Each value expressed in francs is followed by a figure in brackets expressing it as a percentage of the total valuation.

The first thing the table shows is the relative fluctuation of the various items: livestock did not always account for the bulk of the assets; sometimes it was crops. Nor is there any strict connection between the total valuation and the relative structure of the three items. To put it another way: the wealthiest inventories did not consist of a fortune in cereals or livestock, any more than the poorest had one item that was always preponderant. Such differences must be attributed to the particular structure of each farm at that moment in the family's life cycle. Our having eliminated the extremes is not enough to ensure perfect

* A piece of furniture that served the triple function of seat, step up into a box bed, and storage chest [Translator].

Table 25. *Comparative table of the valuations of 46 post-mortem inventories (in francs)*

	Age of head of household	Assets	Debts	Cash	Valuation	Livestock	Cereals and potatoes	Movables
1829	48	4,492.00	106.00		4,386.00	1,453.00 (33)*	882.00 (20)	361.74 (8.2)
1831–1840	53	1,138.50	52.50	117.25	968.75	135.00 (13.9)	318.00 (32)	112.65 (11.6)
	45	1,142.40			1,142.40	699.00 (61.2)	?	107.50 (9)
	49	593.00	12.00	83.00	498.00	66.00 (13.2)	132.00 (26.5)	124.15 (24.9)
1841–1850	41	1,385.85		157.78	1,228.10	346.00 (28.2)	198.50 (16)	132.00 (10.7)
	45	3,113.75	744.00	180.00	2,189.75	644.50 (29.5)	353.00 (16)	147.00 (6.7)
	66	683.70	330.00		353.70	45.00 (12.7)	?	?
	51	1,528.70	120.00		1,408.40	658.00 (46.8)	438.00 (31)	68.00 (4.8)
	58	6,134.10	3,293.30	20.00	2,820.80	902.00 (32)	796.00 (28.2)	206.25 (7.3)
	49	1,941.40			1,577.40	528.00 (33.5)	343.00 (21.7)	183.00 (11.6)
	67	1,932.45			1,932.45	891.00 (46.1)	390.00 (20)	45.25 (2.3)
	67	5,050.00	3,725.25	750.00	574.75	81.00 (14.1)	27.50 (4.7)	96.00 (16)
1851–1860	54	1,654.95			1,654.95	506.00 (30.6)	462.00 (27.9)	180.35 (10.8)
	51	14,823.20	10,301.60	125.00	4,396.60	1,845.00 (42)	1,016.60 (23)	214.50 (4.8)
	44	3,989.50			3,989.50	1,766.00 (44.3)	1,296.00 (32.4)	300.00 (7.5)
	45	1,498.95			1,498.95	482.50 (32)	605.00 (40.3)	75.00 (5)
	50	3,608.20			3,608.20	1,961.00 (54.3)	480.50 (13.3)	153.25 (4.2)
	40	5,379.65	1,250.50		4,129.15	2,282.00 (55.2)	294.00 (7.1)	309.25 (7.4)
	51	3,423.50	306.50	1,595.00	1,522.00	406.00 (26.7)	150.00 (9.8)	242.00 (15.9)
	66	1,514.50			1,514.50	580.00 (38.3)	256.00 (16.9)	186.25 (12.2)
	66	1,071.40		164.65	906.75	144.00 (15.9)	32.00 (3.5)	146.00 (16.1)
	46	12,892.80		4,166.60	8,726.20	2,546.00 (29.2)	3,745.25 (42.9)	482.00 (5.5)
	45	303.05			303.05	139.50 (46)	?	48.75 (16)
	44	1,137.65	900.00		237.65	75.00 (31.6)	40.00 (16.8)	35.00 (14.7)
1861–1870	45	23,084.00	9,710.65	80.00	13,293.35	1,353.50 (10.2)	1,298.50 (9.7)	212.00 (1.5)
	59	3,460.10			3,460.10	1,253.00 (36.2)	1,504.00 (43.4)	221.25 (6.3)

	69	45,184.35	44,515.00		669.35	251.00 (37.5)	54.00 (8)	134.80 (20)
	51	3,648.50		690.00	2,958.55	1,120.00 (37.8)	695.50 (23.4)	210.50 (7)
	60	5,988.75	4,080.00		1,908.75	762.00 (39.9)	666.00 (34.9)	50.25 (2.6)
	45	8,869.65			8,869.65	2,446.00 (27.6)	3,516.00 (39.6)	649.00 (7.3)
	62	2,199.80			2,199.80	932.00 (42.3)	607.50 (27.6)	73.00 (3.3)
	44	2,406.15			2,406.15	512.00 (21.3)	595.50 (24.7)	123.00 (5.1)
	48	11,714.75	2,057.00	4,976.25	4,681.50	2,249.00 (48)	1,043.00 (22)	467.00 (10)
	44	10,755.00	5,479.75		5,275.25	1,436.00 (27.2)	1,373.00 (26)	239.50 (4.5)
	50	1,020.00	900.00		120.00	66.00 (55)		42.00 (35)
1871–1880	52	1,344.00			1,344.00	475.00 (35.3)	513.00 (38)	123.25 (9.1)
	56	364.25			364.25	125.00 (34.3)	262.00 (72)	24.00 (6.5)
1881–1890	62	1,901.00		500.00	1,401.00	546.00 (38.9)	309.00 (22)	202.00 (14.4)
	43	27,306.75	23,641.75	550.00	3,115.00	990.00 (31.8)	1,213.00 (39)	207.00 (6.6)
1891–1900	57	2,471.10	2,000.00		474.10	141.50 (29.8)	76.00 (16)	63.50 (13)
	52	1,549.75	131.25	20.00	1,398.50	835.00 (59.7)	324.00 (23)	184.00 (13)
1901–1910	51	18,777.35	16,507.35	40.00	2,230.00	1,460.00 (65.5)	260.00 (11.6)	190.00 (8.5)
	42	5,292.00		90.00	5,202.00	2,200.00 (42.3)	1,500.00 (28.8)	390.00 (7.4)
	41	11,848.04	11,578.04	70.00	220.00	0.00	?	128.00 (58)
	55	2,088.00		2,088.00	1,745.00	40.00 (2.2)	?	107.00 (6.1)
1911–1920	44	18,170.55	13,395.55	500.00	4,275.00	2,275.00 (53.2)	407.00 (9.5)	424.00 (9.9)

*The figures in brackets represent the item expressed as a percentage of the total valuation.

comparability. These variations also tend to show the margin of manoeuvre that each farmer had at his disposal, since he was free to distribute his assets as he thought best. The main constant relationship between the items is the way in which they form a hierarchy; as a rule livestock attained a higher estimated value than did crops, and both these items had entries higher than the one relating to domestic furniture. Mean distribution is as follows for the three items:

Livestock	30 to 50 per cent
Crops	10 to 40 per cent
Furniture	5 to 20 per cent

with respective averages of 37 per cent, 24 per cent, and 4.8 per cent. So there does appear to have been a structural relationship between the value of livestock (and hence its importance) and that of agricultural production, the two items being connected by the surface area of the farm in question.

While the spread under the item agricultural production might be from 1 to 70 between an inventory from a poor household and one from a wealthy household, that under the item domestic furniture was very much smaller, ranging from 1 to 20 at the two extremes. Once again the spread is attributable to the relevant point in the family life cycle and to the extent of adult co-residence. We know that every couple owned a wardrobe; indeed, it was a symbol of the married state. The difference in value under the item furniture often stemmed from the number of wardrobes, which was itself linked to the number of couples living under one roof. Comparing, for each inventory, wardrobes and beds with the number of persons living on the farm, we are struck by the number of wardrobes: invariably more than one and sometimes as many as five. They also varied in the number of doors they contained (between two and five). The large number of wardrobes is explained by their many functions. They were used for storing linen but also personal papers and cash; and clearly they doubled as larders, since that was where the butter was kept.

Beds were similarly numerous. The inventories distinguish between 'bedstead', 'bed with accoutrements' (meaning with mattress, bolster, and bedclothes), or 'bed with bench'. Adding these three types of furniture together invariably gives a figure lower than the number of persons sharing the dwelling. The number of beds increases only in the inventories of well-to-do households. When Jeanne Le Lay died at Kervouec in 1864, she left eight children between the ages of eighteen and one. Seven 'accoutred beds' thus accommodated parents, children, and four servants. Around the same period, in Jean Tanneau's house at Trevinou six adults, four children, and three servants occupied five *lits à banc-coffre*. At Kerveltré, on what is by no means one of the poorer farms, six adults and one child shared only two beds.

The average number of beds was between three and four, whereas the number of persons sharing the same dwelling was between six and fifteen, with variations dictated by the family life cycle and the standard of living enjoyed. We know, incidentally, that better-off households constituted the majority.

An interesting detail regarding sleeping arrangements is the small number of cots in what was after all a highly prolific society. Even in families with very young children, we hardly ever find a record of a cot. There are two explanations for this apparent anomaly: either the notary did not enter this item, which was not regarded as forming part of the joint estate, or there was no cot, women being in the habit of keeping new-born children that they were still breast-feeding near them.[39] The child might be placed in the parents' bed, though according to some accounts babies were laid in the bottom of the wardrobe with the door left open. Once weaned, children slept in the same bed as their brothers and sisters until puberty, when the sexes were separated.

The communal room was thus lined with one-storeyed or two-storeyed box beds and wardrobes. Another piece of furniture always present was a table, described as *coulante* or *coulissante* (sliding) or *à manger* (dining), The tabletop also served as a kneading surface and either slid off or lifted up to reveal a compartment in which dough could be left to rise. Always associated with the kneading table was the bench seat, with or without arm rests. This collective seat also represented a particular way of sitting with the buttocks well back to keep the knees clear of the belly of the table.[40] The flat table and the chair, when they came in, also represented a new way of holding the body, erect and individually. Chairs figure in the inventories only very irregularly; people continued to sit collectively as they lived and slept collectively. Flour bins and kneading troughs are also present up until 1870, while dressers and sideboards appear only sporadically. Grandfather clocks – an infallible indication of wealth in the mid nineteenth century with a value sometimes amounting to sixty francs – were common in the early twentieth century, even among the poorest households.

Looking at people's furniture thus reveals a uniformity of life style running right through the economic hierarchies. As far as comfort was concerned, the wealthy peasant lived in much the same conditions as the poor. The difference came out in other areas, namely in everything to do with farming methods.

A relatively uniform life style

Analysis of the rest of people's domestic equipment fills out the picture of a traditional way of life both in the matter of eating habits and with regard to the care of household linen. Again there was very little difference between levels of affluence.

The basic equipment hardly varies from one inventory to another. All mention a trivet, a pan for making *crêpes* and *galettes*, described as being kept near the fireplace, as well as a cauldron, still called a *bassin d'airain* or 'bronze pan', and a cooking pot. This was the equipment required to prepare the gruels and pancakes that constituted people's basic fare. Some inventories recorded a number of cooking pots (from three to six), indicating either a more varied cuisine or more people living in the house. Other items mentioned include bowls, spoons, earthenware vessels, milk pans, ladles, plates, and soup tureens; less frequently mentioned are knives, forks, a *castelle* or *ar hastell* (a box in which bread dough was placed to rise), salt containers, and salad bowls. There would be little point in expressing all this household equipment statistically. Presumably a degree of variation arises out of the descriptions themselves, and there is no guarantee that they were either exhaustive or comparable. At most we may note the presence in the wealthiest households of trenchers accompanying the bowls (though not invariably) or of a bread basket associated with the sliding table. A coffee pot is mentioned in 1885; elsewhere there is a record of glasses.

Bowls and spoons tended to be equal in number, ranging from four to fifteen; they were individual utensils. Individual knives, on the other hand, were unusual; instead there was a large kitchen knife for cutting bread or bacon.

Lastly, the kitchen equipment listed in the post-mortem inventories includes what are often large numbers of churns and milk cans. Picturesque descriptions of kitchen and dairy utensils and individual enumeration of each object, together with its valuation, disappeared around 1890, when notaries began to adopt the formula *la batterie de cuisine et les ustensiles de laiterie* and to give a lump-sum valuation.

Quantities of household linen (sheets, feather beds, tablecloths) varied more substantially from dwelling to dwelling. Here it was not the number of co-residents that determined the amount but the affluence of the household. The wealthiest had more than thirty pairs of sheets (counting those that were in the wardrobe as well as those on the beds) and more than ten tablecloths, used to dress the table or tables on the occasion of family celebrations, wedding banquets, or *pardons*. One farm even owned an embroidered sheet used for laying out the dead.

And finally there are mentions of stocks of yarn, skeins of hemp yarn and tow valued in metres, kilograms, bundles, remnants, or individually, showing that up until 1870 fabrics continued to be manufactured in the home. There were few spinning wheels and only a small number of reels for turning skeins into bundles.

The inventories also list such washing equipment as steam coppers and grinders for *ludu* or ashes (or was this ash as used as a fertiliser?).

As regards clothing, the inventories, for all their wealth of description, rather let us down. *Hardes*, as a person's personal effects were called, were seldom either listed in detail or valued, for they were the separate property of the deceased. Occasionally there is mention of 'personal effects of the deceased that will be his exclusive property, the widow having retained her own in compensation'. Clothing was thus a special case and never formed part of the joint estate constituted at marriage. As far as this category of property was concerned, the age of the deceased was of prime importance. Old and young did not have the same wardrobe, either in extent or in value. Hence the big differences in valuations of clothing that bear no systematic relationship to levels of affluence. The wealthiest households did not have a notably more valuable wardrobe than did the less well-off. The range of valuations extended from fifteen to over 300 francs, with the average lying around 80–100 francs.

Jean-Marie Le Berre, the publican in the *bourg* who died in 1848 at the age of forty-seven, must have cut a fine figure in his great buffalo-hide belt (estimated at five francs) and his two hats. In an otherwise very modest inventory, his clothing was valued at the considerable sum of 96.25 francs. He had no fewer than sixteen shirts, three pairs of breeches in *nozé*, linen, and cloth, two pairs of trousers. eight *pourpoints* of jackets, six blue and black waistcoats, and a pair of gaiters.[41] The description we have of Breton clothing in the early nineteenth century tallies with this wardrobe: 'It consists of very wide breeches, a jacket, and several waistcoats, with a broad leather belt underneath. The number of waistcoats is in proportion to the affluence of the inhabitants. The wealthiest wear up to five, those underneath being longer in such a way that all five can be seen.'[42]

In 1866 Corentin Le Donge died at the age of forty-four. He left an impressive wardrobe: four jackets and five waistcoats, forty-one shirts, eighteen pairs of trousers ('including three old ones'), two hats, three pairs of stockings, and a pair of shoes, amounting to a total value of 288.75 francs.[43]

As far as women are concerned, we are able to compare the post-mortem inventory of Jeanne Le Donge, who died in 1847 at the age of twenty-two, with that of Marie-Anne Gloaguen, who passed away in 1864, aged forty-eight.

The former was recorded as owning a cape, thirteen skirts of which some were 'old' or 'shabby' (the rest were of linen, tow linen, hemp, *nozé*, or cloth, and a number were described as underskirts), eight aprons, including one of black silk, twelve *justins* (a tight-fitting bodice) and four waistcoats (their colours are described: black and red, white and red, white, white on black, and so on), one jacket, eight petticoats, thirteen blouses, two pairs of slippers, three pairs of stockings, one pocket handkerchief and one mirror, eight ribbons, twenty-seven *coiffes*, eight small undercaps, and fifteen mob caps, the whole valued at 332.05 francs. Jeanne Le Donge had received a splendid and unusually extensive wardrobe at the time of her marriage.[44]

Marie-Anne Gloaguen's wardrobe was more typical. The inventory drawn up after her death put a value of sixty-eight francs on two skirts, four aprons, two *cotillons* or better-quality skirts, one waistcoat, six blouses, a number of bonnets and mob-caps, one pair of stockings, and one pair of sabots.[45]

Contrary to the image often given of rural society, which posits dress as an index of social class, it does not appear from these inventories that social hierarchies found any direct expression in a distinct wardrobe. Clothing was not, any more than furniture or domestic equipment, the object of the substantial social and emotional investment that many observers have sought to tell us was the case. Well-to-do and less well-to-do shared the same culture, any difference being perhaps evinced more in the quantity of linen and clothing than in their quality. Moreover, for festivals and weddings in particular it was not unusual – many of those interviewed said this – for a woman to borrow a waistcoat from a relative or neighbour. There was no direct connection, in other words, between appearance and level of affluence.

Similarly, diets and methods of heating differ little from one inventory to another. In this damp climate, heating was provided by open fires. There were often two hearths – one at each end of the house – and that was where the trivets and pots and pans were inventoried, all cooking being done over the open fire.

Many inventories valued the wood pile. Firewood appears never to have been short, except on the sandy coastal strip. The cutting of firewood, particularly on the embankments between fields, was strictly regulated in leasing agreements, which distinguished between wood belonging to the landlord and the wood that the domanial tenant might cut for his own use. There is also mention of 'blocks of peat' that people went to cut in the étangs. Houses were poorly heated, to judge by modern standards (but then the human body is capable of adapting to a wide range of temperatures). They were also very dimly lit. Only half the inventories mention candlesticks, and only one a quantity of candles. Probably these were overlooked, but there was certainly very little household illumination. People took advantage of the long spring and summer evenings to work hard, and in the winter they went to bed early. There is the occasional mention of a lantern, used for going out to the stable at night.

The inventories also give us a picture of what people ate, quite apart from what we can learn from the cooking utensils recorded. The relatively common presence of a gun, for example, indicates that game was consumed. Inventoried under the heading 'food supplies' there is nearly always a 'steep-tub with bacon' and one or more cakes of fat; more rarely we find a 'pot of lard'. Despite the inevitable presence of churns and other dairy utensils, only rarely is there any mention of butter. When it was recorded, it was in large quantitities of ten or twenty-four kilograms suggestive of a commercial enterprise. Also mentioned very occasionally are leguminous plants (mainly dried peas) and onions. Potatoes formed the staple diet, together with bacon; they alternated with preparations based on cereals, milk, and butter such as gruels, pancakes, and biscuits (*galettes*).

As regards drink, there are very erratic mentions of casks and barrels. The wealthiest households might own up to eleven barrels, but usually there was only one. Does this mean that cider-drinking was general? Probably it was, because the inventories often mention the apple presses used in the manufacture of that beverage.

As well as cider and butter the household also made its own bread, though the objects used in its manufacture are mentioned very much less systematically than those used to make pancakes. The occasional inventory mentions a bakehouse with its associated utensils: peels and markers to identify the products of different households.

Bread consumption was not as common as in other parts of France, where it formed the basis of the diet. Here it was rivalled by the pancakes and gruels that represented an original form of cereal consumption.

A mode of production based on manual labour

The post-mortem inventory offers us a glimpse of a life style that was also a mode of production. Farm implements, furniture, domestic equipment, and costume were common to all the inventories. The homogeneity of the tools used to till the soil and plant, sow, and reap the crops was also quite remarkable.

Preparing the ground and spreading seaweed or 'mixed manure' involved the use of shovels, forks, pitchforks, and muck rakes (*crocs à fumier*, also called *crocs à framboas* or *framboise*). Hoes and picks were also common, as were spades and harrows, all these implements being used to aerate the soil preparatory to sowing. From the very first inventories we find mention of metal or wooden ploughs; in 1864 the wealthiest household in our sample owned a Dombasle plough.* Where an inventory makes no mention of a plough it is difficult to be sure why. Either the deceased will have already dispersed his farm implements, or he possessed none, working as sub-tenant of a larger farmer who lent him his plough and team of oxen or horses.

Harvesting was mainly by sickle. Scythes are rarely mentioned, and always in the singular, whereas the inventories record quantities of sickles or reaping hooks (six, eight, eighteen, even twenty-one in one instance). These would have been distributed among the team of harvesters that came to the farm to bring the harvest in.

The grain was removed from the ear with the aid of a threshing sheet and a horse-driven mill. Most of the inventories mention several threshing sheets and numerous sieves. There is only one mention of a flail. The horse-driven threshing-mill is not mentioned explicitly; its existence is assumed from contemporary descriptions, detailed analyses of agricultural equipment, and the evident ubiquity of horses.

Potatoes required a great deal of manual labour. They were often put in the ground by hand with the aid of a special rake; they were hoed and weeded by hand; and they needed to be earthed up, also by hand. The crop was then lifted with a hand-held implement – hook, hoe, or fork.

In this production cell in which agriculture and stock-rearing went hand in hand it may seem surprising that there is so little mention of dunghills and strawstacks. These two important elements in the cycle of production had a special status. Without being regarded as real estate, they belonged more to the landowner than to the domanial tenant. Were the latter to be dismissed, he would be compensated 'for his edifices, surfaces, straw, and manure that may not be sold or transferred elsewhere'. The notary never inventoried the strawstack, which was located in its own open yard, any more than he did the fruit trees in the orchard.

Tools and equipment might be virtually identical from one farm to another, but the same was not true of the livestock, the amount of which would depend on the size of the farm. All the inventories list cows, which varied in number from one to eight, the largest numbers

* An invention of the French agronomist Mathieu de Dombasle (1777–1843) [Translator].

belonging to the principal butter producers. A farm with three, four, or five cows not only had enough milk for its own use; it could also produce calves, either for the table or (more commonly) for sale.

Horses (from one or two up to six) are almost always mentioned. Between two and six oxen (with pairs often referred to as 'the two plough-oxen') are regularly present. Steers and heifers (from two to ten) also formed part of the livestock on most farms. Pigs (between one and five, with a single exceptional case of fourteen) were raised on the farm to provide bacon. With three exceptions, there were no sheep; widely reared in the eighteenth century, they disappeared in the nineteenth. The inventories also record a few goats and beehives. Very rarely is there any mention of poultry; ducks occur only once, hens seven times, and geese five times. But does this necessarily mean poultry were not kept? It is more likely that their value was small and the notary's time too precious for him to go chasing round the farmyard attempting to count them. Lastly, we find three bulls belonging to households whose inventories came to substantial totals. But if there was little difference in the kinds of animal kept as between one holding and the next, certain farms were distinguished by the numbers of cows, horses, ploughing-oxen, heifers, and steers owned.

An analysis of livestock numbers reveals two types of farm: one with around three of these animals, and another, wealthier type with eighteen or twenty. This disparity in numbers had to do with the hierarchical system of inter-farm co-operation. The poorer farms assisted the wealthier ones in performing the major operations of the farming calendar, while the wealthier farms lent the poorer ones horses or oxen to do their own ploughing or harvesting.

The animals used for working in the fields and for transport also had to be fed. Apart from the hay garnered each year, their fodder consisted of bruised gorse, and most of our inventories include a 'hammer for bruising gorse'. The plant was placed in large granite troughs. These later became watering-troughs, but their primary function had been to hold the boughs of gorse while they were being bruised – an essential operation if the plant was to be rendered digestible by the animals. The marks of the hammer can still be seen on the edges of these stone troughs today.

Lastly, the inventories listed all the tools, implements, and other objects that can be classed under the heading of means of transport: three or four yokes for the 'ploughing-oxen', a number of collars, various traces and shafts for harnessing horses. Vehicles used for transport included various handcarts, some with shafts, some long, some short, some enclosed, some coupled to the *charrette*, a cart that was harnessed to an animal. A farm might have two, three, or four of these light vehicles or wheelbarrows, but there would be only one *charrette*. Frequent mentions of pipe-boxes and wheel rims are an indication of the extent to which the 'do-it-yourself' mentality reigned on the farm. Rarely was the cartwright called in; when a wheel needed repairing, a group of farmers would get together on a mutual-aid basis. These vehicles were used for transporting various loads (hay, harvested wheat) and possibly, in the first half of the nineteenth century, people too. The famous *char à bancs* or horse-drawn charabanc devised for this purpose is not mentioned regularly before 1880. Prior to that date the most usual ways of getting about must have been on foot or on horseback.

To sum up, this brief examination of a number of post-mortem inventories shows the simplicity of farming methods based essentially on manual labour. There was no mechanisation as yet, and the implements available were mere auxiliaries to human or animal muscle power. In the final analysis it was livestock in relation to surface area farmed that distinguished one household from another.

Marie-Louise Le Berre died in 1859, aged forty-four. A widow with two children by her first marriage, she had remarried and borne a son to her second husband, Pierre Berrou. The household had comprised the husband and wife, three small children of two marriages, a sister of the husband (she is described as 'lame'), and three servants. Three adults, three children, and three servants – not in fact very many, given the average size of household living under one roof.

Her assets amounted to 3,989.50 francs. That was in fact the total valuation, there being neither credits nor cash. By contrast the succession was burdened with debts arising out of various financial transactions. These included family transactions pursuant to the purchase of reparative rights from and eviction of brothers and sisters of Marie-Louise Le Berre's first husband and transactions with the notary in connection with purchasing the freehold of Kergonan. Though he had no right of ownership over the farm, the widowed Pierre Berrou continued to work it for a long time and was later to marry again. Here we see the full interest of the post-mortem inventory, recording as it did assets belonging to minors.

Let us join the notary, Maître Ronac'h, as at ten o'clock on the morning of 29 November 1859 he steps inside the house on Kergonan farm to draw up Marie-Louise's post-mortem inventory.[46] Three beds complete with bedding, a kneading-trough, a wardrobe with four doors, some cakes of fat, three more wardrobes, a sideboard and a dresser with bowls and spoons, two brass pans, another for making pancakes, a large cooking-pot, the 'sliding' table and a seat with back surmounted by a cupboard are valued one after another. The notary does not draw the usual distinction between the 'rising end' and the 'setting end' of the house, but it is clear that his description and valuation make a tour of the room, eventually reaching a heterogenous group of objects situated near the door: jars and earthenware vessels, axes, sickles, wheelbarrows (short and long), two old wheels, and a plough with attachments. Probably the notary stepped outside the house at this point and valued the plough from the doorstep. In the *hangar* (shed) – often termed *grange* (barn) in other inventories – he finds various farm implements: yokes, hoes, shovels, hooks, small pitchforks, a saw, a pick. He also finds the ropes used to harness the vehicles. The shed also contains certain household stores: carrots, parsnips, a 1,400 kilogram heap of small potatoes and a 2,000 kilogram heap of large potatoes (which he values at 300 francs – a tenth of the total valuation), and six bundles of hemp. Here, too, he finds the threshing equipment, a rope loom, and two ladders. The notary then opens the door of the laundry, which besides laundry equipment houses various other items: harrow, casks, cider press, and hemp for hackling are ranged alongside or above five heifers and a young bullock.

The cowshed houses eight oxen, including the two black ploughing-oxen (worth 390 francs by themselves). The pigsty contains a sow and three hogs and the stable four horses, two of which account for 300 francs together. The notary continues his tour with a visit to an open field adjoining the next hamlet where seven cows are grazing.[47] In what he terms the *placitre du village*, the piece of ground common to the farms of the hamlet, the notary lists three stacks of logs, some hemp brakes, oven utensils (a peel, some markers), and a grindstone. He returns to the house, where an upstairs granary contains the cereal stocks: 3,500 kg of barley, 1,500 kg of wheat, 150 kg of oats. Here, too, are the threshing sheets and that vital instrument for cereal trading, a balance with a set of weights. This upper floor of the dwelling house also contains the larder, at its emptiest at this time of year. It was not until the end of November that the pig was killed. Finally, the notary opens 'the deceased woman's wardrobe', one of the five on his list. In it he finds six feather mattresses, ten sheets, six tablecloths, and forty hanks of tow yarn and thirteen of hemp yarn (valued at nearly 90 francs). The dead woman's personal effects, inventoried last of all, are valued separately at a mean figure of 71.50 francs.

The affluence of this household was based on its livestock and its cereal and potato reserves. The absence of any reference to carts, harnesses, and yokes is surprising. Were they overlooked?

Here by way of comparison is another inventory, taken around the same time, in the middling sum of 1,295.70 francs. Jacques Guegaden died in December 1855 at the age of thirty-nine on a small farm at Le Steud, which he had leased for 180 francs per annum. The need for an inventory cannot have arisen immediately, because it was almost a year after the event that the notary was summoned to draw up the document (a costly operation, obviously representing a difficult decision for small households to make). What happened was that the widow, Marie-Jeanne Le Brun, remarried. Her new husband was Félix Blayau, the shoesmith and publican, and the date of their wedding was 9 October 1856 – two days after Maître Kernilis and his colleague had visited Le Steud to establish an 'accurate and faithful valuation of all the movables and other personal objects generally and severally belonging to the joint estate that existed between the said Marie-Jeanne Le Brun and the late Jacques Guegaden'.[48] The purpose of the inventory was to protect the rights of the couple's four children.

On this farm the notary did not bother to indicate the positions occupied by the objects described. Nevertheless, a relative order in their enumeration enables us to trace his progress round the property. First comes the furniture, the sliding table, a bench, two box beds (for the two parents and four children) fully equipped with sheets, feather mattresses, and bolsters, three wardrobes (one 'of poor quality'), and four wooden stools. Then come the kitchen utensils: two cooking-pots and two cauldrons, two churns and other dairy equipment including a milk strainer and pan, two trivets and a pancake pan, dishes, plates, jars and bottles, and lastly the knife and a table basket.

All question of value aside, the basic equipment is very much the same as that described in the Kergonan inventory.

The notary must then have entered a different building – though there is no mention of the fact – where he found five sickles, together with the hammer and anvil used for sharpening the blades, four sieves and two bolters, a reel, two muck rakes, some picks, a quantity of prepared timber for building a wheelbarrow (the household's sole means of transport), and a ladder. This building also contained the cereal reserves (9.1 hectolitres of wheat, 11.2 hectolitres of barley, and 4.8 hectolitres of oats) and a heap of potatoes, which accounted for more than a third of the total valuation. Still in the same building, the notary counted the household linen (tablecloths and napkins) and six grain sacks. It is easy to imagine these items stored away in a loft, since they were used only occasionally – the tablecloths for big meals on feast days and the grain sacks at threshing time. Besides the cereal reserves and the linen, the notary also found the hemp reserves, here in the form of sheaves of hemp (which would have needed to be steeped and hackled before the fibres could be woven), and two large beehives.

The next thing mentioned is the woodpile, which the notary saw as he crossed the yard towards the building that housed the animals. The livestock, comprising two pigs, three cows, and two young *bêtes à cornes* (probably young oxen), accounted for another third of the total valuation. Back in the house, the notary opened the deceased's wardrobe, where he found five more tablecloths and three napkins as well as sheets, bolsters, skeins of yarn, and a remnant of linen cloth. His clothes had already been sold for 120 francs.

The little farm at Le Steud, which was not one of the smallest in the hamlet, differed from the large farm at Kergonan in the size of its cereal reserves and livestock. The ratio was about one to four, as the following table of comparative values shows:

Estimated value of	Kergonan	Le Steud
Livestock	FR 1,760	FR 436
Cereals and potatoes	FR 1,296	FR 314
Movables	FR 299	FR 90

Confusion of productive and social activities

The inventories list large numbers of very simple, hand-held tools representing extensions of human muscle-power. These relate to moments in the cycle of farming operations that called for a labour force larger than the household could provide and needing to be recruited on the basis of neighbourhood kinship.

A number of operations in the cycle of production could not be performed by the domestic group alone but required some collective organisation of labour. They included harvesting, threshing, the renewal of threshing-floors, and burn-beating of fields. The custom of renewing the threshing-floor – *al leur nevez* or *aire neuve* – appears to have been as typical of South Bigouden as the tall *coiffes* worn by its womenfolk. Here work and pleasure were wholly fused into one; they were two facets of a single activity. For it was through people dancing on it that the new threshing-floor was flattened and smoothed. Jacques Cambry described such an occasion:

A ceremonial tour of the threshing-floor was made, with bagpipe and oboe, the principle instruments of the region, leading the way. The master of the house was followed by his friends, carrying on the end of their sticks gifts to reduce the expense of the party; women bearing milk, butter, and sheep [*sic*; joints of mutton, perhaps] brought up the rear. All present gave themselves up to the architricline, the ruler of the feast; they sat down at table, where cider, wine, and food of all kinds were lavished upon the guests. The dancers trod down the threshing-floor as they went, beating time with their feet but exerting more pressure than for ordinary dancing.[49]

Writing a hundred years later, at the end of the nineteenth century, Gabriel de Ritalongi significantly likens the threshing-floor party to a wedding with its associated festivities, the work aspect being largely passed over in silence:

Another local ceremony is that called 'new threshing-floor', which takes place before the harvest. Several days before the party the farmer who is giving it makes the following preparations: the yard in front of the farm where the grain will be threshed – in other words, the threshing-floor – is got ready; the holes are filled with stones, and the entire area is levelled and sanded and covered with a layer of earth and a very light mix bound with cow-dung. The invitations have been sent out a long time in advance for a certain day; the bagpipes have been booked and supplies of food laid on. Here, as at weddings, each guest contributes some food, which is then sold by the recipient. Come the appointed day, the meal is served, followed by dancing on the still damp earth of the threshing-floor, and it is the dancers' feet that have to harden and flatten it. On these occasions everyone may go along and join in the dancing; the farmer is in fact honoured when there is a crowd of onlookers, and certain farms have a reputation for their 'new threshing-floors', which are followed in the same way as pilgrimages.[50]

Many jobs were organised along the lines of the *aire neuve*. Other *grandes journées de travail* were devoted to the grubbing operations necessary for burn-beating a field (*Dervez Bras*), to transporting stones (*Dervez charrea mein*), to house-building, and to road maintenance. Once or twice a year the neighbours in a particular locality would all get together with the smith and the wheelwright to build or repair their carts.

The same confusion of productive and social activities can be seen in the peasant attitude to fairs and markets. Here is Marie T., talking about her grandfather:

Every Thursday he and his wife went to 'send off' the little calves, butter, eggs, and poultry at the covered market in Pont-l'Abbé, near the church in Delessert Square. A dealer from Quimper used to come there. They called him *Bonnet Du*, which means 'black cap'. When they went to Pont-l'Abbé they arranged to meet a butcher to sell him the calf. They went to the fair in Quimper on 15 April and 2 May to sell the one and two-year-old colts. The thirteenth of each month there was a livestock fair in Plonéour. And in Saint-Jean a butter merchant used to visit, outside the LeLoc'h Hotel.

In addition to their economic function, such encounters also served festive and commercial functions. Marriages, inheritances, and farm-leases – often three aspects of the same phenomenon – were discussed. Away from their village, people nevertheless felt 'at home' – among folk of the same region who dressed and spoke like themselves. The lure of the big fairs struck another observer, Jean-François Broumiche, writing in the early 1830s:

> The whole population of the canton loves fairs, markets, dances, and local pilgrimages; they flock to the amusements available on such occasions; the Quimper fairs above all hold a great attraction for them. These folk travel huge distances for no other purpose than that of covering twelve to fifteen leagues on foot in the day* and sitting in the fairground for hours on end or visiting the taverns improvised in tents where men, women, and children abandon themselves without restraint to the pleasure of drinking adulterated wine or a heavy, nauseous cider, almost invariably emerging drunk to set out on the long road home, arriving back at their hamlets utterly done up by the journey. But they are happy; they have seen Quimper Fair and admired the horses exhibited there for sale: that provides them with an inexhaustible topic of conversation with which to while away the long winter evenings by the family fireside.[51]

Broumiche's facetious description highlights the peculiar mentality that prompted the peasant to take part in half economic and half festive collective activities.

Apart from the weekly and yearly religious festivities (Sunday mass and the *pardons*), going into town to market or to a fair gave peasants their only chance of getting away from the farm. Such occasions served not only to realise commercial transactions and exchange information about farm leases and possible marriages; they also provided a break in the monotonous daily round of toil. Small wonder, then, that alcohol abuse occurred with the regularity of a ritual to wrap the men, women, and young adults of the household in the same clouds of intoxication. We know that alcohol consumption in Finistère rose from 2.6 litres of pure alcohol per inhabitant in 1841–1850 to 5.11 litres in 1890–1896,[52] but abuse was sporadic and observed clear-cut rules. Wine was too expensive for regular consumption, and people drank only water. Men used to get drunk three times a year for several days on end. The rest of the time they remained relatively sober. Women dispelled the fatigue of washday by taking the occasional 'back from the wash-house', as it was called – *eur banne d'ar ger d'eus ar poull* – a blend of the 'two saints', Saint-Rémy and Saint-Raphaël.

So we ought really to revise our view of the drunkenness so often deplored by travellers in the light of this confusion of productive activity with social activity, a confusion that was typical of the organisation of peasant life in the nineteenth century.

The local economy was anything but static, then. From 1800 to around 1880 heath was cleared and enclosed. Farmers gave up rearing small

* The distance from Saint-Jean-Trolimon to Quimper and back is about 36 kilometres [Martine Segalen].

black sheep and began to cultivate potatoes instead. On the other hand farming made little progress towards becoming more intensive, and the characteristics of agricultural production were only superficially favourable to the local economy. Notwithstanding an upward trend in cereal prices, the economic equilibrium of farms remained fragile because farm rents were rising even faster. Only landowners benefited from the economic climate. For the rest, the general context of village life evolved only very slowly; roads remained wretched and schools poor, and the greater part of people's strength and resources was still channelled in the direction of an all-powerful church.

Demographic pressure militated against any kind of technological innovation and was a major contributor towards maintaining a mode of production based on the one hand on manual labour and on the other on a confusion of family with working relationships. It also helped to block mobility. The hierarchies outlined at the end of the eighteenth century became diluted; henceforth no family group stood out by its wealth; the period of capital accumulation was quite over.

Relations between economic circumstances and family circumstances were multifarious. The traditional organisation of production, as revealed by post-mortem inventories, continued to be based on the family group. Each member of the household was both a member of the family and a worker whose physical strength and health were vital to the survival of the farm as a going concern. The documents also show that there was little difference in the way people lived and worked as between the richest and the poorest; everyone laboured on the basis of a symbiotic collective organisation. The social and cultural unity of the region was profound.

This analysis confirms the hypothesis of a social organisation of production based on a relative confusion of the two spheres of economics and kinship and distinguished by the persistence of a mental outlook not yet oriented towards profit-making. The social use of fairs and markets illustrates this very well.

The composition of extended households is thus explained by a further group of factors coming on top of the cultural causes linked to the mode of property transmission. Economic factors turn out to give a good account of Franklin Mendels' hypothesis, which predicts an extended household employing adult labour wherever you have a combination of a type of agricultural *métayage* (with which the *domaine congéable* has something in common), low wages (the case here, since labour was plentiful), and an uncommercial approach.[53]

Up until the 1880s South Bigouden presented a model of an organisation of production that both suited the geographical conditions and was at the same time distantly connected with the macro-economic system. The twofold aim of production – to supply not only the household's own needs but also goods for sale in regional markets – was what gave that economic organisation its originality. Neither traditional nor primitive, it offered a solution to the demographic expansion with which it was faced. After the 1880s it began to move in fresh directions. As industrialisation became established, agriculture became progressively less important.

8

Tradition and modernity: the years 1880–1980

Inserting a hiatus into a chronological account of social developments is often problematic. Nothing changed abruptly in 1880, and the domestic organisation of production, as analysed in chapter 7, continued until 1950. However, towards the end of the nineteenth century certain signs of altered social behaviour – the timid appearance of contraception, for example – did become apparent. And above all Bigouden rediscovered the sea as a source of wealth and focus of intense activity that drew an expanding population down to the coast. The year 1880 marked the beginning of a series of complex, non-linear economic and social transformations. Crises in fishing and agriculture occurring at different points in time were followed by large-scale emigration. In the 1970s Bigouden started to look for the new economic and social order that seems to be emerging today.

Those transformations are not easy to describe. Rather than study statistics, the ethnologist prefers to talk to individuals, hearing how Bigoudens themselves experienced the changes at the local level. Their accounts – which, like any other source, need to be examined critically – lend themselves well to an analysis of social change in that those interviewed related their contemporary circumstances to their former situation in the years that interest us, namely the late nineteenth and the beginning of the twentieth centuries. But first-hand accounts provide only a subjective, partial vision of change. The best way to give them objectivity is to organise them by subject-matter in the light of the documentary record. Here we shall be listening mainly to the accounts of a farmer and two blacksmiths whose concern to describe technological changes was perhaps less susceptible to the distortions of memory. These former votaries of iron and fire become, in the 1930s, votaries of the machine, which they helped to introduce to the region.

Running through this analysis like Ariadne's thread is the question of how economic changes related to social changes.

The slow pace of technological change

The domestic and agricultural equipment described in post-mortem inventories before the 1870s was very similar to that found in inventories taken a century earlier in the Le Porzay and Quimper regions.[1] Until 1870, as we have seen, the inventories of rich and poor households hardly differed in terms of their equipment. No particular tools suggested that farming methods varied at all according to people's standard of living. After that date some differentiation did slowly become apparent, emerging very clearly between well-to-do farms and others. The wealthiest farmers progressively introduced mechanisation. The post-mortem inventory of Kervouec farm listed not only a Dombasle plough but also a threshing-machine worth 500 francs, a blower (50 francs), and a chaff-cutter (105 francs). That document was drawn up in 1864 on the death of the wife of the head of the household, and there is no reason not to suppose that these initial mechanical aids had been introduced at an earlier date – as in Le Porzay, where the first mention of a blower was in 1854.[2] Chaff-cutters and blowers were subsequently present in the inventories of well-to-do households from 1880 onwards. Cream separators appeared around 1900.

Transport improved from the 1860s with the arrival of the horse-drawn charabanc. There was not one at Kervouec in 1864, but there was one on Louis Le Loc'h's farm in 1865. When his wife died, the inventory listed a charabanc valued at forty-five francs, and when Louis Le Loc'h himself died seven years later the farm was using another, larger one valued at eighty francs as well as a team of horses worth thirty francs. Here again Bigouden and Le Porzay both embraced a technological advance simultaneously.[3] From 1870 onwards every well-to-do farm had a charabanc.

Machines might mean a slight saving in terms of human effort, but they did not alter the social conditions of labour. On the contrary, they were at the centre of neighbourhood and kinship mutual-aid networks and were often borrowed between several farms that did their threshing together.

'*Il faut être sans pitié de sa carcasse*' ('You have to really flog yourself') is a remark often on the lips of Hervé C., a farmer born on 1 July 1897. An educated man and former trade unionist, he practises what he preaches as, now in his eighties, he turns the heavy straw bales. The owner of a twelve hectare farm – considered the 'proper' size in the

1950s – he knows farming inside out. When interviewed, he readily contrasted the time of his youth with his subsequent experience:

At the beginning of the century, your farm equipment consisted of one or two hand ploughs, one or a number of carts, depending on the size of the property, one or two harrows, a roller, a tool for earthing-up, a hoe, sometimes a press for cider-making – the Breton national drink at the time (anyone who didn't have a press found an obliging neighbour where he could go and make his cider) – a charabanc (to which you harnessed the best trotter out of the horses on the farm to go to market, to the fair, to weddings, to mass – people went on foot more often than in the charabanc – or to go to funerals or visit relatives who lived a long way away), sometimes a seeder, very occasionally a mechanical reaper, but in fact they were never satisfactory, a winnower for cleaning the grain after threshing. Brabant ploughs – the reversible kind – didn't exist in 1900; they came on the market around 1907–1908 and were very soon in general use. Reaper-binders appeared around the same time but took a while to come into general use because of the price (900 old francs in 1913). There weren't ten reaper-binders in the whole of Pont-l'Abbé canton before 1914. There were also machines for threshing cereals, consisting of a mill with a thresher attached, the whole thing driven by six, seven, or eight horses – that's actual, flesh-and-blood horses, not horsepower. Potato lifters put in a timid appearance before 1914 but very quickly came into general use after the war, as did mixed reapers (grass and cereals). Tedders were unknown before 1914. They came on the market about 1930, not much before.

So here we have a first-hand account of a farmer's own experience of using the farm equipment described in the inventories. He outlines a timetable of the introduction of agricultural machinery of a kind that continued to call for teamwork and intense physical activity. Consequently it did not alter the traditional conditions of production, which brought together neighbours and kin to do the work on the farm.

Jean-Louis D., born in 1904, was a representative of the Landerneau agricultural co-operative. In his youth he used to help with the harvest, and he has very clear memories of those pre-machine days:

You had to get up early in those days – before five, if possible. There were ten or twelve of us, all with scythes, for a field of four, five, six hectares. Where we lived they were all small farms of five or six hectares, so obviously they couldn't go in for any big machines. So we manufactured small machines, little harrows and little hoes as well to earth up with an implement we made ourselves. At the beginning there was no machinery; everything was done by hand, if you like, harnessed to horses. I remember seeing the first threshers when I was twelve or thirteen – this was long before the war. The machine was on a farm belonging to one of my uncles at U., on the Plobannalec road. That was a splendid sight. Before, to thresh wheat you laid the sheaves out on the ground. First you'd give the threshing-floor a sweep. You piled all the sheaves of wheat one on top of another round in a circle ten or twelve metres across, and you got two horses to tread it. A touch of the whip and bang bang, round they went. After the horses had been over them the sheaves were spread out and given a flailing. This is going back a bit now, but I saw it done several years running – just across the road from where we lived. There was a farmer who did that every year rather than join in and help his mates. He preferred to do it all on his own rather than spend eight or ten days going round with the threshing team, where several of them helped one another out. A small thresher wasn't anything very much, you see. There was a platform, and the fellows got up and undid the sheaves, taking care not to get too close to the platform so as not to interfere with the feeder.

Hervé C. provides a description that, despite its concern for technicalities, really brings the atmosphere of the threshing-floor to life:[4]

At the beginning of the century cereal threshing was done in farmyards with the aid of threshing mills operated by six, seven, eight horses. Those mills used iron rods – transmission rods, they were called – to operate a thresher, which just did the threshing; the cleaning of the grain was done with a winnower once threshing was completed. The horses tramped round and round in the farmyards from seven o'clock until dark and maybe for an hour or two after nightfall when it was a question of finishing a job off or if the weather looked threatening. The people, men and women, shared the same accommodation as the horses. Threshing was very hard work at that time – so much so that you were dead with fatigue and sleep when it was good weather and you were threshing six days a week. Able-bodied women helped too. Their job was to shake the straw about with pitchforks in order to separate the grain from the straw. It was tiring work when you were at it all day long.

At threshing time the only person left on the farm would be the farmer's wife. She too had her work cut out for her. She had to look after the small children, if there were any, and that was quite often the case because there used to be far more children on farms than there are now. She also had to bring in the cows for evening milking, feed the calves, and look after the pigs that were kept on every farm. What is more, a hard-working farmer's wife would absolutely insist that all the farm work was done before the threshing team returned home in order that her servant or her daughter should not have to go to bed any later than half past ten or eleven, because it was back in harness again at four o'clock next morning.

During threshing the gear was stopped about every hour and a half to give men and horses a breather and to grease and oil the wheels of the machine. The stops were only brief ones; even so, the youngsters (boys and girls) who accounted for the vast majority on the threshing teams used to take the opportunity for a bit of ragging (as youngsters will). That helped them to forget for a while how tired and sleepy they were. But the minute the thresher started to rumble again, back they all went to their posts. The most capable of the women was given the job of preparing the swaths[5] of straw that the strong men on the team then hoisted with long poles – 2.6 metres, they used to measure – to build the ricks, which were fifteen, sometimes twenty metres long. This job, which was hard work anyway, got harder and harder as the rick grew higher, and when there was a wind blowing it was dangerous, too. There weren't many accidents, though – the men were sensible and took care.

The woman who'd been detailed to make up the swaths had to be careful she did the job properly, otherwise she came in for some good-natured joshing from the men. The least arduous of the women's jobs was the one that consisted in stopping the straw clogging the outlet. That was called the dosser's slot, *toul ar fainéant*, and it was given to either the youngest or the oldest member of the female team. This good woman used to get the grain pouring all over her feet and ankles, because at the time I'm talking about there was no question of lumbering yourself with footwear in a farmyard where threshing was being done. Everyone worked barefoot, they'd been so used to working barefoot all summer. Also you never used to get women wearing trousers; they didn't come in until much later. In fact, if a girl had let herself be seen in trousers at that time she'd never have found a husband, no matter how badly she wanted to get married. You've got to admit, though, that trousers are a lot more practical than skirts.

After the straw had been shaken up, two or three men carried the mixed grain and chaff into the barns with the aid of sheets, and when the threshing was done it was gone over again with a winnower, which by a process of blowing separated the grain from the chaff and gave you the commercial grain. The winnower used to be turned by hand.

To make up a proper threshing team with the normal sort of output, sometimes five, six, or seven farms had to get together. It all depended on the size of the farms and the availability of manpower and horses. For things to go smoothly you needed a minimum of eighteen people and nine horses – seven to pull the thresher (eight horses was better) and two horses harnessed to two carts to bring in the sheaves. At a pinch you could operate with four or five horses on the threshing gear, but you couldn't do serious threshing like that.

There weren't any farm sheds on Bigouden farms before the 14–18 war; straw and hay were stored in ricks outside near the stable and cowsheds. Those ricks were sometimes seven or eight metres high.

Threshing using a horse-driven mill went on until about 1930, when the horse was replaced by the internal-combustion engine, which used to drive the thresher with the aid of a belt. A few years later winnowing-machines came in, and just before the 39–45 war big threshing-machines that did both threshing and blowing of cereals and gave you fully commercial grain out of the end of the plant.

This work of communal threshing and mutual aid by groups of farms was accepted willingly by countryfolk, despite its being such hard work. Anyway, they couldn't have done any differently at the time.

So the early years of the twentieth century saw a slow process of mechanisation coming in to back up human labour but not, as was to happen with the second wave of mechanisation in the 1950s, to replace it.

Nor were there any changes in methods of manuring fields. On the farm of Pierre Jean B. at Le Steud they went out three times a week to Saint-Pierre, near Penmarc'h, to gather seaweed, getting up at four in the morning and returning at noon. This was then mixed with dung or gorse to improve the fertility of the soil.

As in the nineteenth century and earlier, the technical constraints of agricultural production necessitated some collective organisation of labour. Hervé C. tells us at first hand how many farms, representing what human and animal resources, needed to get together. These teams used to comprise both farms of the same size and smaller farms that were served in their turn in accordance with the system (already described) of integration of labour between tenant and sub-tenant.

Jean-Louis D. confirmed this:

People helped one another from farm to farm. If there were seven farms to be done, that was seven days' work – minimum. On the big farms they used to do two and a half days and on the others one day. I was a lad, and in summer everyone lent a hand for eight, nine, ten days – for nothing; it was free. I used to pitch in myself, generously dispensing the plonk. On my parents' farm [a very small one] there were two cartloads to be threshed – about fifty sheaves.

The technology employed was based on the driving power of the horse, and horses became increasingly important in the early years of the twentieth century. Their number was in proportion to the size of the farm. A farm of three or four hectares (seven–ten acres) would have only one horse. However, two were needed for ploughing. As Jean-Louis D. put it, 'You did a deal with another outfit about your size.' From eight to ten hectares, there would be two horses. He recalled the non-stop work at the forge when he was a young apprentice blacksmith with his uncle at Pendreff. Of all the craft-tradesmen present in the village from the beginning of the nineteenth century, the smith held his own most successfully and even expanded until the advent of large-scale mechanisation. Jean-Louis's grandfather and uncle were both blacksmiths. He and his young friends were fascinated by grandfather U.'s forge at Pendreff:

I lived on the left and my grandfather lived on the right, so I only had to cross the road, and that was where I took my evening stroll. When the iron boiled, drops of it fell like water. Grandfather would shout *Tann Ruz* [red fire]! – as much as to say: Mind out, kids! I went there with friends who used to ask me: Take us to the forge, Jean-Louis.

Later on Jean-Louis made horseshoes, either new ones or more often what were called *lopins*, made from old, salvaged shoes:

Grandfather had six or seven men, skilled workers, and the big thing every day, day after day, was making horseshoes. We made them in the evening because during the day there were other jobs that needed doing, whereas evenings everyone could concentrate on making shoes. Grandfather made *lopins* out of two old shoes, because it was too costly buying new ones and he would try and salvage as much as possible. He used to weld them.

The evidence of the Saint-Jean-Trolimon blacksmith, Laurent F., confirmed this ancient practice. When he first worked, with a Combrit farrier, they used to shoe about 150 horses. Some were shod every three months, others more frequently.

We knew the horses – knew them all by name. When the animal came in again you knew what you were dealing with. And you knew the size. It's like with shoes: horseshoes have sizes numbered 4, 3, 2, 1 and 0. They aren't all the same, you know. Some horses have great big hooves. The work horses used to have big hooves. You looked them over, just like the shoemaker. You want a pair of shoes? What size. It's the same with horseshoes.

There were still 250 horses in Saint-Jean-Trolimon in 1952. A single blacksmith-farrier could not possibly have looked after them all, especially since besides shoeing horses he repaired all kinds of farm equipment and also, in collaboration with the cartwright, bound wheels. In the 1930s the inhabitants of Saint-Jean-Trolimon, depending on where in the village they lived and on their particular kinship network, could go either to the village blacksmith, or to Le Méjou-Roz, or to Tachen-ar-Groës, or to Beuzec, or to Lestrellou, or to Ty-Boutic. They could even take their custom to Jean-Louis D.'s grandfather at Pendreff.

Up until the 1950s the slow progress of mechanisation did not alter the traditional structures of production. This also continued to be characterised by its versatility. The inventories show cattle-rearing to have been, apparently, always associated with agriculture. The poorest farms, situated on the edge of the sandy coastal strip, were the most versatile of all. But only oral investigation is capable of reconstructing the way in which this particular economy operated.

As domanial tenants of the commune, the farmers of the *palue* were forced into competition with one another every nine years, when their leases were put up for sale by auction. The bidding took place *à la bougie*: a candle was lit, and the last bid made before the candle went out was the successful one. There was no mercy between rivals, as the following story indicates:

A fellow from the coastal strip had painted *Labous* [the bird] on his boat. His lease went up for auction and his bid failed. The man who got it went to cock a snook at the evicted tenant in the bar at La Torche, telling him: I've just been and laid an egg in the little bird's nest.

The men who farmed the sandy coastal strip, whether in Saint-Jean-Trolimon or its neighbours Plomeur and Tréguennec, needed to be resourceful to make up for the poor quality of the soil they had to cultivate. In the 1920s Michel C., who had been born at Pen Ar Voës in Saint-Jean-Trolimon, worked a five-hectare holding at Feunteun-an-Dorchenn in Plomeur. He had four or five cows that, like all the animals of the coastal strip, suffered from a peculiar disease called *Klevet ar palud*, which came from eating too much sand. They could only be kept on the *palue* for a year, after which they were sent to other farms, where their stomachs could recover. Michel C.'s main job was working in one of the factories in Saint-Guénolé, but with the help of his wife and children he grew potatoes, carrots, and peas on his land. His children gathered seaweed. In addition to arable and livestock farming and organising seaweed gathering, this factory-worker-cum-farmer also did a bit of fishing. He had a small boat and used a seine net to catch bass and mullet, which he sold at a fairly handsome profit to the restaurants of Le Guilvinec. He also, like all the farmers of the coastal strip, extracted sand, which went into the manufacture of the cement used to rebuild so many old houses in the early years of this century. Michel C. in fact took versatility to the point of being a part-time builder, constructing his own house with stones taken from La Torche. That was how the last remaining prehistoric monuments disappeared from the region; the stones were recycled for building. The only surviving evidence of numerous dolmens that were still standing before the First World War is in contemporary picture postcards or photographs taken by amateurs in search of the picturesque.

So the arid land of the coastal strip concealed greater treasures than might at first appear. In particular seaweed torn loose by storms gave rise to a well-organised business. The job of gathering it was given to children, who had to get up at three in the morning. But it was also collected by farmers travelling in horse-drawn carts from the heart of Bigouden – places like Peumerit and Pluguffan. The men were able to sleep on the way.

Oral evidence tells of a system of sharing the seaweed 'harvest', in accordance with mutually agreed rules, among the inhabitants of the part of Plomeur that borders on the coastal strip and is known as La Torche. Heaps were made of the seaweed gathered, and each farmer put his mark on a pebble. A referee was sworn in, known as the 'keeper of the seaweed', and he shuffled these tokens in a bag and then threw them to decide which heap should belong to whom. Everyone accepted the judgement of fate.

As well as serving as manure for the fields, seaweed was used in the manufacture of soda. It was the spring seaweed, rich in iodine, that was employed for this purpose. Called *Teil ruz*, because of its red colour, it was dried or well burned in special ovens. Such ovens operated all along the coast of Audierne Bay, and chemists came to assess the iodine content for use in the manufacture of blocks of soda in Audierne or Kerity.

So the farmers of the coastal strip were even more versatile than the inland farmers. They were tillers of the soil, fishermen, builders, factory workers, and seaweed gatherers. They combined all these activities in an ardous life made up of days of endless toil, relying on the vigour of all those – men and women, adults and children – who constituted the domestic group.

However, while the technological conditions of agricultural production remained relatively unchanged, land ownership and the prices of farm products were evolving. Right from the beginning of the twentieth century landowners started selling farms freehold to their occupiers. This launched a process whereby farms became split into smaller units, a phenomenon unknown until then since holdings had passed from hand to hand down lines of noble or middle-class proprietors who then leased them out in their entirety. Often a farm would be too large for a particular purchaser's means and he would be forced to break it up and immediately resell part of it. Moreover, the partible system was applied rigorously. While they might stop short of dividing a holding into as many lots as there were children, people did not hesitate to create two or three farms where there had been one before (see chapter 3). As a result, the gap that had existed up until 1900 between large farms of twenty or so hectares and tiny farms in the coastal strip or farther inland now narrowed as the average size of holdings showed a clearly diminishing tendency.

This process of parcelling out is recent and has been relatively sudden: in two generations there were no farms left above twenty-five hectares. Moreover, agricultural purchasing power, still relatively high before the First World War, began its inexorable decline from 1922 onwards. Social structures slowly evolved, with day labourers and servants disappearing from the census records. Even in 1901 some 25 per cent of farmers employed servants; by 1931 the figure had dropped to 10 per cent. A fresh economic rationality came in – marked particularly by the disappearance of the *aires neuves*, the festivities associated with remaking the threshing-floor.

The crisis in agriculture ought to have resulted in a mass exodus from the countryside at the end of the nineteenth and the beginning of the twentieth centuries. However, this was deferred for thirty years by the growth of fishing and related activities.

The expansion of the fishing industry

Fishing of any importance had ceased in Penmarc'h around 1660. Its vigorous resumption towards the end of the nineteenth century was due to improvements in canning techniques that made it possible to conserve and commercialise the catch. More and more canning factories (called *friteries* because, among other operations, the headed and gutted sardines were fried in boiling oil for two minutes)[6] were built, at first along the coast but subsequently inland as well. Around 1906 there were seven canneries in Saint-Guénolé, two in Kerity-Penmarc'h, five in Le Guilvinec, two in Lesconil, one in Loctudy, and two in L'Ile Tudy. Some were set up by local entrepreneurs; some belonged to nationwide companies such as Saupiquet or Cassegrain. Their numbers continued to increase until the 1930s, when the whole southern coast from Saint-Guénolé round to L'Ile Tudy was engaged in sardine-fishing and numbered twenty-four fish canneries employing 1,800 people, two ice-manufacturing plants, ten boat-building yards, two factories processing the waste from the mackerel and sardines, and so on.[7] Canneries were set up in Pont-l'Abbé and Plonéour, and there was even one in Saint-Jean-Trolimon. Fishing and the canning industry attracted the poorer farmers. This influx of population to the coast in turn led to a building boom, and many sons of day labourers became builders around that time. In this way the sea gave rise to a period of intense activity based on a new technology (canning) and financed by a new system of production, namely capitalism. However, the change in social relations was slower than when industrialisation takes root in an urban context, mingling populations. Here it was still the same people working in a different economic environment.

Another, technologically more traditional activity grew up alongside this new economic activity. It had to do with the busy commercial relations between the coast and Pont-l'Abbé, terminus of the railway that freighted the products of the fishing industry towards the centre of France.

Jean-Louis D. gives an account of this traffic seen from the viewpoint of the blacksmith. His grandfather's forge at Pendreff was situated at a major crossroads on the route used by the carters who plied between Le Guilvinec and Pont-l'Abbé.

> Every day, day after day, the horses used to come from Le Guilvinec. A horseshoe lasts ten to fifteen days, but the horses had to be reshod often because the horn was not strong enough; if you let it get thin it broke and the nails no longer held. You had to keep a close watch on that. From Le Guilvinec to Pont-l'Abbé is about twelve kilometres, so a horse galloping the whole way to get there just in time for the train wore its shoes out very quickly. If you got to the station with a full load and the train had gone, that was it. You had to go back, put the fish in the ice house, and bring it again the next day. You had to get there in time, because the station master liked his train to leave on the dot. Sometimes he would make it wait a few minutes for some special mate of his.

The fishmongers delivered to the little Pont-l'Abbé train at a certain time, then it was transferred to the main line, and every day it was the same old story. And do you know something odd? The horses were that used to it, when he reached the station road the horse would do an about-turn and go into reverse. Why, he'd practically go off to his stable by himself.

As far as the blacksmith was concerned, the fish transport not only increased the amount of shoeing that had to be done for the horses making the trip every day; it also meant more work binding wheels. Jean-Louis D. again:

At Le Guilvinec all the fish carters had charabancs with wheels measuring 1.20 mm or 1.30 m. Every week those wheels, which were made of wood, needed binding with iron. So off grandfather would go to Le Guilvinec. He did a lot of work at Le Guilvinec, my granddad.

This repeated binding was necessary because after a lot of use the wood gave slightly; then the iron expanded and

the whole thing started to jerk around. When that happened the iron tyre had to be cut and the two ends joined up again. That was done with special things we called plates. They were heated up – not quite boiling so as not to burn. My grandfather was doing that all the time, every day, day after day. The actual [iron] tyres were made outside. The tyres were prepared, then the measurements were taken with a hand roller, all the way round; you counted the number of turns. That was a business, that was! And we didn't have many tools in those days, either.

So the fishing and canning industries created jobs and fresh outlets for what at the beginning of the twentieth century was a steadily expanding population. Indirectly – by relieving the pressure on the land – the development of these activities had an influence on agriculture. But it also directly affected what was grown. Looking for something to keep them going all the year round, the canneries turned to peas as providing a summertime activity capable of complementing the business of canning fish, which kept their (mostly female) workforce busiest during the winter months. The resultant huge demand explains why most of the farms in Bigouden began growing peas in the early years of the century. It was a clear sign that attitudes had changed. The potato had come up against what outside observers called the habits and resistances of the peasant. Pea-growing, on the other hand, spread very rapidly. Old people living on the sandy coastal strip recall what a laborious job it was. You planted a handful of peas in a wet, cold hole in the ground and covered it with a handful of manure. Often strong winds would tear up a whole field of young plants at once, the flying clods of earth carrying off all the farmer's hopes of harvesting a crop.

The development of pea-growing had two aims: to make farms

profitable and to provide full-time employment for the factory workers (mainly women, as we have seen). Nevertheless, the Bigouden economy remained a fragile organism, both on the agricultural side and on the fishing side.

During the decade 1890–1900 it was the potato blight. Then the complaints of the farmers were abruptly swamped by the enormity of the crisis that hit the fishing industry, a crisis made all the more catastrophic by the fact that fishing and canning together constituted the sole activity of the coastal population.

In the wake of a terrible tidal wave the fish deserted the waters frequented by the fishermen of Penmarc'h and Le Guilvinec. Storms made it impossible for the boats to go out, for at the beginning of the century fishing was still done on a small scale, using row-boats. With no fish, the men had no resources, and when all the activities associated with the canning industry ceased, the women and girls lost their jobs too. According to the letters of priests at the time, 3,500 parishioners in Penmarc'h were living on charity. The parish priest wrote to his bishop on 19 March 1904: 'Fishing is completely non-existent.' Religious communities served meals every day. Cooking on so-called 'economy stoves', they prepared between 850 and 900 litres of *bouillon* a day, using concentrated soup, together with 140 pounds of bread.[8] The nuns also taught people how to do a particular lace stitch called *picot*,[9] which men, women, and children proceeded to turn out until their normal activities resumed – and went on doing subsequently as an extra activity on the side.

Oral accounts amply confirmed this poverty. People of sixty or seventy and over can remember fishermen's families coming inland to beg. At K., a large farm that raised twelve cows and had four servants up until 1914, the former farmer's wife recalled the beggar-women who came to weed the fields of wheat or potatoes, keeping the weeds in a sack to give to their own cow. They used to come for the day with one of their children and a cow, because that way the three of them got fed. In the evening these wretched women, exhausted and under-nourished, would stay for the *mern vihan* (the 'little meal', or tea) washed down with *lambic* (a kind of apple brandy). Afterwards the cart had to be harnessed to take them home, half-drunk. Certain times of the year were set aside for ritual mendicancy, when anyone in a position to give was obliged to react positively. On Shrove Tuesday and at Christmas, in exchange for a prayer the beggar-women would be given potatoes, some sweet bread, a little milk, and a few coins for the children clinging to their skirts. Similarly, between All Saints' day and All Souls' day it was the custom on the wealthier farms to leave a few coins and a few pancakes on the sill of the main window. Singers went round the farms carrying a little bell, which they rang before reciting the prayers for the dead. By the end of the century people no longer put out pancakes but just money for the singer, called *arc'hant ar grampoennic* or 'little-pancake money'.[10] The wives of the Le Guilvinec fishermen were given butter, which they melted down in order to preserve it until May. But poverty was not confined to the coast. People still recall – half sympathetically, half scornfully – the figure of the *receveur buraliste*, the licensed tobacconist. An aged bachelor born in 1883, Pierre R. or *Per Reo* as he was nicknamed ('Frozen Peter') came of a line of day labourers who had had jobs all over the village of Saint-Jean-Trolimon. People said of him:

There was poverty for you. He used to eat great pieces of bread and butter with *kig sal* [salt pork], and you could see his Adam's apple going up and down with each mouthful, like an old soak's.

In the first decade of this century the deliberations of the Saint-Jean-Trolimon parish council were largely concerned with drawing up a list of the poor. Most of those receiving medical assistance lived on the coastal strip. An average of some fifteen households (out of 200 household recorded in the census) were permanently supported by council assistance. In the mid 1920s a number of major social conflicts broke out in the canning industry over working conditions, notably over seasonal employment and the very low level of wages.[11]

The transformation of the *bourg*

During this final period of population expansion in Saint-Jean-Trolimon, at the same time as agriculture was slowly being transformed the physical appearance of the village underwent a change. There was a gradual concentration of population as time went on. In 1893 Paul du Chatellier, who often came to Saint-Jean-Trolimon when he was excavating the site of Kerveltré, described the *bourg* as 'insignificant, consisting of barely twenty houses'.[12] The body of artisans remained stable, though expanding to include a baker from 1896, a butcher from 1921, and a 'tradesman' operating as a retailer of drinks and tobacco and as a middleman for the butter dealer who came from Pont-l'Abbé to collect what the farmers had brought in.

In 1911 the *bourg* comprised twenty-two houses, which as in the past accommodated agricultural day labourers and various artisans: an innkeeper, a sabot maker, a roadmender, a shoemaker, a joiner, a cartwright, several tailors, and the schoolteacher. With the exception of the last, these artisans were still linked by a network of blood and marriage ties, as the following chapter will show. Immediately after the First World War a small factory was set up, making special sabots. The man who launched this little local industry with considerable enterprise and ingenuity was Pierre Jean Le P., the son-in-law of Widow D. who ran the bar and tobacconist's shop set back from the square in front of the church. His shoes were a great success; warm and comfortable, they insulated the foot against the damp, cold surface of the mud floor, and they were also very strong. The business provided jobs for about fifteen people in workshops built in the church square, once the cottages that stood there had been demolished. Many women also came to collect the inside slippers that could be partly assembled at home. It was the kind of proto-industry that in other regions had typified earlier centuries.

Alongside this small factory, more and more businesses were set up in the *bourg*. A notable newcomer in 1921 was a large restaurant catering for weddings and banquets. The people of Saint-Jean-Trolimon particularly remember the *patronne*. After losing her husband in 1930, 'Mar'Louis'' S. continued to rule her establishment with a rod of iron. She ran a grocery, a butcher's shop, and a bar, and with the help of her cooks used to serve the extended wedding banquets after people stopped having them at home on the farm. In those days, when the clergy used publicly to condemn what they saw as excessive enjoyment, someone like Mar'Louis' was bound to meet with their disapproval. It was a source of great chagrin to the priest that one of the crosses of the *bourg* stood beside the restaurant, so that pilgrimage processions had to go right past the forbidden premises. The monument was actually known as *Croas Mar'Louis'S.*, and they made Mar'Louis' pay to have it moved.

The structure of the *bourg* was to some extent governed by these two poles – the *barbutun* or *café-tabac* at one end and the café-restaurant at the other. Then, 500 yards away near a crossroads marked by another pilgrimage cross, not far from the presbytery, the hôtel de la Croix was built, giving the village a third commercial pole. The period from the 1920s up until 1960 saw a big increase in the number of businesses, both in the *bourg* and out in the countryside, serving groups of farms. The tiniest hamlet had its café, which also sold bread and groceries; later it operated as an insurance agent as well as performing other functions. From the *bourg* out to the hôtel de la Croix a string of little houses began to go up, replacing the old mud-and-straw cottages. It began to become a real *bourg*, in fact, especially when a new school was built in 1921 and a third café opened in the old school building at the heart of Saint-Jean-Trolimon.

Finally in 1926 the Le B. canning factory arrived, providing a source of economic activity not only for Saint-Jean but also for Plomeur. A butcher from Plonéour built a pea cannery, modelled on the canneries of Pont-l'Abbé and the coast. There was a big increase in pea-growing in the commune, with every farm producing a crop, particularly those on the sandy coastal strip. The cannery not only helped to keep the population at a healthy level until the 1950s; it also provided local employment. In the peak months of June, July, August, and September, following the harvest, the factory employed between 100 and 120 people. The majority of the seasonal workers were women, but there were also permanent jobs for twenty men. During the season it was possible to work twenty hours a day, because when the factory had finished canning the local crop it brought in peas from central Brittany, notably from Briec, where the harvest occurred later. For the remainder of the year the staff were employed on a half-time basis. Various attempts were made to diversify production and so keep the factory going all the year round. It turned out pork pâté, green beans from the centre of Finistère, and diced mixed vegetables. The tinned vegetables and potted meat produced under the 'Petits pois du Calvaire' label, coupled with the manufacture of shoes, sustained a whole working population. People well remember those long hours of work in the summer when the weather was so hot. When the itinerant strawberry sellers arrived in the square, work stopped for a few moments while everyone rushed out to buy quantities of the red fruit to quench their thirst. Their memories conjure up the animated picture that the *bourg*, nowadays so empty, used to present daily.

The area traditionally described as the *bourg* in censuses from the beginning of the nineteenth century was in the process of becoming one – but for an ephemeral period of a mere thirty years or so, before the job scene became restructured on fresh foundations. The modern era was slow in arriving because the roads that might have put an end to the isolation of the more remote hamlets made little progress. In the 1920s only three of Saint-Jean-Trolimon's roads were metalled: the Route du Méjou running west past the cross and the presbytery, the Plomeur-Plonéour road, and the road leading to Pont-l'Abbé. All the access roads to farms had to be maintained by the farmers themselves putting in so many days' labour as a sort of tax in kind.

Around the same time the first bicycles appeared. Before the First World War there were no more than two or three bicycles in Saint-Jean-Trolimon; after it they spread rapidly, giving greater autonomy to everyone but especially to the young.

The efforts of the local council were also directed towards education. This enjoyed a great boost in the 1880s, and with the passage of the

Jules Ferry laws large numbers of children flocked to the school in the *bourg*, where as we have seen a new school was built in 1921, and to the one at Kerbascol. In the 1920s, when the population was at its peak, there were 100 pupils at the school in the *bourg* and 160 at Kerbascol. The children of farms in the coastal strip, when they could go to school at all, went either to Kerbascol or to a nearby school in the commune of Plomeur. The way children were distributed among the schools reinforced the geographical and social divisions between rich and poor, freeholders and tenants, and inland and coastal-strip farms. On top of that there was the division into 'reds' (children who attended a secular school) and 'whites' (those who attended a church school) so well described by Pierre Jakez Helias in his memoirs. In Saint-Jean-Trolimon there was no church school, but there was one in Plomeur and another in Plonéour, and of course in Pont-l'Abbé St. Gabriel's school for boys and the Carmelite girls' school educated generations of youngsters. The red/white quarrel was still raging in the period under discussion here; Catherine told how, when her father was elected a member of the local council on a red list, her mother was afraid the children would no longer be allowed to attend the convent school!

When youngsters from different schools met, there was often bitter fighting. At Tronoën the children of the Saint-Jean-Trolimon *palue* would fight those from the Plomeur *palue*, the girls helping with ammunition in the form of stones that they carried in their pinafores. These battles between schoolchildren were echoed in scraps between the youths and young men from the *bourg*, those from Le Steud, Tréganné, and Kerioret, and those from Tronoën. They prefigured the confrontational tensions that were later harnessed in the form of football matches.

The church still exercised a considerable hold at the beginning of the twentieth century, as we have seen from the misadventures of our widowed restaurant-owner and the rivalry between red and white schools. This is a difficult phenomenon for anyone who did not experience it to imagine. It meant the priest thundering from the pulpit against 'those who are not doing their duty' (they were practising contraception), or entering a house, seeing a novel on the table, and throwing it into the fire (he would offer a few pence by way of compensation), or excommunicating a child's parents for making him a pair of trousers out of red cloth and sending him thus attired to Easter mass.

Priests made ready use of excommunication, in fact, and it was a weapon their parishioners feared more than any other. So people were assiduous in their churchgoing. The official report of the 1909 pastoral visitation did mention a 'deplorable attendance at vespers' but noted 'between 650 and 700 parishioners making their Easter communion'. To the question: 'Is the catechism taught in the home and do people read the lives of the saints?' the answer was: 'Yes, in a fairly large number of instances.'[13] Pastoral missions were still frequent at that period. There was one in 1920, during which forty-eight *Buhez ar Sant* (*Lives of the Saints*, one of the few Breton books to which peasants had access) were distributed. A further mission of worship was held in 1924 with 661 acts of communion (296 men and 365 women).[14] In Bigouden, unlike other parts of France, religion was not purely a matter of women; men, too, attended services and carried the insignia in procession on the occasion of *pardons*. Churchwardens continued to organise services and take the collection. Rolland A. recounted how, after the harvest, the churchwarden went round collecting the grain due to the parish priest – some ten or twelve kilograms per farm. This annual tour of the parish in a horse-drawn charabanc went on until 1950. When in 1953 he went round on a bicycle, people started giving money instead.

In addition to these technological, economic, and psychological changes, the years 1880–1930 were marked by the appearance of new social divisions. The old opposition between rich and poor farmers was replaced by fresh contrasts: farmer/fishermen, farmer/worker, farmer/shopkeeper. Giving an account of the rise of these divisions is thus no anecdotal matter but lies at the heart of the subject under investigation, for they help to explain marriage practices. All of these economic and social changes intensified in the period beginning in 1930.

The mechanisation of farming

After the First World War the population continued to increase – almost as if to compensate for the murderous blood-letting of the four-year conflict.[15] Then in 1931 a reversal of the demographic trend became observable in all the communes of South Bigouden. The decline in the population from that date was accelerated by extensive emigration. Certain groups of farmers had already left the region in the 1920s, immediately after the war, to try their luck either elsewhere in Brittany – in Morbihan, the neighbouring *département* down the coast to the south-east, particularly around Sarzeau, or to the north in the Châteaulin basin – or in the Laigle region of Normandy, or in the more distant *départements* of the Yonne or even the Dordogne.

The pressure on the land was not relieved, however, particularly since large numbers of tenant farmers continued to purchase their farms. Their children, still very numerous, could not all hope to settle on holdings. At the same time the craft trades were closing ranks. From now on the only way to become a builder was by marrying a builder's daughter; you served your apprenticeship on the job. The class of fishermen was also now established, and access to it was strictly by birth or marriage.

The canneries found themselves experiencing the economic and social difficulties that we have just been looking at, and in 1934 the whole Bigouden region suffered the grave economic crisis depicted by Youenn Drezen in his Breton novel *Itron Varia Garmez* (translated into French as *Notre-Dame Bigouden*).[16] The children of farming families who could not find a place on the land turned to the craft trades or to jobs in public administration. Two changes in popular attitudes lent impetus to this tendency: the growing importance accorded to education as offering an escape from straitened circumstances, and female pressure to get out of farming.

This enforced opening-up to the outside world implied a rejection of traditional culture and a thirst for modernisation that brought elements of rupture in their wake. The teaching given to children was extremely loaded in value terms, notably the teaching of French. The memoirs of Pierre Hélias show with great clarity how Breton culture was dealt a blow by the school-imposed abandonment of the language, but they also bring out the deep desire of rural communities to reject traditional values. We see the rupture occurring in the minds of the children of farming families; farming was denigrated as a way of life, and parents themselves discouraged their sons and daughters from settling on farms that were in any case far too small to be viable. In the 1950s a number of farmers' sons left the village (many of them joining the *gendarmerie* or state police, which in France comes under the army). Women played a key role in effecting these changes. As Auguste S. put it:

> People looked at all the possible ways of getting out of farming because there weren't enough farms, and then from about 1930 the dream of every country lass was to marry a state policeman or a petty officer in the navy; they had a great time compared to girls who stayed on the farm.

This female rejection of farm work, expressed as much by mothers as by daughters with their mothers' encouragement, was a common phenomenon in regions of small mixed farming where people's life-style between the wars was markedly different from that of townspeople.[17] The result was an increase in agricultural celibacy, with workers' daughters refusing to marry farmers and the daughters of farmers seeking to marry men outside farming. Several of the women interviewed in Saint-Jean-Trolimon had accepted the proposals of farmers' sons as soon as they became builders or took a job at the *mairie*.

Agricultural purchasing power continued to decline in relation to workingmen's wages. The 1930s saw the arrival of the major process of mechanisation that was to substitute the engine for the draught horse. The blacksmith Jean-Louis D., Pont-l'Abbé representative of the Landerneau co-operative, saw it happen:

The mechanical sheaf-binders used to come from Chicago in crates made of pitch pine. That's a red wood, and there wasn't a single split in it. Lots of the crates were still good for making small pieces of furniture. Sometimes a joiner was glad to buy it, because there were still plenty of things you could make with such clean wood. The binder came all packed and had to be assembled bit by bit. You put the big wheel – the drive wheel – upright in the centre. The mechanical sheaf-binder reached Pont-l'Abbé in, say, 1930. There was no manufacturer locally. The biggest main dealer was at Rueil; otherwise they were manufactured in Vesoul. After the Chicago binders we took another make: Tollier of Rennes. The machine was all assembled then and made in France.

Jean-Louis D. also sold a great many small milking machines – known as Diabolos – as well as potato-lifters.

When I came to Pont-l'Abbé in 1926, lifters had already been around for ten years or so. They were rod-driven machines with a disc that threw the row a metre, a metre twenty-five. You could lift the whole field in one go, but the horse used to trample on the potatoes. In strong sunlight you had to do a row and then pick up straight away, otherwise the potatoes started to go green. As soon as one was slightly affected it was no good any more. It tasted bad, it wasn't clean – wasn't commercial, if you like. New potatoes were even more delicate and were picked up straight away, as you went along, as soon as they were lifted. I sold an enormous number of potato crushers. I also sold masses of screw conveyors, and press jacks went like hot cakes.

A fresh shock wave came with the Second World War. Many farmers imprisoned in Germany became aware of the different levels of equipment between their farms and those beyond the Rhine. On their return they were ready to welcome major mechanisation. Immediately after the war carts with pneumatic tyres came into general use:

I made carts with tyres from American stocks – 750 by 20, they were. You had to hurry if you wanted to buy them, though. At the beginning we found some quite cheaply, but later, when we were doing more and more carts, it became a problem getting hold of tyres. I managed to make some contacts. There was a chap in Landerneau who had bought up a tremendous stock, so we went up there to buy them. I had two men on carts the whole time. We built the bed with a cartwright or someone else, and then we had to find some axles. All the farmers bought carts with rubber tyres instead of the old iron-bound wheels. The carts were still horse-drawn – there was no question of tractors yet.

Laurent F. explained that the beds of these carts, of which he, too, manufactured large quantities, were a lot wider than those of carts mounted on traditional wheels. This meant that they were able to carry more sacks of potatoes or a larger load of hay. Far from his business being harmed by this technological change, the blacksmith found himself with a great deal of work, adapting the mountings and rails to the new dimensions. These carts 'used to last a chap's lifetime'.

The advent of the tractor dates from the 1950s. Once it had arrived, it spread very rapidly. Jean-Louis D., who used to sell them, clearly recalled what was to prove one of the biggest events in farming in the last thirty years:

Twenty-five years last April [he said in August 1976] I've been doing Renault tractors. The first tractor I sold was at Quimper Fair on 15 April 1951. That was a very big fair. The whole region around here – virtually the whole of Finistère used to go to Quimper. Folk even came from Châteaulin and Pleyben. It was amazing the people you used to see there. It was the fair for horses, young heifers, and piglets. People exhibited there and went over every day. The fair lasted two days, and the co-operative [the Landerneau co-operative, of which he was the representative for the Pont-l'Abbé region] had a stand promoting Renault. There were also MacCormick, Someca, Leizin, and all the big Quimper dealers. Renault were just starting out then. There weren't a lot of tractors about, and people used to come and look, discuss prices, ask questions. There were dealers who used to shoot a line or tell whoppers as long as your arm. I was the only one in the region doing Renault. When you've once hit on a good make you don't change. We never had any big problems with repairs, and you could always get spares when you needed them. At that time the tractor cost one million five, one million six [old francs, of course].

The first tractor was sold to a farm in Plonéour – a man called Daniel who farmed fifteen or sixteen hectares. He had three horses. He could easily get rid of two, and with the proceeds he was able to buy a tractor.

You even got tractors on farms of four or five hectares around here, gradually. The old existing equipment was no good at all. Farmers had to buy plough, harrow, spreader – everything new. Buying a tractor wasn't the end of it! You had to have the right equipment for it. There was no way of adapting the old equipment. Take a roller, for instance. You put the shaft in the middle with a couple of bolts and had the tractor pull it. But often the equipment was too small, because the roller was usually a metre forty or fifty or sixty and a tractor was capable of pulling a metre ninety and even wider if there was space for it on the farm. When you're selling a tractor you first sell the plough and the fertiliser spreader. When we sold a tractor we were pleased because a tractor brought other sales of all the machinery that went behind it. It wasn't just the tractor that was interesting; it was the little implements that went with it: ploughs, seeder, roller – all the things for the different jobs on the farm.

There could hardly be a better description of the process set in train by the appearance of the tractor as it plunged farmers into the infernal spiral of constant indebtedness. Before a new item of equipment was either worn out, paid for, or written off another version of it – newer, bigger, more powerful – had come along to render the previous one obsolete.

So it was that, as farms lost their horses and their men and women, machines invaded the territory and proceeded rapidly to transform the traditional structures of production. The family cell might continue, henceforth without paid help, to work a tiny farm, but the economics of it now became dissociated from kinship systems and bowed to the logic of capitalism.

Despite the introduction of pea cultivation, the fields of the coastal strip started to fall into disuse. As early as 1934 the commune had to agree to a reduction of 10 per cent to its domanial tenants on certain lots and 25 per cent on others. In 1947 the renewal of leases ceased to be settled at auction and was done by agreement. In 1956 it was abandoned altogether.

The income that the commune had derived from leasing out the *palues* was more than made up for by the revenue from sand extraction, an activity that expanded in the post-war years. Previously the cartloads of sand had been used only to build houses for the inhabitants of Saint-Jean-Trolimon. Taking their cue from the Germans during the Occupation, the extractors now turned it into a profitable business and started exploiting the dunes on a relatively large scale. Until the 1960s the royalties paid were small, but they have since increased to the point where financially the commune is quite comfortably off.

The population decline that had begun inexorably in 1936 accelerated, particularly between 1954 and 1962. It affected the farms of the coastal strip as much as the inland farms. The percentage of the population involved in farming steadily diminished as the centre of gravity of the population shifted towards the *bourg*, where the new building tended to be concentrated along the Plomeur–Plonéour and Tronoën roads and between the *bourg* and the hôtel de la Croix. School rolls fell. The school at Kerbascol was closed in 1970; the Plomeur school that had taken the children from the coastal strip followed it around 1975. In 1977 the school in the *bourg* numbered no more than forty-seven pupils, while a stable proportion attended the free school in Plomeur.

In 1974, out of 232 heads of household only 40.8 per cent were employed in agriculture. Of those, 85 per cent were owners and 2.4 per cent domanial tenants, and most of them combined the roles of owner and tenant, working not only their own land but also that of relatives or neighbours, leased to them under verbal agreements. The scale of holdings was very small, with only six farmers working more than twenty hectares and the rest making do with smaller properties.

In joining in the dominant mode of production, the farmers of Bigouden suffered all the negative aspects of change without managing to pick up the rhythm of modernisation quickly enough. Constantly shackled by the debt spiral, they were too late in opening their pig sheds and their milking parlours. Altogether, in Saint-Jean-Trolimon as in the region as a whole, agriculture has failed to find its niche in the contemporary world, contrary to what has happened in other parts of Brittany.[18] What makes this all the more regrettable is that Bigouden is probably one of the most fertile regions of Brittany, as the past clearly showed.

The average age of the population, particularly the farming population, is very high. In 1974 more than 120 inhabitants out of 660 were over sixty years old. Farmers apart, the 1974 population also comprised 30 per cent retired people, 8 per cent factory workers, and 5 per cent construction industry craftsmen (mainly builders). The canning factory had closed its doors in 1968. The slipper factory was still in operation (it closed in 1988), and a small joinery business employed five or six workmen. Astonishingly for anyone not familiar with the Breton environment, this tiny population included a large number of tradesmen who had set up shop between the wars, most of them with several irons in the fire, as it were (bars also selling bread, meat, or hardware, for example). Nine of them were located in the *bourg*, one at Tronoën, and another at Kerbascol.

This commercial structure is wholly anachronistic. It no longer corresponds to the needs of the population, which with more and more private cars at its disposal is able to shop at the supermarkets now springing up around Pont-l'Abbé. The result is that the village shops are closing one after another as the old men or women running them decide to 'pack it in'. As each one goes, its surviving clientele loses the opportunity for a daily chat; the exchange of information that each shop helped to promote now no longer takes place. Instead of the animated scene they once presented, the former *café-commerces* offer only a blank façade, and shelves that used to be laden with merchandise now hold nothing but pots of flowers.

In the years 1970 to 1975 Saint-Jean-Trolimon seems to have experienced another kind of social death. Paradoxically it was marked by the building of new houses, but as second homes. Either these have a purely artificial existence, being occupied for no more than three months of the year at most, or they were built to house old people on their retirement – to await their owns deaths in.

The mechanisation of farming upset the social relations based on the tradition of mutual aid by making physical strength superfluous, and farmers have not succeeded in finding a fresh basis for association. In the 1960s a new co-operative venture (CUMA or 'Coopération d'utilisation de matériel agricole') was unsuccessful. It was doomed to failure, in fact, twenty-five farmers having together acquired two binders and a baler. For one thing there were not enough machines, which led to friction. But above all people were not ready psychologically to accept a form of co-ownership that had no precedent in the traditional structures of production. The CUMA broke up rapidly. Two others were set up between farmers working land in the north of the commune and those in the south. Today's farmers reject CUMA-type associations in favour of forms of multiple ownership of agricultural machinery among selected partners over a wider geographical area. Whereas farms of different sizes traditionally got together on a neighbourhood basis, farmers today look for other farms of the same size and kind as their own with which to embark on joint equipment purchases. Because few farmers have adopted modernisation measures, whether large-scale cattle-breeding or pig-breeding, those who have are obliged to look beyond the narrow relationship of the immediate neighbourhood to find others like themselves. The result is that mutual-aid relationships nowadays tend to be spread over a large area instead of being confined to the neighbourhood, as used to be the case.

Multiple ownership of agricultural machinery has proved no substitute for the traditional forms of co-operation. Threshing used to be done on the premises, and several farms would be mobilised for the task, parents, children, neighbours, servants, and animals all working together in the cordial atmosphere described by Hervé C. in the passage quoted earlier in this chapter. With modern machinery, even if eight farms have got together there are never more than three people in the field at a time. Face-to-face human contact has been replaced by a telephone call to find out whether the machine is free on such and such a day. There is no meal for the co-owners, just a few figures jotted down over a cup of coffee and a look at the rota to see whose turn it is next.

Professional institutions structured at a level extending beyond the commune (Landerneau Insurance, department-wide family-welfare associations, the Bigouden pork producers' group, and so on) help to define the farmer's place within the region to which he belongs. Mutual-aid relationships of a more everyday kind are still possible wherever immediate neighbours continue to farm in spite of their advanced age. But they are tinged with uncertainty when all the farms round about are worked by men already in their seventies whose children will not step into their shoes. Mutual aid in cases of misfortune still operates, if somewhat shakily where the invalid farmer is out of action for any length of time. It takes the form of a solidarity reaching out over time and space and involving the whole social group. It is not a bilateral relationship between two households who keep an account of loans and borrowings; it is a pool of potential help, supplied by anyone in a position to give it to anyone in need of it. When Alain P. broke his ankle, for example, the neighbours took turns to plough and sow his fields. Lisette S. helped a woman neighbour in her dying days, ten years after the woman had helped her to look after her aged, senile parents. When Marcel B. celebrated the marriage of his two children a neighbour, Jean R., milked his cows; Marcel B. will pay off his mutual-aid debt sometime in the future – possibly to a different neighbour.

On the farms that modernisation has passed by, mutual aid is organised in two contrasting ways. Either the farm turns to a threshing contractor, who carries the eradication of human relationships from farming to the extreme in that the crop is harvested and threshed without the owner even being informed, which leaves him with an acute sense of loss. Or on the other hand the bulk of the work is done by the aged parents together with their children, who though they have got out of farming have not left the district; they come back at weekends to drive the tractor or bring in the hay. A fresh town–countryside link is thus established, as well as a direct relationship between parents and children, though the latter is fairly self-contained in contrast to the broader kinship solidarities of the past.

Following a pattern peculiar to the development of agriculture in France, we find women have been very definitely removed from work relations. Closely associated with work in the fields before and immediately after the war, they were progressively relieved of such tasks by the advance of mechanisation. Nowadays their main job is feeding the animals on the farm. Deprived of the contacts and discussions as well as of the responsibilities they once enjoyed, women have become marginalised as far as farm work is concerned without finding a corresponding niche elsewhere. They have fallen back on housekeeping and looking after their children.

The new spatial structuration of farmers' collective activities tends to alienate them as far as the commune is concerned. Largely absent from the council, for example, they are consequently excluded from such forms of village social life as hunting and football. The people who embody the dynamic peasantry with family roots going back far into the past are actually rejected by their fellow villagers in favour of the representatives of the numerically dominant class of artisans and wageworkers.

The period 1930–1970 brought some abrupt changes. For the first time in its history the population of Bigouden was forced to emigrate. Those who stayed behind found it hard to adjust because of the succession of crises that affected fishing, the canning industry, and farming.

In the 1980s the Bigouden economy has found a fresh equilibrium, and with the demographic cycle on the upturn again social relations are also being restructured on fresh foundations.

The economic turnaround of the 1980s

The booming fishing industry constitutes the principal economic asset of the region. Despite the closure of almost all the canneries, it fuels a great deal of activity both up and down the line. The South Bigouden ports of Saint-Guénolé, Penmarc'h, Kerity, Le Guilvinec, Lesconil, and Loctudy together have an annual catch of nearly 30,000 tons of fish. The industry supports 2,000 fishermen as well as the firms that buy and sell the fish, unship the catches, build and paint the boats, service the engines, and all the rest of it.

The expansion of Le Guilvinec had the effect of pushing the population northward, with the result that Plomeur experienced something of a building boom. The growth of the coastal communes has led to a redrawing of the cantonal boundaries. Instead of being included in the canton of Pont-l'Abbé, as formerly, the five communes of Penmarc'h, Le Guilvinec, Treffiagat-Lechiagat, Plobannalec-Lesconil, and Loctudy now make up a separate canton with nearly 20,000 inhabitants. The canton of Pont-l'Abbé is now confined to the communes of Tréguennec, Tréméoc, Saint-Jean-Trolimon, Plomeur, L'Ile Tudy, and Pont-l'Abbé itself, with a population of around 13,000.

Despite this separation the two cantons can only live in symbiosis, the boom in the fishing industry taking up the slack from farming, which is still in a critical condition. In the decade 1970–1980 almost one-third of farms in the two cantons disappeared, while the average age of those working them rose. Young people hesitate to take up agriculture because of the amount of indebtedness involved in making the investment needed to launch a new farm.

The third pole of economic activity in Bigouden is tourism, a somewhat diffuse industry at the moment but one that seems to hold out some hope for the region, even though it contains the seed of fresh struggles for living-space. Unauthorised campers and wind-surfing enthusiasts are spoiling the dunes and are the despair of the ornithologists who come to watch the birds nesting in the vast étangs fed by the streams that cross the coastal strip. There is now an organisation trying to make sure that these magnificent stretches of water are protected.[19]

The secondary and tertiary sectors are busy, but industry is much reduced. The economy of South Bigouden is still fragile; in October 1982 there were nearly 1,500 persons in the region looking for work, 900 of them women.[20] However, after the population exodus of the years 1930–1950, the situation is now reversed. This change is very noticeable in Saint-Jean-Trolimon.

Spending ten years in a place, you get a better picture of the way it is evolving. The trends that marked the years 1930–1970 might have continued inexorably, bringing the commune to depopulation and a lingering social death. Not a bit of it. Suddenly – not even gradually – the situation has turned right round; the population has risen again (735 inhabitants at the last census), and the average age is coming down.

In 1976 and 1977 the number of building licences issued was about fifteen a year, most of them for second homes or retirement homes. In 1978, 1979, and the early part of 1980 more than sixty building licences were issued, and 90 per cent of those were for principal residences; forty-five houses were built by young couples from the village (that is to say, whose parents or parents-in-law lived there) or from the immediate vicinity, the rest being strangers to the village, to the region, or even to Brittany. Since the law forbids residential building in zones set aside for agriculture, most of those houses are situated in the *bourg*, which is becoming larger and larger. There are now two *lotissements* or developments. The private one comprises twelve houses, all built in the last two years, whereas it had been on the market unsuccessfully for about eight years before that. The second belongs to the commune and is at the other end of the *bourg*. Launched in 1979, it consists of ten houses around the newly erected community centre, the *Maison pour tous*. A third development, either private or public, in now being considered.

So what we are seeing is a complete reversal of the situation that obtained around 1975, when such building as was going on – second and retirement homes – did not imply any genuinely village-based economic activity. Today it is young families that are moving in, with young children. The school, which had been threatened with closure, opened a new class in 1980.

How is one to account for this sudden phenomenon, and of what is it symptomatic? An initial explanation can be advanced at a purely material level. Saint-Jean-Trolimon is situated four kilometres from Pont-l'Abbé (and eighteen kilometres from Quimper), and the price of land in the village is half what it is in Pont-l'Abbé. Furthermore, despite the closure of the small regional canning or manufacturing business, the building trades have enjoyed a period of relative expansion, and young people have been able to find work in them. An entire age group that might have chosen to leave the village, as young people did in the 1950s and 1960s, has decided to stay, while others are coming back. Outsiders working in the region have taken advantage of what to them were relatively inexpensive plots, and there are people from Le Guilvinec who were no longer able to find homes in their own commune.

The appearance of the *bourg* is changing as the old cafés in the centre continue to close or have adopted the outward and visible signs of modernity. The hôtel de la Croix, for example, has ripped out its fine, austere façade to make room for a mini supermarket and a modern-style café. The restaurant for weddings has adorned its walls with its splendid name – 'Le refuge' – painted in huge letters. Opposite the *mairie* the royalties from sand extraction have paid for the construction of a car park and community centre. A new cemetery has been opened as you leave the commune on the Pont-l'Abbé road, and the tombs are gradually being moved from around the church, where they have rested since time immemorial. They will be replaced by another car park.

The crisis in farming is being resolved only very slowly, and the restructuring that each year seems inevitable happens on an *ad hoc* basis with no overall plan. At the time of writing a threshing contractor from Plomeur is working a number of fields belonging to elderly farmers, when those field are not lying fallow. The farmers refuse to let to the few young people who could use the land. 'Let is as good as sold', they say, meaning that they would lose all power over their own property. In 1981 a young farmer and his wife moved in and started an intensive piggery, the third in the commune. As for the farmers now turning fifty who have modernised their farms, they can see one of their children possibly succeeding them.

The ever-increasing concentration of dwellings in and around the *bourg* is leading to fresh social divisions. Before there was a gulf between those who lived in the coastal strip and those who lived in the *bourg*; now there is one between *Nordistes* (the inhabitants of the upper part of the *bourg*) and *Sudistes* (the inhabitants of the lower part), who mutually accuse one another of being proud and stand-offish. The fact that such antitheses, not to say rivalries, are able to find expression today is an indication of social health. One has the feeling that a fresh mode of social relations is coming in, of relations that transcend conflicts relating to the commune and cross the traditional dividing-lines between town and county.

Among the new inhabitants of the commune there are two distinct types: those who have stayed and those who have consciously decided to return. The latter's decision is mainly connected with the possibility of finding a job locally. The reason why young people left in such numbers in the 1950s was that the cities – Nantes, Rennes, Paris – offered them openings, notably in the *gendarmerie*. Family ties were not overstretched by such emigrations, and these men (now in their fifties) have built second homes to which they regularly return. They immediately find their way back into the traditional networks of social life centered on football and hunting. 'It's great to see how much village spirit [*esprit de clocher*] they still have, the people who've bought land here.'

However, living in Saint-Jean-Trolimon does not necessarily imply that one works there. The dissociation between domestic group and production has been followed by a dissociation between residence and place of work. Suddenly it is not longer clear where one belongs. There are no more villagers on the one hand and townspeople on the other; people work in town and live in the country. Craftsmen – those involved in the construction industry, for example – no longer confine their activities to South Bigouden but work on sites over much of southern Brittany (Cornouaille). For some, work and social relations are centered on the town, while the country is a pleasant place where they managed to find satisfactory accommodation. For others it is quite the reverse. They have as little to do with the town as possible. They work there, but their whole social life is concentrated in the country, in a place with strong emotional connotations that they refer to as *chez moi* or *à la maison* – even if the *maison* in question is their parents'. As in the past, the boundaries of the commune are too narrow for most social activities not to spill over them. The new range of the social relations of the farming community extends beyond those boundaries, and the same is true for most socio-professional groups. People work outside the commune and do their shopping outside the commune; to visit the hairdresser or the dentist or the doctor they go to Plomeur or Pont-l'Abbé. However, the geographical context of social relations has not really changed; it is still South Bigouden, and in that respect there is great continuity with earlier behaviour patterns. But of course people's social world has expanded considerably. They have been away, travelled, sometimes worked outside the region or even outside Brittany. It would be quite wrong to give an impression of South Bigouden as being backward-looking; on the contrary, openness to the outside world has been very much a feature of the last few decades.

Viewed as a whole, the region continues to represent a social entity that has its relevance. The geographical area of choice of spouse, for example, has not increased substantially. It exceeds the narrow bounds of the commune, but it is still pretty much confined to South Bigouden. The young people get to know one another in the schools and colleges of Pont-l'Abbé, which take most of the village youth, and they still marry within the same geographical area. No one will be surprised, then, at the unchanging significance of the mutual acquaintanceship that continues even nowadays to mould social relations.

Altogether the region seems to have experienced an economic upswing over the last fifty years, a rising standard of living without the accompaniment of violent industrialisation. What has been lost in terms of jobs has been made up for in quality of life. The abandonment of the traditional culture – particularly the language – has left social relations fundamentally unchanged; they still revolve around kinship as the source of emotional and material mutual aid. That is why the next chapter, which deals with the relations between kinship and economic and social spheres since the beginning of the eighteenth century, has no difficulty in incorporating the contemporary period.

9

The importance of being kin

How did the economic and social changes described in the foregoing chapters relate to changes in matrimonial practice, identified particularly in terms of a reduced level of relinking within kindreds?

In the eighteenth century and for much of the nineteenth the organisation of production was based on the domestic group, obeyed a non-productivist logic, and depended on human labour. That and the slow development of farming know-how help to account for the overlapping of the sphere of production and the sphere of the family considered from the narrow viewpoint of the co-residential group. So what purpose was served by relinking? Surely its declining incidence suggests a separation of the circumstances of kinship from social circumstances?

It is customary nowadays to state that the two spheres are independent. With regard to South Bigouden, however, there is a case for questioning the rift that supposedly occurred between them and examining the existence and nature of possible inter-relationships. Doing so involves calling upon a variety of evidence: archives, reconstituted genealogies, interviews, and participant observation. And as we approach the present, written evidence increasingly gives way to oral. Social expressions of kinship are in fact more often traceable through interviews than through archive documents (no document will list the members of a threshing team, for instance).

Because of this shift from the past to the present, the subject-matter tackled in this chapter also involves juxtaposing problematics and a vocabulary that do not fit easily together. Ethnologists deal with kinship relations and are chiefly interested in their structure; sociologists look at them in the wider, less well-defined context of social relations. A person's social relations, whether formalised in associations or the informal product of interaction between individuals, are to be observed particularly in regard to work, the family, and leisure.[1] In this sense there is certainly continuity between the eighteenth and nineteenth centuries and the situation today. Moreover, the concept of the individual's social relations* encourages us to become aware of social and cultural differences that are usually not much in evidence in the societies studied by ethnologists. This is a valuable dimension as regards studying the present-day world in that new kinds of socio-professional differentiation are beginning to emerge together with new relationships between the urban and rural contexts.

Conditions of production and kinship relations

Those who did not expect to succeed their fathers or fathers-in-law tried to discover which leases were about to become free. They might inquire among their kindred, or they might turn to the notary or to the lord of the manor, perhaps to ask whether what was regarded as a large farm (in excess of twenty hectares) might be divided in two. It is difficult to evaluate the respective degrees of importance of the two courses of action. On the other hand various examples make it possible to establish that peasants who held reparative rights, as many did, sub-leased them to members of their kindred. As far as the eighteenth century is concerned, this can be shown by comparing information culled from the table of lay leases[2] with the genealogical files; for the nineteenth century it emerges from the reconstituted genealogies. What then appear are several groups of kindreds in which kinship and economic relationships overlap.

The Cosquer–Guirriec example (see figure 26) illustrates this interlocking of leasing and marriage, this merging of the economic and kinship spheres. Maudé Cosquer Sr. leased to Louis Guirriec, who was married to his daughter-in-law's aunt. At the same time Louis Guirriec leased to Maudé Cosquer Jr., the husband of his niece by marriage, and the loop appeared to close when Maudé Jr. leased to a distant cousin who was married to a cousin of his wife. Marriage relinking (because in this triangular relationship there was no consanguine marriage) was here underlined by a triangular relationship of an economic nature. It is pointless trying to establish whether the marital tie or the economic tie was pre-eminent. The leasing of reparative rights and the contracting of a marriage were two aspects of a single relationship. One might precede the other in time, but there is no knowing which was causally anterior. It was simply a family matter.

The case of the leases held by François Quittot (see p.187 and fig. 27)

* French sociologists speak of *sociabilité* [Translator].

makes an interesting study. Initially, in the 1760s, he ceded them to distant relatives. Then around 1790 he gave them in direct line of descent. This arrangement is typical of the general development that occurred as a result of demographic pressure. A person who held several farm leases was able to give distant relatives some of the benefit because his direct descendants were already looked after, as it were. As the number of leases held by each individual diminished (with more

Fig. 26. Circulation of leases within a kindred.

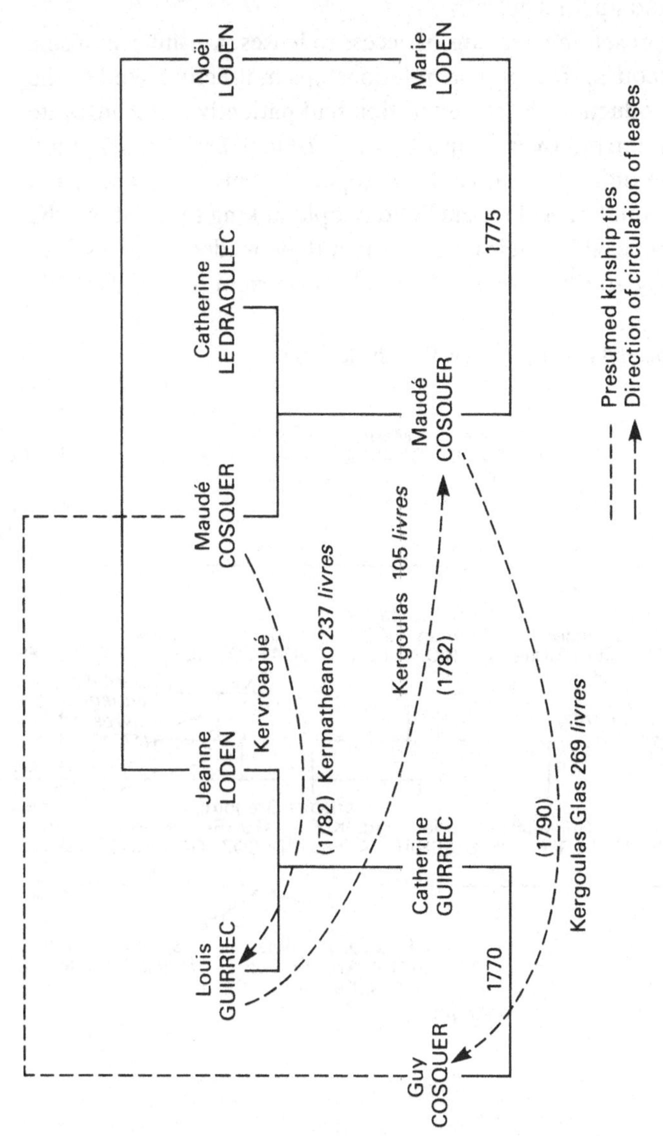

children surviving), there ceased to be a surplus and were only enough for direct descendants to benefit. By the decade 1850–1860 there were not even enough for that.

The nineteenth century also offers a number of examples of leases being transferred between distant consanguines or affines or of the land worked by a domanial tenant being purchased by a distant consanguine or affine with more money. There is no suggestion that all economic transactions were settled in this way; all this is meant to show is the importance of the kindred as source of information about leases available and land up for sale.

As well as representing a means of access to leases, kinship was also a person's first recourse in economic relationships indirectly linked to the conditions of production. Each generation had patiently to reconstitute the rights of ownership over a farm that inheritance kept breaking up. Loans had to be obtained that could be repaid out of the income from farming. Who would provide them? Old people talking today about the problems experienced by their parents when they needed to get hold of money invariably speak of family loans. Post-mortem inventories cor-

Fig. 27. Economic and marriage ties within the kindred.

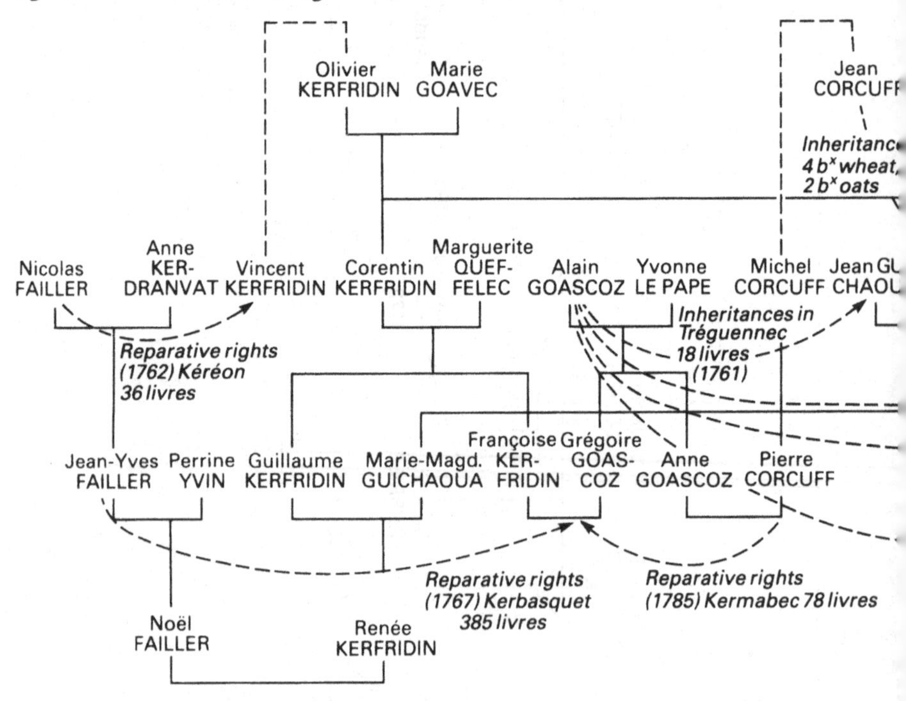

roborate these accounts. At the same time, however, they show how often people appealed to the notary.

Indeed, when it came to getting hold of money people were forced to resort to a variety of strategies. Here are some examples of the importance of the kindred in an economy in which hard cash played a major part.

Louis Loch died in 1872 with no direct descendants. His heirs – nephews and great-nephews – are known. The assets of the estate consisted essentially of credits totalling 55,685 francs. A quarter of that total was accounted for by loans made by Louis Loch to his sister and to his nephews and great-nephews in respect of sums ranging from 300 to 1,800 francs. Of all his rightful claimants eleven borrowed money from him, notably a branch of his wife's nephews. Was it because these children had stayed closer to him through the medium of their kinship, or was it because they had remained in Saint-Jean Trolimon while the others had gone off to Penmarc'h and Pont-l'Abbé? Geographical proximity probably counted for something – but not necessarily, because a sum of money was also lent to a great-great-nephew living in Combrit.

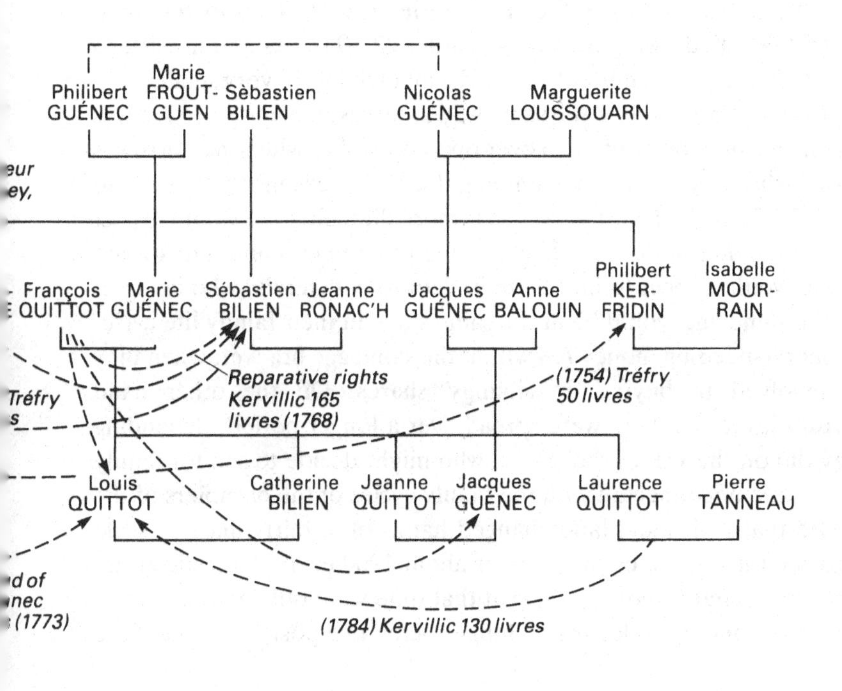

In any case loans to relatives and notarial loans overlapped, because the former were not verbal; they were based on notarial agreements, with careful accounts being kept of the interest on them. Sometimes the notary made up the sum himself. So it is difficult to tell whether the borrower had first gone to his relative, who then urged him to formalise the situation in a written document, or whether, having approached the notary, he had been encouraged by the latter to enter into debt with a relative who had money on deposit with him.

Louis Loch also lent money to non-relatives, those credits representing something like a quarter of his assets (12,270 francs loaned to non-relatives out of 55,685 francs). The rest was loaned to farmers, tradesmen, and even Vice-Admiral de La Grandière, who bought all his landed property at Kerbascol from Louis Loch for 36,000 francs – of which 18,000 francs were still owing at Loch's death.

On the other hand those farmers whom we know from their post-mortem inventories to have been in debt drew on both sources, family and notary. Corentin Le Maréchal, for example, died owing 3,632 francs 'in the cabinet of Maître Emile Arnoult for divers instruments in principal, interests, costs, and fees';[3] Pierre-François Stephan owed 2,320 francs,[4] and so on. Other inventories reveal debts to relatives: Félix Trébern died owing his wife's brother 22.50 francs and his sister's husband 30 francs.[5] Admittedly the sums involved were very small.

Loans, then, operated in the same way as leases; the appeal to kin was not exclusive but constituted an ever open possiblity when the borrower did not wish to ask the notary. Pinning down the reasons that governed the choice of one or the other course is more difficult. One would expect to find the young in debt to their parents or to uncles and aunts more than one would expect to find them in debt to collaterals, who being of much the same age would be at the same stage in their family life cycle. Most people needing money fell within the same age bracket, when they were involved in buying up siblings' shares. On the other hand opportunities to buy land were spread over a longer period, depending as they did on the will of the lessor, who might decide to sell for family reasons quite unconnected with the family cycles of the occupiers of the farm. Be that as it may, land changed hands on a fairly massive scale around the turn of the century. After about 1900 every domestic group was in debt. What is more, people at that time were obliged to resort to institutional funding, relatives no longer being in a position to meet the need.

Jean-Louis D. confirmed this:

Between the wars lots of people became landowners. They used to take out a loan with the notary. He was the man who knew about everything because all the business deals passed through his hands. So they'd go along and ask whether there wasn't a farm for sale, or, if their farm was too small, they'd be after one of ten or maybe twelve hectares if they could get it.

In cases of insolvency, bankruptcy, and public sale, kin were among those present, buying up farm implements, furniture, clothing, livestock, and reserves. This could even be seen as a way of settling family debts, with relatives acquiring things at a lower price than their inventory valuation, or as a symbolic continuation of the line expressed as a reluctance to allow the objects of someone's domestic existence to become dispersed.

However, ties of kinship did not operate to the exclusion of other connections, any more than in the case of taking out loans, and neighbours also turned up at such sales to bid for and acquire a variety of things.

Take the case of Louis Quittot, a deceased bankrupt farmer whose possessions went on sale on 22 January 1841, seven days after the inventory.[6] He was in debt to – among others – his two sons-in-law, Jean Nicolas and Pierre Carrer, to his brother-in-law, who was godfather and surrogate guardian to his children, and to Pierre Carrer's father. Nicolas and Carrer, among others, bid for and bought his possessions. Pierre Carrer, who was in the process of setting up house, purchased the largest number of items: a wardrobe, a bedstead, a bed with bedding, a large cooking pot, a medium-sized cooking pot, a pan for cooking *crêpes*, a large cart, two iron-bound ploughs, a beehive, a white horse, two oxen, a young bullock, two more young plough oxen, and lastly some barley and oats. He spent about 500 francs or a little under one-third of the total yield of the sale (FR 1,734). Jean Nicolas was

Fig. 28. Example of kin buying up property following the death of a relative. The case of Louis Quittot of Kerstrat.

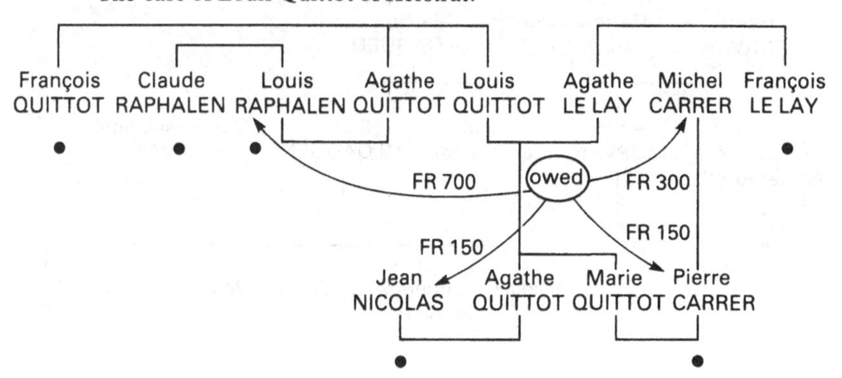

already installed. He bought a dresser, a wardrobe, some tablecloths, a chopping-board and two knives, some threshing flails, a black horse with its tack, two young bullocks, some barley, and some oats. Between them Carrer and Nicolas bought a fat pig and the seed corn standing in the fields. Other relatives who dropped in on one of the three days it took to sell off everything included Louis Raphalen (he bought only two plates), Claude Raphalen (a bronze basin), Louis's brother François Quittot (a heifer and two bundles of tow), and François Le Lay (130 litres of barley). The remaining furniture, livestock, and so on went to farmers or to various dealers from Plonéour.

The same sort of thing happened at the post-mortem sale at Keryoret in 1851.[7] The principal buyer was the third son, who was setting up house and purchased the main implements and pieces of furniture, the plough-oxen, and so on. Two further relatives appeared who lived nearby: a distant cousin, Henri Le Borgne, who occupied another farm in the same hamlet, and a first cousin who lived at Tronoën. The publican in the *bourg* bought the cake of fat, and various small farmers from the sandy coastal strip bought stacks of firewood (they were always on the lookout for fuel).

For countryfolk, then, kinship, taken in the sense of a group of people related by blood and marriage, usually members of adjacent generations but often reaching out a long way collaterally, played a key role in regard to access to land, whether it was a question of purchasing reparative rights, acquiring leases or loans, or providing mutual aid in the event of insolvency.

Fig. 29. Example of kin buying up property following the death of a relative. The case of Nicolas Gloaguen of Keryoret.

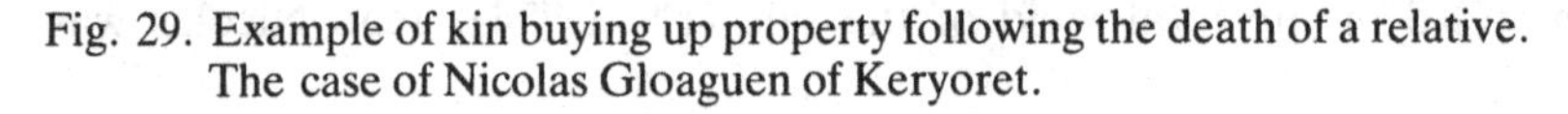

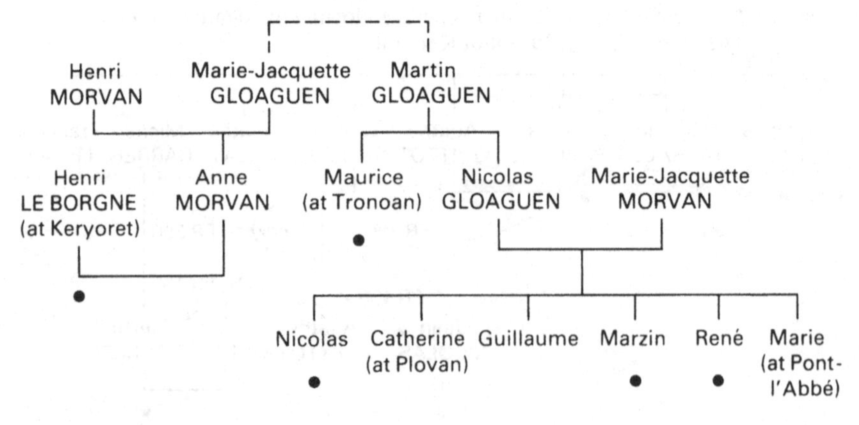

In the early nineteenth century, when ownership of reparative rights, which had been concentrated in the hands of a few lines, became more diluted, it was through the medium of farm leases that the relationship between the kinship and economic spheres was maintained. Should the perenniality of this relationship in its many forms be seen as a throwback to Celtic culture? It is a fact that Irish kinship is characterised by an unwritten law to the effect that, in the absence of a direct heir, a person's property is sold or leased to his nearest kinsman rather than being allowed to pass into the hands of a stranger.[8]

But Gérard Delille found similar behaviour in southern Italy in the fifteenth and sixteenth centuries.[9] Is it therefore a stage in an evolving system that passes from a period in which the kinship and economic spheres coincide to one in which they become dissociated?

Peasants in nineteenth-century South Bigouden, while trying to keep leases circulating within family networks, nevertheless found themselves in a competitive situation *vis-à-vis* a landowner whose sole guide was considerations of profit. Two ways of thinking met and clashed here: that of the peasant (who tended to a greater or lesser degree to confuse the kinship and economic spheres) and that of the landed proprietor (who kept them separate). From that clash it was the latter way of thinking that emerged victorious.

As for artisans and tradesmen, who began to increase in number in the 1920s, the conditions of production were twofold: acquiring a tool (in the shape of premises, be it forge or café) and acquiring technical know-how. Sometimes that know-how constituted the person's entire capital, as with the tailor or the builder. Particularly the building trade required a minimum of capital investment, which is why so many children of day labourers turned to building around the turn of the century.

Artisans, like farmers, passed their professional skills on within the family, and as with farmers the jobs available were handed down from father to son or son-in-law. There was a limit, after all, to the number of forges that could open. Each had its own catchment area, and no blacksmith took kindly to a competitor setting up on his patch and trying to poach his clientele. Some increase was possible, of course, as the population grew and more horses and carts came into use. But the number of forges, like the number of farms, remained a fairly fixed quantity.

So it was through marriage that one gained access to this relatively closed artisan class. The story of how Saint-Jean-Trolimon's last blacksmith set up in business throws light on the role of the kindred in the transfer of know-how and in access to the means of production.

Yves R., blacksmith in Saint-Jean from 1929 to 1942, had taken over the business from his wife's cousin. Laurent F., an orphan from Combrit, heard from a passing commercial traveller that Yves R. was looking for an apprentice – what was known as a *mousse*. Yves died during the war, and Laurent went on working for his employer's widow Catherine, who kept things going as best she could. After the war, in 1947, Laurent set up on his own. As he said,

It wasn't easy, starting up here in Saint-Jean. I found it hard being in competition with my old boss's wife, you know what I mean? She'd told me once, after I'd got set up, that she wanted me back because Georges [Laurent's replacement] hadn't proved up to it. I'd gone into business with my cousin, Le M. He was a tinsmith, more into zincware for the building trade. He made guttering and all the things they used on roofs in those days like cornices and dormers. That was his trade, tinsmith, and I was the blacksmith. We'd signed a lease with M., who was married to a cousin of my wife's. It was after that that things changed. What happened, you see – I said to myself, M. is going to marry off his daughters, and if one of the husbands is a blacksmith . . . Well, you never know, you've got to face facts. And that's very nearly how it turned out. M.'s eldest daughter married a plumber and zinc roofer – Emile C., his name was. There was a bit of meadowland in the middle of the village that I bought and built on in 1952.

The interview shows that Laurent F. had secured his position by first making use of professional contacts and subsequently enlisting the help of a cousin of his wife, who was established in the village. He knew, however, that someone with a closer family tie to M. could well supplant him, and that is indeed what happened when the daughter's husband went into business with his father-in-law.

Note that work ties gave rise to marriage ties; Laurent's wife's brother, Jean B., married the daughter of his former boss's wife, Catherine. As in the case of farm leases preceding marriages between the lessors and lessees of a piece of land or a *domaine congéable*, a marriage was here contracted through the medium of an employer/employee relationship.

For Jean-Louis D., forge and family were a single entity:

There were lots of blacksmiths in the family; there were mechanics, cycle dealers, what have you. Now at Pendreff crossroads, on the right, there was an uncle, the youngest of the U. family – out of fourteen in all. My grandfather was a blacksmith at Pendreff, too. My grandfather died around fifty or sixty – not very old. There were sons so they took his place. One of the sons moved the forge up to where the roads divide at Pendreff. It was better for business than down at the lower corner on the Le Guilvinec road. That's where I started at twelve or thirteen. Another U. brother, the father of the man who runs the Caravelle [a restaurant in Plomeur], was a cycle dealer in Pont-l'Abbé. It was that uncle I replaced at Peumerit [where he was apprenticed].

Shopkeepers were often former artisans turned entrepreneurs who had embarked on a new line of business in the early years of the century. Subsequently their numbers became restricted, as happened with cart-wrights and blacksmiths. Henceforth access to commerce was by descent or marriage only. With regard to farmers, a certain social gap inhibited marriages; a shopkeeper would gladly marry a farmer's daughter if her dowry helped him finance the working capital of the business, but a shopkeeper's daughter would never marry a farmer. One effect of a shopkeeper's contracting such a marriage was of course to turn a kindred into a clientele.[10] In Saint-Jean-Trolimon all the shop-keepers of the *bourg* are linked in a network of affinity or consanguinity that is itself connected with the kinship systems of the farming community.

Examining the conditions of production leads one to wonder about the existence of the kind of independence of the economy that may be mooted in connection with exotic societies. it would be absurd to deny a distinction between the economic and kinship spheres. A whole area of economic ties and of the economic structure escapes this mode of social relations. Locally, however, the economy and kinship are linked at several levels. If the one is swallowed up in the other to create social hierarchies, rich and poor are not marked off from each other by different life styles. Kindreds are based on economic relations, with wealthy farmers enjoying kinship ties with the less well-off. Without justifying domination, kinship can make it easier.

The fusion between kinship and economic spheres is also apparent in the ritual and political organisation of society.

Ritual uses of kinship

As Maurice Godelier writes:

> To account for the fact that, within a given society, a type of family organisation operates as unit of production and/or unit of consumption or does not so operate, either not at all or only partially as such, it is necessary to go beyond these visible aspects of kinship relations and to examine the social conditions of production, the mode or modes of production of the material means of social existence. Kinship relations must operate as production relations, governing the respective rights of groups and individuals over the conditions of production and over the products of their labour. And it is because they operate as production relations that they regulate politico-religious activities as a whole and also provide an ideological framework at the core of symbolic practice.[11]

Baptismal sponsorship provides a good illustration of the polyvalence of the social manifestations of kinship. Simultaneously symbolic, ritual, and practical, the roles of godfather and godmother were extremely important in a society that death rendered fragile until the end of the nineteenth century. The godparents gave the child a first name – very often their own, although family tradition was also taken into account. The ritual tie was expressed in the appellations *tad bouern* (godfather) and *mamm bouern* (godmother), while these addressed their godsons and goddaughters as *filhor* and *filhores* respectively. Called by a special name by their godchildren and singling those children out among their nephews and nieces, godfather and godmother played a special emotional role. It was to them that one turned in a family crisis. When godchildren met their godparents on the occasion of the summer *pardon* processions, they expected to be given a coin. Godparents played a key role in the marriage ritual and in fact stood in for the parents during the wedding celebrations. Up until the First World War it was their job to lead their godchild to the altar. So important was this ritual that if the godfather or godmother was prevented from taking part one of their sons or daughters performed it in their stead.

The ritual functions of godparenthood were underlined by the ever-present possibility of one or both parents dying. The godfather acted as surrogate guardian in respect of all official documents concerning the child's estate. Occasionally he was made the orphan's actual guardian.

When Françoise Le Berre, aged thirteen, and Ambroise Tanneau, seventeen, went to the appropriate ecclesiastical court (the *tribunal de l'Officialité*) in Quimper on 14 April 1784 to ask for a dispensation for their forthcoming marriage (they were 3–4 cousins), the girl explained that she was an orphan, that Ambroise's father, her great-uncle, was at the same time her godfather and guardian, and that 'this marriage will get round legal discussions regarding the account that [Ambroise's] father owes to the said le Berre'.[12]

So godparenthood was not merely a ritual and spiritual link; it implied an idea of protection. Day labourers used sometimes to ask the farmer who employed them to stand as godfather to one of their children (underlying this practice there is the notion of clientship).

A great opportunity for a child, particularly in the second half of the nineteenth century when competition for farms was fierce, was to have a godparent whose marriage was infertile or all of whose children had died young. This accounted for the privileged situation of a number of households that succeeded in reconcentrating what the inheritance system had dispersed. In 1855, for example, Anne Le Lay was lucky enough to be both goddaughter and niece to the wealthy Louis Loch, who on signature of her marriage contract gave her, over and above her share of the estate, the sum of 5,000 francs.[13]

The adoption of a nephew by a godfather and his wife was not without its problems when the wife subsequently gave birth to an heir. Take the case of a farm that shall be called Kerlenn, where the proprietor Pierre B. and his wife were without children from 1880 to 1913. They sub-leased the farm to Yves, Pierre's nephew and godson. When Yves died, his young widow remarried and continued to work the farm in the belief that it would be hers by right on the death of her first husband's uncle. However, Pierre had himself been widowed, and in 1913 he got married again to a woman who bore him four children – four heirs. Legal proceedings were necessary to compel Yves's widow and her husband to vacate the farm, and the family sustained such emotional damage that the present-day descendants of those involved are still not talking to one another.

At the end of the last century grandparents were chosen as god-parents for the first children, the two lines being treated equally. According to this ancient model, the first child had its mother's father as godfather and its father's mother as godmother. For the second child it was the other way around. Subsequent choices fell on the child's uncles and aunts or their spouses, usually in order of birth and in such a way that, in each couple, the husband was godfather or the wife godmother to one child in each of the families of their siblings. In this way the family ties woven by descent and marriage were constantly reinforced by ties of godparenthood.

In the event of remarriage following widowhood, which remained very frequent up until the end of the nineteenth century, godfather and godmother were chosen from among the brothers and sisters of the deceased spouse when a child was born of the second marriage or from among its half-brothers and half-sisters. Certain patterns of god-parenthood show how links were forged down the generations in order to preserve relationships that collaterality had a tendency to stretch.

The godparental tie is still very strong today. As before, it tends to tighten collateral links that are the more distended for people having moved away from the village. Remembering godchildren's birthdays and attending their first communions are godparental obligations. Rather than to grandparents, parents now turn to their own brothers and sisters to fill these roles, often choosing those who have left the region. Age differences are used to cross-buttress godparental ties, as when Jean R. stands godfather to his nephew, who in turn stands godfather to Jean R.'s youngest son.

To the individual, then, the kindred of which he or she forms the centre is seen as a flexible network providing work and emotional and material support. In this sense it is possible to speak of the kindred as possessing real power. It is a different, less diffuse sort of power that finds expression in politics.

Kindreds and political power

As we saw with respect to the late eighteenth century, political power at the local level was held by the kindred of the ancestors of the genealogies investigated. A study of the mayors and councillors of Saint-Jean-Trolimon during the nineteenth century reveals a similarly flagrant degree of overlap between kinship and political power. If other men appeared who did not spring from the well-to-do kindreds, it was because they embodied new economic relationships. Examining the village from the political point of view helps us to understand the social diversification of local society.

A comparison of genealogies, marriage networks, and members of the Saint-Jean-Trolimon *conseil municipal* confirms the classic process whereby power is kept in the hands of the wealthiest classes.[14] In contrast to other communes in the region, where a resident lord of the manor appropriated the *mairie*, municipal responsibilities in Saint-Jean-Trolimon were discharged by the peasants themselves; with the exception of three brief periods (the early nineteenth century, 1838–1851, and 1977–1983) no notable stood against them. Every mayor since the beginning of the nineteenth century has been related to his predecessors or successors by blood or marriage, and their councils, if we take a cross-section of their composition at a number of dates, appear to have been recruited from within the same kindreds, who represented so many political clienteles.

Every mayor of Saint-Jean-Trolimon up until 1935 and then from 1948 to 1964 was a farmer, whether freehold or tenant, and every one stemmed from the lines that have been identified as constituting the wealthiest stratum of the farming class. In order to be mayor a man had to be able to read and write, speak French, and have a certain amount of education. Pierre-Jean Tanneau had his *brevet* or school certificate.

The typical Bigouden conflict between 'white' and 'red' coloured the politics of Saint-Jean-Trolimon too.[15] The white mayors were close relatives backed by a solid kindred. The red mayors embodied a break with local tradition, an irruption from outside. Alour Le Cossec, for example, had been born in Plobannalec, where he was related by marriage to the great R.-Quiniou-Calvez kindred. However, marriage ties can be traced from one to the other, those ties sometimes post-dating the mayoral mandate, possibly with the object of establishing a fresh kindred/clientele (for example, the marriages of Michel C. and Corentine C. or of Pierre-Louis V. and Anne-Marie J.).

Table 26. *Mayors of Saint-Jean-Trolimon, 1808–1983*

	Place of residence	Date
M. de Penguily-Merle		1808
Louis Loch	Kerbascol	1830
Jean Desban	Le Steud	1838–1851
Jean Le Garrec	Kerstrad	1851–1878
Alour Le Cossec (red)	Kergreac'h	1878–1888
Guillaume Volant	Kerbascol	1888–1892
Jean-Louis Le Garrec (white)	Gorré-Beuzec	1892–1903
Pierre-Jean Tanneau (white)	Kervouec	1903–1919
Jean-Marie J. (red)	Kervouec	1919–1935
Jean Le R. (red)	*bourg*	1935–1947
Pierre B. (white)	Kerstrad	1948–1964
Jacques Le R. (red)	*bourg*	1965–1977
Dr. Yves J.	Botégao	1977–1983
André C.	*bourg*	1983

The mayors had not necessarily been born in the commune, as we have seen. It is even interesting to note where they lived. They were residents alternately of the enclave – 'the most populous and wealthy [part of the commune]', as mayor Louis Loch said in 1835 – and of farms closer to the *bourg*. No mayor ever came from the sandy coastal strip.

Up until 1935, then, the same kindreds occupied the Saint-Jean-Trolimon *mairie*. Politics, kinship, and marriage were closely interlinked. Although the inheritance system did much to level out social differences, political clientship was associated with the symbol of ancient prestige that a particular line continued to bear. In contrast to what Claude Karnoouh observed in the Lorraine village of Grand-Failly, relinkings and consanguine marriages were not incompatible with the formation of political clienteles. In Grand-Failly it was at the cost of an impoverishment and a dispersal of inheritance that a person diversified and increased his electorate, whereas endogamous lines concentrated ownership among first cousins but in so doing limited their potential political clientele.[16] Here it was not the same mechanism that operated. Family marriage policies never had the function of consolidating or increasing peasant ownership of land; moreover, the birth rate was high enough for the dominant kindreds to find sufficient voters, the real turnaround coming only when the newly elected councillors depended on the poorest section of the population (see table 26).

The 1935 election did indeed mark the irruption into politics of a new economic order. The new mayor, Jean Le R., was the proprietor of the canning factory in the *bourg*, which provided jobs for the workers and a market for the many *palue* farmers whose pea crop it purchased.

Previously the split between whites and reds had run through the same farming lines; now it was intensified by professional differences. Jean Le R.'s appointment as mayor was in any case hotly disputed. It is said that he got the job only on age, because after three days' debate the twelve-man council was split down the middle between him and his opponent (a farmer related by marriage to the major kindreds). As with his predecessors, marrying his daughter to the son of one of the village shopkeepers was to be, for him, a way of integrating.

The period from 1935 to the present has seen a struggle between the kinship and economic spheres, with political power passing back and forth between the old kindreds and the cannery proprietors (father and son). When the cannery closed in 1968, the economic relationship between the mayor and his electors gave way to the standard sort of relationship of alliances that had by then become established with certain farmers and with the shopkeepers and artisans of the *bourg*.

If we make the effort to trace the members of the council from 1836 to

Fig. 30. Kinship and marriage ties among mayors, 1830–1983.

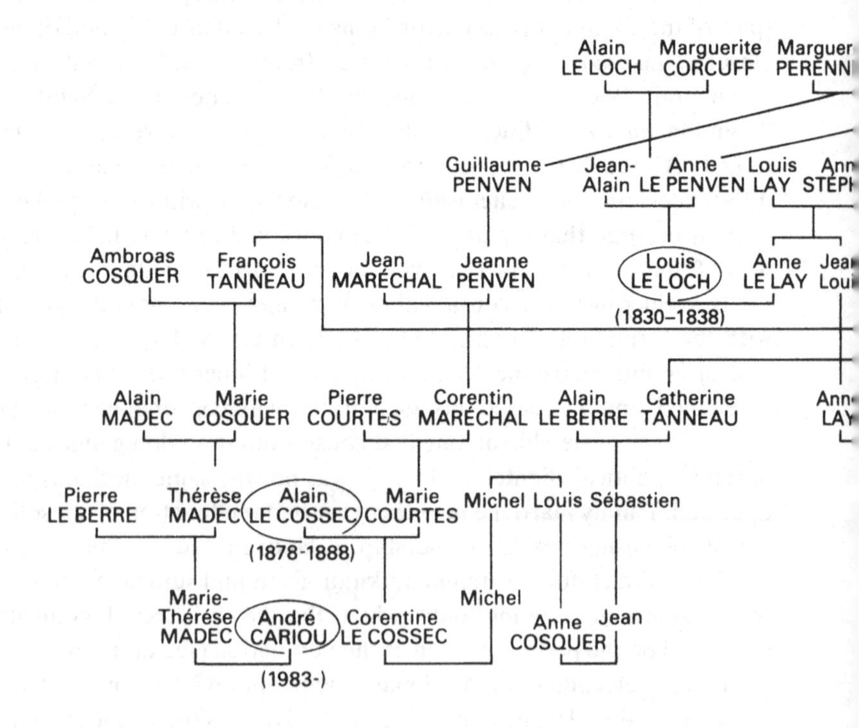

today, we find many of them falling into the same kindred systems. There is a quite remarkable stability, whatever the political colour of the *mairie*. Just as between red and white mayors, there are so many ties between red and white council members that it is impossible to tell their colour from the family they belong to.

The kindreds that kept their hand on power did not, as might be thought, cover the whole population of the village. Not surprisingly, their members were both wealthy and long-established residents, well known to everyone. The councillors sitting in 1856, for example, for the most part held a very stable mandate. They came from much the same areas of the commune as the mayors. Nor was the enclave forgotten, with every council including inhabitants of Gorré-Beuzec, Kerbascol, and Kernahu. The people of the sandy coastal strip were kept completely out of it as being too poor, too illiterate, too different, and too remote. Not a single member of the council ever came from the *palue*, even when the population of that part of the commune was at its highest (see table 27).

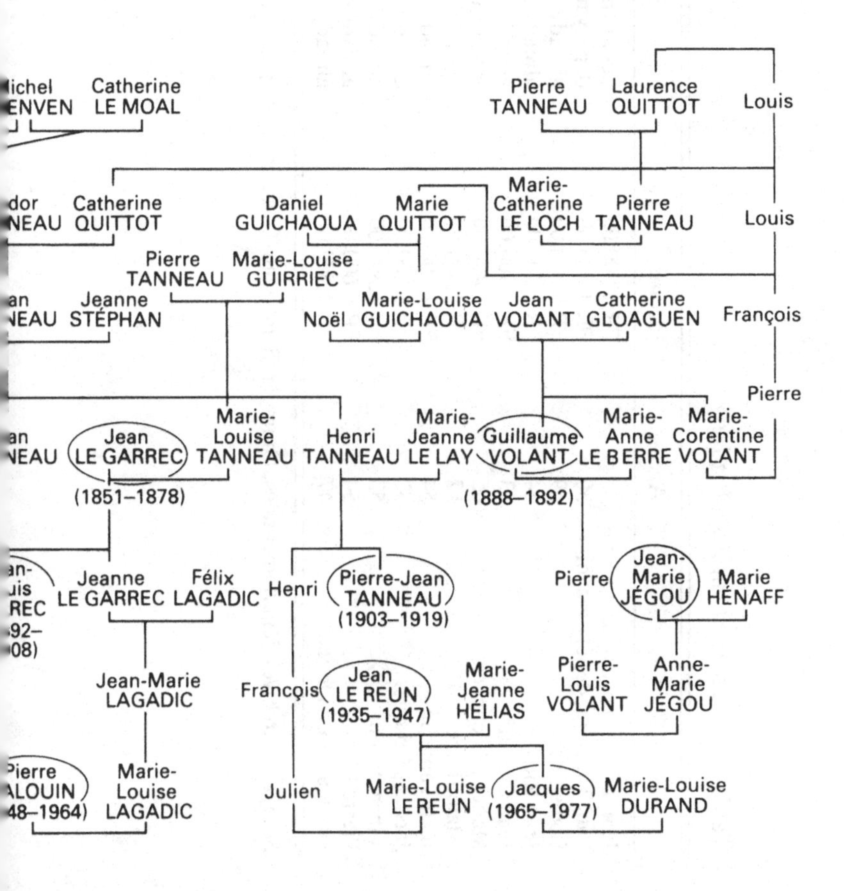

Table 27. *The conseil municipal in 1865*

	Age	Income		Place of residence
Jean Le Garrec, mayor		FR 4,000.00	deputy mayor 1846–1851, mayor since then	Kervouec
Paul Tanneau, deputy mayor	30	FR 4,000.00	deputy mayor since 1864	Kerbascol
Louis Loch	69	FR 3,000.00	already on council for 41 years	Kerbascol
Jean Riou	64	FR 1,200.00	already on council for 15 years	Kerbascol
Jean Corcuff	63	FR 900.00	already on council for 48 years	Kernahu
Maurice Gloaguen	65	FR 400.00	already on council for 18 years	Tronoën
Alain Berre	57	FR 4,500.00	already on council for 27 years	Kerinval
Jean Tanneau	54	FR 1,500.00	already on council for 24 years	Trevinou
Jacques Cosquer	57	FR 3,000.00	already on council for 10 years	Kerfilin
Vincent Coïc	53	FR 1,300.00	already on council for 12 years	*bourg*
Pierre Berrou	54	FR 600.00	already on council for 5 years	Gorré-Beuzec
Guillaume Le Donge	47	FR 2,500.00	already on council for 18 months	Tréganné

Source: Archives départementales, Quimper, Saint-Jean-Troilimon, no. 33.

One is struck by the great stability of the kindreds right down to the present. Not only are council members related to one another at each point in time; they are related to their predecessors and to their successors.

At the council elections of 1983, the team was once again recruited on the same foundations of mutual acquaintanceship. Between ten of the fourteen councillors it is possible to trace links that are not merely the fruit of genealogical research but are well known to the persons concerned and indeed account for their positions in the administrative organs of the commune.

Given all the technological, economic, and social upheavals described in the foregoing chapters, will kinship go on preserving its vitality and importance in fresh forms?

Do economic circumstances nowadays tend to be dissociated from family circumstances?

Kinship and conditions of production are largely dissociated today. This even applies to farmers, who have stopped borrowing money from relatives and now borrow from union or professional organisations, which have imposed a new kind of logic: profitability. Domestic group and labour unit no longer coincide, and incomes have become individualised. For all these reasons it has become customary to assert that kinship systems have lost their social uses.

On the contrary, South Bigouden offers a wealth of examples of the vigour of kinship systems and of their use by most socio-professional categories other than farmers. Networks of kinship ties and mutual acquaintance run through schools, trade unions, administrative departments, and so on. A patient admitted to Pont-l'Abbé hospital, for example, may be sure of finding a number of people he or she knows among the nursing staff (the doctors come from outside the region) or among the patients, whether distant relatives, friends' relatives, or friends of relatives.

This is why looking for a job becomes a highly personalised affair when one is local and kinship comes into play. Networks of social relations, whether among family or friends, carry a great deal of information about the state of the labour market – information with regard to both quantity and quality. These networks intersect. The proprietor of the largest building-materials supplier sits on the committee of the football club. He often recruits staff there, and he is in a position to supply others with very precise information about the needs of local tradesmen. The president of the football club is the manager of the local branch of the *Crédit agricole* bank. Bigouden-born, he knows

70 per cent of his clients personally; as a result, the granting of loans is done on a rather different basis than in an urban branch, whose clients are often anonymous.

Self-employed artisans often find their clientele in the kindred. Local-government employees experience interference at the professional level from kinship or neighbourly ties seeking to supplant the administrative relationship. People prefer to consult Dany R. (of the Douarnenez branch of the ANPE or 'Agence nationale pour l'emploi', the government job-finding agency) or Jacqueline P. (who works for the inland revenue in Pont-l'Abbé) at home rather than in the institutional surroundings of their offices. Young people looking for a summer job will turn first to a relative, who will give them priority when taking on extra staff.

The collective forms of labour by which certain farming operations were carried out in the old days – and which brought together relatives and neighbours – find an echo today in the practice of collaborative house-building.

Many young people intending to settle in the village build their own houses. They reactivate the old forms of family mutual aid in the context of modern economic arrangements – never failing, for example, to obtain bank loans to finance the purchase of materials.

Building a house by mutual aid within the family involves the need for a measure of professional qualification in the building trades as well as a major investment of labour by the young owner personally. Since every family network nowadays includes at least one or two representatives of the various branches of the construction industry, mutual aid is entirely possible in terms of both advice and actual help. Building houses for other people eventually gives a man the know-how to build one for himself, but it means putting in long hours of toil in the evenings (after a day's work) and every weekend for many months. The young man who used to play football before or in the early days of his marriage will have to give it up. Building one's own house is an adult step; it is a rite of passage cutting the young married man off from the group of bachelors.

Houses built by the family are often houses without an architect. The future owner draws up the plan, which then has to be submitted to the authorities by a registered tradesman. This is a service that a cousin, say, who is in the building trade will perform by signing a plan he has no intention of executing himself. Once the administrative formalities have been dealt with – and this may take some time – the next step is to organise what are still called *corvées*, reminiscent of the old *Dervez Bras* or *grandes journées de travail*.

The family team is made up of close relatives: the fathers and brothers of the couple building the house (house-building is regarded as men's work). However, it is not usually possible to rely exclusively on free help of this kind. Where the team members lack the necessary skills it is necessary to turn to professional craftsmen. But the site boss is the owner of the house, and he will make sure that such intervention is only occasional.

The situation of Bernard and Gérard M. probably offers the archetypal model of family house-building. Not all examples quite come up to it, but it may be taken as a yardstick for the various degrees of mutual aid that the family network is called upon to supply.

The M. brothers had both been married for two years. One of them worked as a plumber and heating engineer and the other was on the payroll of the largest local building firm. Their father was a builder; their mother owned three plots of land in the *bourg* that she had inherited from her father, a small farmer in the days before the First World War. She gave one plot to each of her sons, keeping the third for her daughter, who was still a spinster.

One August, during his holidays, Bernard built his house in the space of a fortnight. His employer, who was also one of the managers of the village football team for which Bernard had played for a long time, lent him a crane. The mutual-aid team consisted of his brother Gérard, his father and father-in-law and an uncle who lived nearby (all three of them builders by trade), and his young cousin as odd-job man. His brother-in-law, who was in the haulage business, fetched all the necessary materials from Briec-de-l'Odet in a single afternoon. During that August the whole team worked flat out to erect the shell of the house (all except the roof frame and roof covering). The rest Bernard completed virtually on his own. The same team, plus Gérard's wife's brothers-in-law, got together again two months later to build Gérard's house. Family, locality, and mutual aid are closely interlinked.

In the case of Guillaume B. the team brought together the same family group: father, father-in-law, and three brothers-in-law. A firm poured the concrete floor; the family took care of the roof, woodwork, insulation, heating, and electricity. The tiling and plumbing were done by a craftsman – a cousin who took longer than he was meant to but whose work was appreciated.

One last example: Louis E., an electrician by trade, poured the concrete floor of his house with the help of his father, his father-in-law, his two brothers-in-law, and an uncle. It took them two and a half days. Then Louis built the external walls and the plasterboard partitions and installed the electricity, plumbing, and heating single-handed. However, he did call in professionals to do the plastering and put the roof on.

A man building with the aid of his network of relatives can choose how much or how little help he wishes to receive. All help accepted presupposes a return in kind, building someone else's house, and after two years of weekends spent erecting his own home a man may be unwilling to commit himself to the cycle of returns; otherwise he may find himself spending all his free time for the next year on family building sites throughout the region. Consequently some prefer to restrict the amount of aid they are prepared to receive rather than have to pay it back. Others, however, will accept a great deal of help.

But one wonders how much they actually save, these services rendered by the family network. The specific reason given for falling back on family mutual aid is economic. 'We shouldn't have been able to afford such a big house', is a remark that is often heard. Undoubtedly the amount of labour invested personally by the future occupant of the house is a big factor in bringing down the building costs. But opinions are divided as regards how economic this kind of mutual aid really is.

House-building on Saturdays and Sundays and during the holidays is not done at the same tempo as paid work. That is precisely what distinguishes this kind of co-operation from 'black' work, which is performed to the same sort of standard of speed as ordinary work. When relatives and friends are building a house together there is less hurry. Often more time is spent chatting than actually working, and in any case the rhythm is more relaxed. Moreover, there are equipment problems: 'You've never got the right tool, so you hop in the van and drive to Criou's in Pont-l'Abbé. They're closed, right, so you go dashing off to Pendreff to try there.' And collaborative house-building undeniably represents a rediscovery of the pleasure of male socialising; under cover of an obvious pretext, the men have a chance to chat as well as to enjoy the odd drink together.

Family house-building is hungry and thirsty work. This is the festive aspect of mutual aid. It has always been present. But it ought really to be added to the building costs. In fact it constitutes another reason why this way of going about things is less economic. A typical female remark will be: 'The amount of food and drink we had to supply them with!' The womenfolk, with no part in the actual work, find themselves recreating the traditional organisation of agricultural operations – say at harvest time. Their job is to bring snacks and drinks for the men on the building site. They prepare meals worthy of those that used to accompany 'major works' in the old days – not just for their brothers and brothers-in-law who have been doing the work but for all the wives, sisters, sisters-in-law, and children who come along with them Sunday after Sunday. Family house-building offers an opportunity for a get-together with brothers and sisters such as will not recur until the children's first communions or some other family celebration.

So it is perhaps not wholly certain that, when all is said and done, building a house with the aid of relatives really is less expensive than having a professional builder do it. The cost is further inflated by the longer construction time – two years, two and a half years sometimes – and the months of paying rent on temporary accommodation that would have been unnecessary had one been able to move in sooner. No one works it out in those terms, though, because for most young Bigoudens family house-building constitutes the norm. Henceforth the young person is involved in a cycle of giving and receiving aid that reaches beyond their parents to incorporate other relatives of the same age living in the region.

Family house-building distinguishes those youngsters who have chosen to settle in the village and who are local in origin from those who are not. The latter occupy the houses on the two developments, while 'natives' are usually scattered on plots of land inherited from their parents. Outsiders *have* to rely on building firms, which supply their houses ready-made; they do not enjoy access to the kinship systems so useful to native Bigoudens.

The usefulness of kindreds

As it did in former times, the kindred today lends identity; it offers a way of locating the individual and, if need be, relieving him or her of the 'outsider' label. To indicate a person, all one has to do is add the place of residence to the first name or surname – saying 'Marie Gorré-Beuzec', for example, or 'Cariou de Kerinval'. The emigrant returning to the village for the holidays or the young person setting up house outside the *bourg* and experiencing assimilation problems is identified by his kindred; people recall that the grandfather of the one worked a farm at K. or that an aunt of the other ran a bar in Pont-l'Abbé. Reference to a person's kindred makes it possible immediately to relocate that individual among persons who are known by their business as by their place of residence.

Attitudes to the reconstituted genealogies differed as between locals who had never left the village and those returning to it as either temporary or permanent residents. As far as the former are concerned, their genealogy is something experienced and therefore of little use to them in written form. For the latter it is frozen at the level of memories that are already dead and as such is highly prized. Folk still living on the family farm, presented with a genealogy that traces some lines back as far as 1670, will show a polite interest in the work involved but not much more. Finding their great grandparents, the people who first acquired the farm and of whom they vaguely remember their own grandparents speaking, elicits some reaction; they want to know where a particular ancestor was born, or they recall having kinship ties with some other line as well. 'We knew we were related to the Tronoën folk but we didn't know how.' The rediscovery of that link is something precious to them, possibly because they wish to revive it, if only symbolically. But when the names recited no longer find an echo in memory, the genealogy becomes a collection devoid of meaning. They do not ask if they may take a copy or have a copy made of it. They are not interested in knowing their ascendant kin. This is because they have no need to establish an identity or find roots; they have always known that they belonged to and were a part of this region, and having the fact confirmed strikes them as merely futile. On the other hand those who had left the district or whose job put a distance between themselves and the village placed a high value on their family tree as furnishing proof of identity and local roots.

However, the integration of kindreds is very much less strong nowadays than it was in the nineteenth century. As is true all over France, one observes a tightening of bonds between parents and married children. Sometimes inter-generational family ties are developed to the point where, residential independence notwithstanding, one finds a kind of symbiosis between the two domestic groups, particularly when a child/grandchild arrives.

As far as these young couples are concerned, the most notable service rendered by their parents has to do with looking after small children. Grandmothers take the place of mothers when children are at the toddler stage or at infant school. The fact is, building the house having necessitated taking out a loan, the young woman now has to work to help repay it. Where it has been possible to build the house on a piece of land near that of the young woman's mother, she (the grandmother) has the baby with her all day and sometimes at night as well during the week, the parents taking it back only for the weekend. When the child reaches school age, grandmothers play as big a role as mothers in taking it to and fetching it from school, keeping the child in the evening until the parents get home from work. Crèche facilities or day nurseries exist that mothers might use, but they do so to a very limited extent, preferring the family solution. Similarly, during the summer holidays children stay with their grandmothers. What this system makes manifest is a desire for cultural continuity. In contrast to what sociologists observe in other social categories in relationships between parents and married children, there is no conflict over the children's education.[17]

The choice of staying or returning and having a house built gives rise to a further choice: whether to bring up one's children in the way one was brought up oneself. Breton-speaking grandmothers make an effort to speak French to their grandchildren, which actually leads – so the children's teachers say – to an impoverishment of their vocabulary.

When the mother's house and the daughter's house are situated close to each other, it becomes grandmother's job to prepare the midday meal, which her daughter and son-in-law will eat with her. In other words, it is accepted that no man may cook for himself or his family; if the young woman is working and has no time to cook lunch, her mother does it for her. The net result is that the lovely house built by mutual aid for the young couple to live in only really comes to life for Sunday lunch and on the odd afternoon during the week when the washing-machine is in use. In these family configurations mutual aid between parental and filial households is developed to the point where an almost symbiotic economy exists between the two. This finds expression particularly in exchanges of produce, as rendered possible by the partnership of the horticultural economy and the deep-freeze.

Take the case of Jeanne P., whose house stands one garden away from her mother's. Jeanne sows a plot of land with cereals every year. She gives the grain to her mother to feed the chickens that her mother keeps on her own land. Jeanne's return on her grain comes in the form of oven-ready chickens, killed, plucked, and drawn by her mother. The gardens of Saint-Jean-Trolimon supply a steady stream of green vegetables and potatoes that will be given and received according to season and depending on what has 'come on early'. These gifts and counter-gifts, in feeding the family members concerned, also foster relations between them. The produce of people's gardens and chicken-runs and the contents of their freezers nourish both stomach and imagination and provide a constant pretext for family meals as well as much matter for conversation.

At the material level everyone is well aware of the economies that such produce means for each individual budget. When supplies are plentiful these exchanges move outside the restricted context of the family and flow into broader channels that have nothing to do with kinship and where no immediate return in kind is expected. A person will give some courgettes to a friend who has a wallpaper shop and lets him have a discount. Someone else will say 'thank you' to a child's teacher with a crate of new potatoes. A third person will give some apples to a fisherman cousin in Le Guilvinec, who will reciprocate with a few kilos of *langoustines*. The cycles of alimentary exchange also draw in brothers and sisters with second homes in the village, thus opening up to the outside world. Even the lady from Paris who descends on the village several times a year to conduct some sort of research and is received with kindly toleration goes away each time bearing a rabbit, a box of leeks or carrots, and several jars of gherkins or pots of apple jelly (she returns with champagne and birthday presents).*

As Claude Macherel suggests, such exchanges may be analysed in accordance with three principles indissociable from a single reality: the position of the partners to the exchange within the social structure, the symbolism of the goods exchanged, and the form of the system of exchanges.[18]

In Saint-Jean-Trolimon and throughout Bigouden it is not just anyone that participates in these gifts and counter-gifts: close relatives, neighbours, workmates, and persons occupying some sort of hierarchical position *vis-à-vis* the giver are involved because something may be expected from them in return – indirectly, perhaps, or at a later date, and not necessarily identical in kind to the thing given. An implicit account is scrupulously observed with regard to what has been given and what received.

The vitality and significance of relations between kin are observable at a more general level in the wider context of the individual's social relations. Fathers and mothers and any married brothers and sisters who have stayed in the immediate vicinity form the object of the Sunday afternooon visit, and ceremonial meals have been emerging for several years. It has become customary among brothers and sisters who have remained in nearby communes to hold a festive meal each year at the house of one of them, often on New Year's Day. However, if the one due to play host has already given a meal in the past year on the occasion of a baptism, birthday, first communion, or marriage, nothing is done. Family meals still observe the traditional calendar of the *pardons*, those festivals that are both religious and parochial (*communal*) and are dedicated to a saint and his or her chapel. As in the old days, the family still gets together for a meal. Take Valentine S. and her two sisters, for example: one plays host on the first Sunday in the year; the second, who lives in Plomeur, receives on the occasion of the *pardon* of La Treminou, which takes place in September and is the main one in that commune; the third, who usually kills her pig some time in February, invites the others to a meal of pork. Also observable is an important expansion of family meals associated with the major rites of passage, together with an increase in the number of relatives involved

* Even the lady's translator and his wife – virtual strangers – came away from a visit to the village with the makings of a copious and delicious meal (as well as the memory of one already enjoyed) [Translator].

each time: at baptisms, first communions, weddings, and funerals. Baptism and first communion become the occasion for family celebrations at which gifts are exchanged (often of quite considerable material value), with formal meals lasting all afternoon. When a marriage takes place the kindred in the broadest sense is involved; even today invitations go out to the uncles, aunts, and married cousins of both bride and groom. A certain conflict is observable nowadays between this traditional model of the wedding with a large number of guests and the more modern idea of the restricted wedding. In the former case each person pays his or her score; in the latter the newlyweds pay for the relatives they invite. The first type of wedding may be attended by 250 people, the second by 100 at most. Formal relationships with the wider kindred cease at marriage. Funeral processions may be followed by more than a thousand people who, though aware of being kin, are unaware of the precise links. Each person experiences one last family obligation in the face of death. No cards are sent out when someone dies, but before reading the paper people look down the deaths column to see whether their presence is required at the funeral of some great-aunt or uncle. The cult of the dead remains firmly established. When a close relative dies, brothers and sisters stand by one another. When the deceased's spouse has gone before the children take turns cleaning the tomb at All Saints' tide. This involves the equivalent of a vigorous spring-clean in line with the latest standards of modern cleanliness. For a brief moment the tomb and the cross surmounting it cease to be sacred objects to be revered from a distance and become elements of a dwelling in need of upkeep. The cross is given a thorough wash and has its shine restored, while the pebbles and other accoutrements of the tomb are taken home and cleaned with whatever product will make them gleam most brightly. This is women's work, and daughters-in-law willingly take their turn to clean the tomb in which their parents-in-law are laid to rest. Every couple will place a pot of chrysanthemums on the tomb on All Souls' Day, and if the deceased had a large number of children his tomb will be covered with flowers. For a week the cemeteries of Bigouden resemble entries in a best-kept garden competition.

What radiates through all these examples is the continued vitality and significance of the social uses of kinship. Nowadays they may present themselves more in the context of the immediate family – that is to say, of emotional and material relationships characterised by a tightening of the bonds between parents and married children in conjunction with a certain loosening of collateral ties (though every effort is made to sustain these at the symbolic level) – but they still have some connection with the sphere of economic relations and belie the alleged rift between the two realms of experience. In principle, what pertains to kinship is private while what pertains to economics is public and organised in accordance with the laws of supply and demand – laws that have nothing to do with the family sphere. If today these economic relationships unexpectedly woven into the meshes of kinship are described as 'informal', 'parallel', 'underground', and so on it is in order to bring out their quasi-illicit character and the family's power of resistance to allowing itself to be enclosed in an entity devoid of all functionality.

It would be wrong to ascribe the maintenance of family systems to some sort of anachronistic form of village life – rather like an island cut off from and forgotten by civilisation and the modern economy with all their constraints. The fact is that South Bigouden is seeing the same sorts of relationship between town and country as can be observed in many other parts of France.

Conclusion

The primary axis of this study was as follows: to isolate a particular model from among the variety of peasant societies, bringing out the inter-relations between demographic, family, and economic circumstances; then to trace the evolution of that model through two and a half centuries in order to restore a dynamic perspective to the generally static analysis of peasant society.

The importance of kinship in peasant societies, of which ethnologists are well aware, prompted a closer study of this social fact and the way in which it developed. The harder the features of that development were to pin down, the more interesting problems it raised. The high degree of mobility of domestic groups, the absence of discrete exchanging units, and the existence of flexible bilateral networks discouraged research into marriage regularities from the start. This second problematic axis, inserted in the first, called for some theoretical reflection on the place of kinship in marriage in peasant societies. It also meant developing a particular method of study.

South Bigouden appears to present an original model within which family realities – viewed from the angle of demography and kinship – and economic realities fit together in a coherent fashion.

In the eighteenth century and in the nineteenth the population was highly mobile within the commune and within a perimeter described by matrimonial migrations. That perimeter clearly bounded the southern communes of Bigouden up to and including Plonéour-Lanvern. Social practice defined a homogeneous area that a number of socio-cultural traits distinguished from the northern canton (North Bigouden), relations between the two being very much more distant than the folklorists – who seek to convey an image of profound regional unity – would have us believe.

This intra-regional mobility was exclusive of any emigration. When they moved, wealthy farmers and poor day labourers all remained within a geographical area that was familiar to them through shared customs and a common language. The two social categories that made up the farming population were both equally mobile, and their frequent movements were connected with the domestic life cycle. That cycle began very early in South Bigouden, where age at marriage was exceptionally low. One result of this was to necessitate inter-generational co-residence. When the generations did split up at one or another point in the family life cycle, young couples took to the roads in search of a farm lease or sub-tenancy.

Mobility was also consequent upon a system of inheritance that had the twofold characteristic of being partitive and bilateral. Transmission of property effectively took place after the disappearance of the older generation and never when the children got married – in contrast to regions practising a single-heir system and even to the situation observed around Plozévet. Dowries, in failing to provide couples with the means to set up house on their own, contributed to the co-residential model by reinforcing young people's reliance on parents. The latter retained all their power as regarded choosing a successor but were at pains to respect the egalitarian principle of partible inheritance, which did away with all differences between older and younger children or between brothers and sisters. The contradiction between the partible system, which fragmented the family patrimony at each generation, and the need to keep the farm of a size to make it economically viable was not difficult to resolve here because peasants were not usually freeholders but by virtue of the peculiar system of the *domaine congéable* owned only the farm buildings and the arable layer of the land.

Following the family life cycle, children co-resided with their parents for a while, then left the parental farm. They might return to it later, succeeding other brothers and sisters who had occupied it for a spell. The notion of the successor was very fluid in this inheritance system as long as peasants did not as yet own their own land. This was something that came about only very late on (towards the end of the nineteenth century). The imprecise nature of the status of successor led brothers and sisters to circulate among themselves compensatory balances (*soultes*) that meant they lost all right to the paternal or maternal property. This had the effect of detaching them emotionally and symbolically from the family farm and encouraging geographical mobility.

Contrary to all expectations, the partible inheritance system reinforced the power of the older generation, which also controlled the marriage system. It was consanguines and affines of older generations that married the young people, who were sometimes still children and who spent several years co-residing with their parents. Mobility, family life cycle, mode of property transmission, and age at marriage all helped to shape matrimonial practice. The latter was found to have been characterised by relinking, a practice that was first identified and described in the village of Minot in Burgundy and that in this instance a computer programme made it possible to measure statistically. Such relinking occurred between affines, two lines exchanging spouses down the generations in such a way that four spouses ended up having two ancestor couples in common. Relinking also occurred through remarriage, the incidence of which was very high because of the high rate of widowhood. Second marriages described either closed configurations within which the children of remarrying widows were joined in marriage or cycles consisting of a series of remarriages closing with a descendant of the first spouse who had started the chain. Relinking appeared as a very frequent form of behaviour overlaying ties of consanguinity with marriage ties. Analysis of the various figures for these relinkings confirmed the profound bilaterality of the system, male and female lines forging relinkings with equal frequency.

Consanguine marriages, on the other hand, supposedly as typical of Bigouden as the towering *coiffe* worn by its womenfolk, were few in number. Moreover, when contracted between distant kin they often coincided with relinkings.

Relinkings and consanguine marriages between distant kin reveal the existence of a genuine preference for choosing to marry a relative – one, however, who was far more often an affine than a consanguine, and far more often a close affine than a distant consanguine. All social categories intermarried by relinking, but some did so more than others. A few kindreds, covering some twenty lines spread geographically throughout South Bigouden, were endogamous and exchanged spouses very busily between the lines that made them up. No cycles or loops can be distinguished closing sequences of marriages on the initial union, but we do find horizontal exchanges between near generations affecting all the lines embraced by the kindred. The wealthier the lines that made up a kindred, the more endogamous that kindred was and the more exchanging it did.

Relinkings represented strategies of protection operated by a social group that sought in the course of the eighteenth century to climb to the top of the social hierarchy. The economic climate was favourable to peasant farmers. The region was extremely fertile and exported cereals at prices that in the late eighteenth century rose steadily. Wealthy peasants stood out clearly at the time from the poverty-stricken mass of tenant farmers and agricultural day labourers. It was possible to put a figure on this group: around 1,200 persons. This was virtually the same as the so-called founder ancestors (that is to say, individuals whose parents are unknown) of the genealogies studied.

Although the land did not belong to the peasants, the rise in cereal prices gave them access to *convenants* or leases of reparative rights under the *domaine congéable* system, and accumulating a number of *convenants* was the key to possible enrichment. It was the peasant owners of several *convenants* that systematically married all their children amongst themselves, relinking not on the basis of immediate geographical proximity but throughout South Bigouden. These highly integrated kindreds combined marriage functions to such an extent that they were often simply the two sides of a single transaction. The overlap between the kinship and economic spheres appears to have been extensive.

The kindred can thus be thought of as a system of coloured lines criss-crossing the whole area of South Bigouden, even if the individual points making it up – domestic groups – were unstable. From Ego's point of view the kindred was a fluctuating group without collective rights or obligations. Seen as a system of vertical lines joined in horizontal relationships through the medium of relinkings, it resembles a corporate group owning assets jointly – in this case *convenants*, information about farm leases available, political rights, and so on.

In certain French peasant societies the family framework is more of the vertical type, organised around a single line. Here the horizontal was dominant, though it performed the same function of social reproduction generation after generation, echoing the marriage practice of the previous generation.

Within kindreds, which were simultaneously systems of mutual aid, information and marriage exchange, and the provision of certain means of production, each domestic group constituted a unit of production, and it was these that bore the brunt of everyday work. Their mode of production was characterised by a very low level of mechanisation; it relied heavily on human strength and on the oral transmission of know-how down the generations. At certain points in the farming year major collective labour gatherings blended work and play, as it were, providing occasions for social contact and cultural creation. *Aires neuves* (when threshing-floors were renewed), harvesting, and threshing relied on neighbourhood mutual aid as well as on the kinship network. Together with *pardons*, fairs, and markets they were as much religious, work-based, or commercial social occasions as moments that broke with everyday routine. As such they were marked by drunkenness, but it would be wrong to regard this as vicious; it was part of a way of life. Alongside a work ethic the Breton character seems to have set great store by an ethic of spiritual escape through the medium of ritual alcohol abuse.

Economics and kinship are shown to have been closely interlinked in South Bigouden when we look at peasant social practice. In establishing this link, however, we must be careful not to push the comparison with primitive societies too far. Like such societies, peasant society is admittedly characterised by a form of abundance and by a work ethic not based on profitability; the confusion of work activity and celebratory activity suggests that here is an economy exhibiting all the structural conditions of under-production. That is as far as the similarity goes. The economic sphere possesses characteristics of its own that are independent of those of the local society and that respond to constraints coming from without: product prices, lease prices, size of regional markets, and the extent to which those markets fluctuate. Peasant society is subject overall to macro-economic constraints. These interfere with the kinship order by overlaying the segmentation of lines with social cleavages.

The foregoing propositions characterise a peasant society studied statically – as one might describe it at the end of the eighteenth century. However, studying this society statically is misleading since it was subject to dynamic forces that helped to change it – slowly at first but then more spectacularly from the 1880s onwards. Demographic, social, and economic causes were so enmeshed as to make it impossible to say which came first.

The extreme youth of brides at marriage was responsible for the marked population increase that first affected the region in the late eighteenth century. One of the highest fertility rates in France was recorded there right up until the 1880s, showing the absence of any kind of contraception. Demographic growth was prevented from becoming explosive by the region's very high mortality (peasant emigration, as we have seen, did not antedate the 1930s). But it had a place amid the many causes of a peasant mobility marked by contrary movements within South Bigouden.

During the eighteenth century peasants moved up from the densely populated southern parishes (Loctudy, Plobannalec), to such less populous northern parishes as Saint-Jean-Trolimon and Plonéour. In the late eighteenth and early nineteenth centuries the poorest sections of the population moved west to occupy the sandy coastal strip. Potato-growing became more common in the 1830s, particularly on these arid stretches of the *palues*. It was in symbiosis with demographic growth since it called for a large input of manual labour – more than cereals – and was on the other hand well suited to feeding an ever-growing population.

The youthfulness of brides and bridegrooms, a sign of parental determination to control marriages in order to protect social hierarchies, contained the seeds of the elimination of those hierarchies. All lines, rich and poor, were eventually hit by the effects of the vigorous population increase, which were reinforced by those of an inheritance system that, despite many distortions in the way in which it was practised, remained deeply egalitarian and ruled out any accumulation of capital. That meant that the wealthiest lines became poorer, generation by generation, and the systematic exchange of spouses within endogamous kindreds lost its point. The practice of relinking continued, but henceforth it involved a very much larger number of lines. Since there was no longer a wealthy peasant category to be protected, that having sunk back into the mass of the peasantry, from the 1850s onwards kindreds were no longer as integrated as they had been; they opened up to other lines as their power over access to land disappeared in the wake of competition from systems external to kinship. From the point of view of the structure of alliances, what was seen was a disorganisation of the system, which broke up into short exchanges.

Demographic developments and the way in which property was transmitted helped to impoverish the peasant population. The economic climate of the nineteenth century reinforced the process. The Revolution did not favour Bigouden society. Peasants were not able to buy up *Biens nationaux* as they were in other parts of France. Repeated crises of production were aggravated by the very pronounced upsurge in the population that occurred in the early part of the nineteenth century. Endemic poverty became rife. The organisation of society, largely dependent on the church, was disturbed by the events of the Revolution. Uncertainty over boundaries (in South Bigouden the boundaries of the new communes were redrawn several times) partly explains why the commune system of local government was so slow to get off the ground.

Land clearances notwithstanding, agriculture remained extensive for a good part of the nineteenth century. There was no major change in the mode of production, which continued to be based on human physical strength and hand-held implements. Labour was supplied in normal times by the domestic group and at the climaxes of the farming calendar by a form of collective organisation. The infrastructure of communes, to judge by that of Saint-Jean-Trolimon, lent itself neither to technological improvements nor to a disenclosing of the region. The roads were unreliable, education was in a state of stagnation, illiteracy had been advancing throughout the eighteenth century and was not yet on the retreat, and religious disputes took up much more of the commune's energy.

Judged from the outside, however, economic circumstances favoured the local economy – notably in that prices had increased. That increase did indeed benefit landowners, but the economic balance of farms, squeezed between cereal prices and land rents, remained extremely fragile, as an analysis of post-mortem inventories reveals.

These documents also throw light on the cultural identity between peasants both rich and poor who shared the same mode of production, the same diet, the same sleeping habits, and the same style of dress. Where it is possible to compare the descriptive content of the inventories with information relating to the domestic group concerned, the analytical value of such documents is considerably enhanced. For example, they reflect the way in which peasant hierarchies were progressively eliminated in the late nineteenth century. The gaps between rich and poor diminished; beggars disappeared, as did farm servants. However, the memory of the past splendour of the relinking lines was not effaced from popular awareness, for they went on being voted into positions of authority all down the nineteenth century. In the collective memory of the village the ancient hierarchies retained their symbolic power.

At the same time new socio-professional categories were appearing. Fishing, which had lost its importance to the region back in the sixteenth century, experienced a fresh boom towards the end of the nineteenth century with the emergence of the canning industry. A great deal of economic activity grew up on the basis of this maritime revival: haulage, construction, boat-building, and so on. Fishermen, factory workers, and tradespeople constituted genuinely differentiated social categories whose economic interests and attitudes of mind were in contrast to those of the peasants. Fishing and its associated activities made it possible to absorb an ever-growing population. The direction of movement was once again reversed and was now from north to south, with the children of the poorest peasants turning to fishing, factory work, or the craft trades. Not that the standard of living of the population was raised at all as a result; in the early years of the twentieth century the new maritime activities suffered recurrent serious crises.

The first fracture in Bigouden society became apparent in the 1880s, when the first signs of a deliberate restriction of births appear in the reconstituted family record cards, relinking marriages ceased to integrate kindreds, and the first inroads of mechanisation began to affect farmers, though without challenging the traditional structures of collective labour. Moreover, peasants started buying the freehold of the land they farmed, which began a process whereby farms became broken up into smaller and smaller holdings.

The inter-war period saw a growth in local enterprise. Canneries, small manufacturing businesses, and shops sprang up in increasing numbers, offering a wider variety of jobs. However, these new activities were unable to check the massive exodus that got under way in the 1930s. In a parallel development, attitudes were transformed by the growth of education and as a result of contact with the first emigrants, who brought back from the towns and cities a different life style, different ideas, and a different way of talking.

The second fracture dates from the end of the Second World War. Many peasant farmers, having been taken prisoner during the war, had seen the effects of the mechanisation of agriculture in Germany at first hand and were ready to try it themselves. One effect of mechanisation was to throw traditional social structures into upheaval. The association of economic and family spheres became less intense as the former obeyed a logic of its own to which farmers, fishermen, and artisans were obliged to adapt.

In recent years the social composition of the commune has become diversified. Alongside farming, now in the hands of elderly peasants waiting for a younger generation to take over, a new socio-professional category is emerging. This consists of blue- and white-collar workers and artisans who work in the region and live in the village. The new injections of population are not very large, however, and Saint-Jean-Trolimon continues to function on the basis of mutual acquaintanceship – everybody knowing everyone else. One consequence is that the separation of economic and family spheres is not yet complete, though they do now interconnect on different foundations that run right through town–country relations.

As far as farmers are concerned, access to ownership is still, as in the past, linked to descent but kindreds are no longer in a position to provide capital, land, or labour. We observe a contraction around the domestic group and the ascendant generation. The age of the collective 'major works' is over. Mechanisation has taken a load off the farmer's back, but it has also deprived him of occasions for celebrating.

The continued power of the kindreds, despite economic and social changes, among artisans and among blue- and white-collar workers is not obvious at first glance. As the ethnologist gets to know the population more and more intimately, the strength of the tie between parents and married children stands out and the multiple functions of networks of consanguinity and affinity become increasingly apparent. Indeed, beyond the symbolic and affective roles to which recent sociology has tended to confine them, they perform genuine parallel functions in the job market. These powers appear in the form of resistances operating at the more or less hidden level of the economy that does not figure in administrative records or official statistics – the 'underground' economy as it is called for that very reason. The strength of the kindreds today is still exercised through the medium of local political power. These communes are mutual-acquaintanceship societies; one's family or that of one's spouse has to have lived there for at least two or three generations before one has any chance of being elected to the council. Votes are cast on the basis of a combination of two things: what one thinks of the individual, and how long his family has lived in the commune.

The dissociation of kinship and economic spheres is, as we have seen, not yet total. The relationship between them rests on different foundations than was the case in the last century. Observation of the way in which they now interact helps us to understand how this rural society has combined tradition and modernity.

A question that remains to be asked is: how specific is the Bigouden model, both with regard to the past and with regard to the present.

The demographic characteristics of the region – low age at marriage, high fertility, high mortality – seem to make it a veritable showcase of behaviours observed in previous centuries in other parts of France. To what are we to attribute them?

Bridal juvenility – one of the first factors in this chain of cause and effect – was typical of peasant society prior to the sixteenth century as far as one can tell from the few investigations bearing on the subject. Why did age at marriage not increase in Bigouden as it did in other parts of Brittany and the rest of France? Possibly this was a result of some violent demographic crisis – a particularly high death rate, say, that left substantial areas of the countryside empty. From this point of view the legendary ravages of the ruffianly La Fontenelle in the late sixteenth century ought perhaps to be taken more seriously. High mortality encouraged young people to occupy the places left vacant, and the habit of marrying young became part of the pattern of cultural behaviour, surviving even after farms had become harder to come by. In so far as it fitted in with the family life cycle, structural co-residence of the generations, and modes of property transmission and landholding, bridal juvenility was able to persist.

A further historical accident conferred on the population certain features that may be regarded as peculiar. This was the fact that the Bigouden peasantry did not gain access to landownership when confiscated property (the *Biens nationaux*) was sold off at the time of the Revolution. Had they done so, the dismantling of estates might have occurred more rapidly and the exodus from the countryside have begun earlier.

It is important, moreover, not to underestimate the isolation of the region, which is bounded on three sides by the sea and had access to its largest town (Quimper) only through the lock chamber of Pont-l'Abbé, as it were. The geography of North Bigouden has always made it more open to outside influences. Local particularism with its cultural traditions and its language has been maintained longer in South than in North Bigouden.

As a social model, however, Bigouden society belongs in the continuum of peasant societies. Its specificity has been falsely highlighted by collective representations coloured by the wearing of a special costume and the use of the Breton language.

The structuration of the domestic group with its temporary periods of co-residence, the domestic organisation of production based generally on the household and occasionally on collective labour operations bringing together kinsfolk and neighbours – both these features characterise many other peasant societies. The partible-inheritance system is not unique to Lower Brittany but may be observed – with variations – in other parts of northern France. In practice it sometimes even came close to the single-heir system. Bigoudens had no hesitation in favouring one successor, and not all fathers on their *oustas* in the south and south-west of France sought to disadvantage their children with the exception of the heir. Relinking is attested to in a number of peasant societies, even ones that seem to be structured longitudinally in patrimonial cycles. And while there is a genuine affinity between the double brother–sister marriage and the single-heir system of property transmission, the fact remains that as a marriage configuration it is also common in the partible system.

In other words, we need to challenge the typologies that seek to contrast the association of partible inheritance and consanguine marriage with that of single-heir inheritance and marriages with cycles. On the contrary, it is important that we stress the continuity observed right through peasant societies, pointing out that they all tend to be egalitarian and that they all, in social circumstances yet to be described, develop marriages between affines.

Furthermore, the traditional social uses of kinship, closely interwoven with economic relationships, are attested to in Italian society during the fifteenth and sixteenth centuries. The modern uses of kinship continue, though in a less transparent way, to associate the economic sphere with the kinship sphere; this too is not peculiar to present-day Bigouden. Those uses cannot, with the boundaries between town and country becoming so fluid, be ascribed to some kind of peasant archaism; nor can they be attributed to a specificity of the region that might be said to be due to its out-of-the-way position ('Finistère', the name of the *departement*, means of course 'end of the earth'). Bigouden has now stepped right into contemporary French society.

The model developed here would thus appear to be suited to giving an account of the evolution of French peasant society, with variants associated with its culture and its history, and to offer a good illustration of a kind of late twentieth century family and social organisation shared with other regions of France.

Notes

Abbreviations:

AD *Archives départementales*, Departmental Archives
AE *Archives ecclésiastiques*, Church Archives
AN *Archives notariales*, notarial archives (various deposits)
ANat *Archives nationales*, National Archives
CM *Contrat de mariage*, marriage contract
IAD *Inventaire après décès*, post-mortem inventory

Introduction

1 The present work is based on a doctoral thesis submitted to the University of Paris (Sorbonne) in January 1984 under the title *Quinze générations de bas-Bretons: Mariage, parentèle et société dans le pays bigouden Sud. 1720–1980.*
2 Pierre Jakez Hélias, 1975; André Burguière, 1975; Jakez Cornou and Pierre-Roland Giot, 1977; Marcellin Caillon and Guy Riou, 1980.
3 Jeanne Favret-Saada, 1977, p. 21.
4 Camille Vallaux, 1907, and Jacques Bernard, 1968.
5 Jacques Cambry, 1835–1838 (year VII).
6 Martine Segalen, Goanv 1983–1984, pp. 5–8.
7 This study could not have been completed without the help of three people in particular: Arlette Schweitz, who coded part of the genealogical file, Marie-Claude Babron, and Philippe Richard, who wrote the computer programmes for processing the genealogies.
8 Surnames, Christian names, and names of hamlets as recorded in archive documents dating from before 1880 are cited unchanged to give the reader something of the flavour of the Bigouden identity, much of which flows from the very special sounds of such names. After that date, or where the genealogy introduced continues up to the present day, I have transposed identities or replaced surnames by initials.

1 A mobile population

1 André Burguière, 1975, pp. 43–5.
2 Jacques Cotty and Gérald Hamon, 1974, p. 45.
3 Pierre-Jean Berrou, Goanv 1983–1984, pp. 11–21.

4 AE Quimper, letter from the *recteur* of Penmarc'h to the bishop of Quimper and Léon.
5 AD Quimper, Maître Mauduit, IAD 8 July 1872.
6 L.-F. Sauvé, 1878, p. 153, proverbs 952 and 953. Canon J.-L. Le Floc'h was kind enough to translate the second proverb into French for me (M.S.). [The English versions are translated from the French; *Tr.*]

2 A population explosion

1 K.W. Wachter, E. Hammel, P. Laslett (eds.), 1978.
2 John Hajnal, 1965.
3 Michel Fleury and Louis Henry, 1965.
4 Etienne Van de Walle, 1978, pp. 287–8.
5 Alain Croix, 1981; Jean Meyer, 1966.
6 Daniel Collet, 1982, p. 89.
7 Ibid., p. 90.
8 Ibid., p. 95.
9 For example, E.A. Wrigley, 1966; Daniel Smith, 1977. Louis Henry contributed methodological elements of response in *Population*, 1982, estimating on the basis of a theoretical calculation that it was appropriate to work on a 'cluster' of twenty communes.
10 'Situation démographique de la France', 1978, *Population*, March-April, p. 309.
11 R. Netting 1981, pp. 132–5.
12 Etienne Van de Walle, 1979, p. 128.
13 Gérard Delille, 1981–1982, p. 167.
14 Alain Croix, 1981, p. 213.
15 'Situation démographique de la France', 1978, *Population*, March-April, p. 305.
16 R. Leprohon, 1972, p. 711.
17 R.M. Smith, 1983, pp. 109–10.
18 'Situation démographique de la France', 1978, *Population*, March-April, p. 310.
19 Jack Goody, 1972, p. 4.
20 Sylvia Yanagisako, 1978, p. 166.
21 Meyer Fortes, 1971.
22 Peter Laslett and Richard Wall, 1972.
23 H.J. Habakkuk, 1974, p. 25.
24 John Hajnal, 1983, p. 69.

3 To each his (or her) share

1 Georges Augustins, 1979, pp. 128–9.
2 Maria Couroucli, 1987.
3 See particularly the works of Georges Augustins, Pierre Bourdieu, Alain Collomp, William Douglass, Pierre Lamaison, Sandra Ott, Anne-Marie Rieu-Gout and Marie-Lise Sauzéon-Broueilh listed in the bibliography.
4 Robert Cresswell, 1969, pp. 438–9.
5 Lutz Berkner, 1976, p. 74.
6 Robin Fox, pp. 122–4; Marie-Claude Pingaud, 1978; Michelle Salitot-Dion, 1977 and 1978; Giovanni Levi, 1976.

7 Jean Yver, 1977, pp. 122–3.
8 Jean Meyer, 1966, pp. 109–11.
9 Alexandre de Brandt, 1901, pp. 184–5.
10 Elicio Colin 1947, p. 63.
11 Alain Le Grand, 1980.
12 Jacques Delroeux, 1979.
13 Michel Izard, 1963, pp. 66–8.
14 AD Quimper, Maître E. Arnoult, 4 E 205/377 to 438, CM 8 November 1861.
15 AD Quimper, Maître Kernilis, 4 E 206/71 to 93.
16 AN, Maître Mauduit, IAD 30 January 1900.
17 AD Quimper, Maître Verrye, 4 E 213/144 to 156, IAD 4 October 1844.
18 AD Quimper, Maître Ch. Kernilis, 4 E 206/94 to 107, IAD 9 December 1842.
19 AD Quimper, Maître Le Deliou, 4 E 205/439 to 482, CM 26 December 1865.
20 AD Quimper, Maître Flamant, 4 E 212/106, CM 10 June 1841.
21 Jack Goody, 1976, p. 28.
22 AD Quimper, Maître Le Moallic, 4 E 162/120, CM 23 July 1854.
23 AD Quimper, Maître Arnoult, 4 E 205/377 to 438, CM 8 November 1861.
24 AD Quimper, Maître Le Deliou, 4 E 205/458, CM 9 December 1869.
25 AD Quimper, Maître Kernilis, 4 E 206/71 to 93, IAD 8 November 1842.
26 AN, Maître Mauduit, *Etat de valeurs supplétif d'inventaire* (addendum to inventory) 3 August 1909.
27 AD Quimper, Maître E. Arnoult, 4 E 205/377 to 438, *donation* (deed of gift) 12 March 1848.
28 AN, Maître Mauduit, *testament* (will) 8 July 1905.
29 Alain Collomp, 1981.
30 Marie-Claude Pinguad, 1978, p. 125.
31 Yves le Gallo, 1980, p.145.
32 Jack Goody, 1976.

4 Regular relinking through affinal marriage

1 Claude Lévi-Strauss, 1968; Françoise Héritier, 1981.
2 Pierre Lamaison and Elizabeth Claverie, 1982; Allain Collomp, 1983.
3 Françoise Zonabend, 1980.
4 Martine Segalen and Philippe Richard, 1986.
5 Tina Jolas, Yvonne Verdier, Françoise Zonabend, 1970. The French term is *renchaînement d'alliance*.
6 Jean Sutter and Léon Tabah, 1948.
7 Ibid, p. 611.
8 Michel Verdon 1973, pp. 269, 272.
9 Pierre Philippe and Jacques Gomila, 1972, pp. 54–9.
10 Sylvie Postel-Vinay, 1981.
11 Jacqueline Vu Tien Khang and André Sevin, 1977.
12 Jacques Cotty and Gérald Hamon, 1974, p. 50.
13 AD Quimper, Série V Dépôt, except for Plobannalec, for which documents were consulted in the church archives.
14 AE Quimper, 1817–1838, 2 L 1.
15 Françoise Héritier, 1981, p. 163.
16 The low level of Bigouden consanguinity was also established by the results of the National Institute of Population Studies (*Institut national d'Etudes*

démographiques or INED) run on the corpus of reconstituted genealogies. The programme works out 'the probability of origin of genes' and examines whether, among the ancestors described as 'founding', some had a larger genetic contribution than others. That kind of concentration is observable in small closed populations. On the contrary, in the Bigouden genealogies the genetic contribution of founders merges in the mass of genes circulating. My thanks are due to Marie-Claude Babron, who took on the job of adapting and running the programme.

17 E. Ernault, 1896–1897, p. 167.
18 Jack Goody, 1983, pp. 184–8.
19 Alain Signor, 1969.
20 Claude Lévi-Strauss, 1968, p. 305.
21 Ibid., p. 306.
22 Ibid.

5 The Bigouden wedding ceremony

1 Michel Izard, 1963.
2 Sylvie Postel-Vinay, 1977, p. 251; in French: *la parenté est sortie*.
3 Joan F. Mira, 1971, pp. 116–17.
4 Jean-Louis Flandrin, 1975.
5 Arnold Van Gennep, 1909.
6 François Lebrun and Martine Segalen, 1980, Introduction to the catalogue *Le mariage en Bretagne*, Rennes, Buhez, p. 28.
7 Alexandre Bouët and Oliver Perrin 1918 (1835–1838).
8 Villermé and Benoiston de Châteauneuf, 1982 (1840), p. 22.
9 Jean-Michel Guilcher, 1971.
10 Ibid., p. 42.
11 Nicole Pellegrin, 1979–1982.
12 Arnold Van Gennep, 1943–1958, I, 1, p. 314.
13 E. Le Doaré, 1896.
14 Alexandre Bouët and Olivier Perrin, 1918.
15 Paul du Châtellier, 1893, and E. Le Doaré, 1896.
16 Armand du Châtellier, 1863, p. 186.
17 Paul du Châtellier, 1893, and E. Le Doaré, 1896.
18 Pierre Hélias. 1975, p. 448.
19 Paul du Châtellier, 1893, and E. Le Doaré, 1896.
20 'Une noce à Loctudy', *L'Impartial du Finistère*, 8 May 1858.
21 Jean-Michel Guilcher, 1973, p. 28.
22 Jacques Cambry, 1835–1838, p. 152.
23 *Le mariage en Bretagne*, 1980, pp. 150–3.
24 Jacques Cambry, 1835–1838, p. 152.
25 Paul du Châtellier, 1893, and E. Le Doaré, 1896.
26 René-Yves Creston, 1974, p. 36.
27 Dan Lailler, 1947.
28 René-Yves Creston, 1974, p. 106.

6 The sea in abeyance

1 Pierre Bourdieu, 1972, p. 129.
2 Marshall Sahlins, 1976, p. 246.
3 Karl Polanyi, 1983, p. 102.

4 Henri Sée, 1906, pp. 371–6.
5 Hugh Clout, 1979.
6 Henri Sée, 1906, p. 237.
7 Armand du Châtellier, 1858.
8 Jean Meyer, 1966, p. 524.
9 Ibid., p. 519.
10 Ibid., p. 457.
11 Villermé and Benoiston de Châteauneuf, 1982, pp. 92–3.
12 Jacques Cambry, 1835–1838, p. 153.
13 Henri Sée, 1906, p. 371.
14 Vincent Le Floc'h, *La vie rurale à Plonivel, paroisse de Cornouaille, 1675–1789, mémoire* for a *diplome d'études supérieres* (DES) in history, written under the direction of Pierre Goubert, Faculté de Rennes, 1965, 322 typed pages. This information is taken from post-mortem inventories. I am grateful to Vincent Le Floc'h for permission to consult his *mémoire*, only part of which has appeared in print (in the *Bulletin de la Société archéologique de Finistère*, 1966).
15 Armand du Châtellier, 1863, p. 181.
16 Auguste Dupuy, 1890, p. 21.
17 Jacques Cambry, 1835–1838, p. 152.
18 J. Savina and D. Bernard, 1927, Vol. 1. p. XXXIV.
19 J. Cornou, 1977, pp. 213–24.
20 J. Lemoine and H. Bourde de La Rogerie, 1902, p. 136.
21 J. Lemoine and H. Bourde de La Rogerie, 1902, p. 157.
22 Jacques Cambry, 1835–1838, p. 159.
23 J, Savina, 1920.
24 Vincent Le Floc'h, 1966, p. 137.
25 Henri Sée, 1906, p. 281.
26 Jean Meyer, 1966, p. 524.
27 Vincent Le Floc'h, 1966, p. 140.
28 Alain Signor, 1969, pp. 31–2.
29 For Brittany as a whole, Jean Meyer speaks of farm rents doubling (p. 1247).
30 Private archives, Mlle Yvette Balouin. A 'heaped' measure (*mesure comble*) differed from a 'full measure' (*mesure raze*) in that the grain came up above the height of the measuring vessel (E. Cognec, 1904, p. 90).
31 Private archives, Jean Calvez.
32 Private archives, Jean Calvez, power of attorney granted by M. Gabriel-Louis Malherbe to *demoiselle* Marie-Béatrix Bernard, proprietress, of Pont-l'Abbé.
33 Private archives, Jean Calvez, *déclaration à domaine* (description of lands), 19 September 1785.
34 Private archives, Jean Calvez, 'Between noble gentleman François Bernard Louis Le Maire, of the town of Quimperlé, and honourable gentleman Pierre Le Perennou', 2 December 1775.
35 Vincent Le Floc'h, 1966, p. 173.
36 Ibid., p. 169.
37 Pierre Keraval, 1954, pp. 65-6.
38 Armand du Châtellier, 1858.
39 AD Ille-et-Vilaine, C. 1632; quoted by P. Flatrès, 1944, pp. 189–90.

40 *Le sieur Ansquer avocat, sieur Audouin procurer de Quimper, étant actuellement en deffense avec Messieurs les Juges près de ceux de Quimper et pour deffendre et soutenir le droit et propriété de la palue de la Magdeleine et de la Torche jusques à la croix nommé croix-an-dour, et pont suividant y adjacent jusques au lac de l'étang de Pest alae et aussie la propriété de lance [l'anse] dit ast bihan, autrement dit pors carn, et la pocession dans laquelle se trouve de tout temps immemorial le di général [conseil municipal] de Plomeur de faire paistre leurs bestiaux en la ditte palue, y mottoyer et faire sécher leur goémon, exclusivement [à l'exclusion] du général de Beuzec Cap Caval qui a la palue particulière ci-nommée la palue de Tronön qui est au nord de celle de la paroisse de Plomeur.* Archives communales of Plomeur, Registre BMS (*baptêmes, mariages, sépultures*), 18 February 1729.
41 AD Quimper, 28 L 33.
42 J. Savina and D. Bernard, 1927, vol. 1, p. XIX.
43 Ibid., p. 33.
44 Ibid., pp. 215–16.
45 Ibid., p. 204.
46 P. Keraval, 1955, pp. 75-6.
47 J. Savina and D. Bernard, 1927, vol. 1, p. 173; quoted by Henri Sée, 1906, p. 407.
48 Ibid., p. 212.
49 Ibid., p. 174.
50 Ibid., p. 216.
51 Alain Signor, 1969, pp. 40–50.
52 Mme Audouyn de Pompéry, 1884.
53 AD Quimper, series L. Du Châtellier archives, château de Kernuz.
54 They experienced certain temporary difficulties during the Revolution. Mme de Pompéry was imprisoned in the *château* of Pont-l'Abbé. So were Guillaume du Haffond of Treffiagat, Marie-Josephe du Boisguehenneuc of Loctudy, François de Pompéry, Mme de Penfeuntenyo of Kerfillin in Saint-Jean-Trolimon, Marie de Marallac'h of Tréourron in Plonéour, Pierre de la Boixière, and Jean-Maurice de Penfeuntenyo, aged 75, who had previously escaped the burning of his château of Loctudy. This was the period when orders were given to smash and remove any remaining crests, coats-of-arms, and fleurs-de-lys on the façades of manor houses, chapels, or other buildings. Masons and roofers were commissioned to carry out this work where proprietors had not done it themselves (*Dialogue*, 23, 21 July 1975).
55 Alain Signor, 1969, pp. 40–1.
56 J. Savina and D. Bernard, 1927, p. 205.
57 ANat Paris, F 20 187, 'Mémoire sur la Statistique du Finistère 1789 et an IX'.
58 AD Quimper 1 to 19 L 28, 'Schedule to be completed in accordance with the law of 13 ventôse, Year II, municipality of Saint-Jean-Trolimon, on its responsiblity within eight days of receipt'. This document was kindly brought to my attention by Françoise Gestin.
59 Abbé Eugène Cignec, 1904, p. 152.
60 ANat Paris, F 20 187, 'Tableau de la Statistique en 1806'.
61 Yves Le Gallo, 1980, p. 176.
62 Ibid. pp. 187–91.

63 A. Chayanov, 1966, pp. 75–6.
64 Goulven Mazeas, 1940.
65 Yves Tanneau, 1958.
66 Jakez Cornou, 1977.
67 Yves Tanneau, 1958.
68 Armand du Châtellier, 1849, pp. 4–5.
69 Maurice Agulhon, Gabriel Désert, Robert Specklin, 1976, pp. 140–1.
70 Louis Ogès, 1949.
71 Maurice Agulhon, Gabriel Désert, Robert Specklin, 1976, p. 228.
72 Armand du Châtellier, 1835–1837, p. 74.
73 *Enquête agricole*, 1868.
74 Maurice Agulhon, Gabriel Désert, Robert Specklin, 1976, p. 197.
75 *Régistre de déliberations du conseil municipal*, 1847.
76 Paul du Châtellier, 1878.
77 Chanoine Abgrall, 1891.
78 Armand du Châtellier, 1835–1837, p. 77.
79 Twenty-nine documents from 1678 to 1882, mainly bills of sale, leases, and exchanges.
80 Twenty-seven documents from 1638 to 1904, comprising bills of sale, receipts, leases, and summary property returns.
81 The farms of Trevinou, Leac'h ar Prat, Kernel, and Kervouec.
82 This was the case with Castellou Peron: the domanial rent was 500 francs in IAD 22 July 1865, (AD Quimper, Maître Flamant, 4 E 212/102 to 144), 700 francs in IAD 1909 (Maître Queinnec).
83 AD Quimper, Maître Le Deliou, 4 E 205/439 to 482, IAD Marie-Jeanne Le Lay, 4 October 1846.
84 Fatou private archives, lease of 20 July 1834.
85 Fatou private archives, lease of 15 April 1835.

7 Frozen hierarchies and social relations

1 Marshall Sahlins, 1976, p. 130.
2 A. Chayanov, 1966, pp. 75–6.
3 Marshall Sahlins, 1976, pp. 118–19.
4 Alexandre Bouët and Olivier Perrin, 1918, pp. 358–9.
5 Jean Meyer, 1966, p. 14.
6 Henri Sée, 1906, p. 46.
7 Alain Signor, 1969, p. 41.
8 Jean Meyer, 1966, p. 661.
9 Ibid., p. 1121.
10 Ibid., 1966, p. 658.
11 Charles Pelras, 1965, pp. 532–3.
12 AD Quimper, 30 C 15/1 to 8,. *Table des baux de toute nature de biens appartenant aux laïcs de 1733 à 1791* ('List of leases on all kinds of property belonging to laymen from 1733 to 1791').
13 Alain Signor, 1969, p. 49.
14 During the search only the names and accounts of South Bigouden peasants were picked out. The document covers the following items in respect of each lease:
- date of lease
- name of lessor

- name of lessee
- name of reporting notary
- date of official check of lease
- price of farm lease in cash
- estimate of share-cropping lease
- quality and quantity of cereals of lease and their value
- nature, quality, and situation of property
- observations

The initial items are completed systematically; the last three are usually vacant. One problem arises immediately: identifying the lessor. The name of the locality being leased out is mentioned, sometimes in specific terms, sometimes only vaguely, but it is most unusual for the place of residence of the lessor to be given, though there are exceptions. Not until the 1781 list does the document specify 'farmer at . . .'.

Because of the high incidence of homonymy already mentioned, it is difficult to be sure, before that date, that the farmer named in the account is in fact the same person as appears in our genealogies. Furthermore, there were members of the urban middle class with the same patronymic as peasants, being descended from common ancestors but having become differentiated during the course of the seventeenth and eighteenth centuries.

15 Armand du Châtellier, 1863, pp. 203–4.
16 Alain Signor, 1969, p. 49.
17 Vincent Le Floc'h, 1966, pp. 107–8.
18 Jacques Cambry, 1835–1838, p. 152.
19 Chanoine Peyron, 1901–1910, p. 392, *acte* (deed) 333, 29, 7.
20 AD Quimper, Saint-Jean-Trolimon, 7 August 1853.
21 AD Quimper, 3 Q 12578/73 to 3 Q 12646/3, deed of 13 August 1859.
22 Archives paroissales Saint-Jean-Trolimon, 12 January 1886.
23 AD Quimper, Saint-Jean-Trolimon, no. 36.
24 AD Quimper, Maître F. Kernilis, 4 E 206/94 to 107, *vente* (sale) 22 January 1811.
25 AD Quimper, Maître Charles Kernilis, 4 E 206/94 to 107, IAD 2 August 1851.
26 AD Quimper, Maître Emile Arnoult, 4 E 205/377 to 438, IAD 27 July 1848.
27 AD Quimper, Saint-Jean-Trolimon, no. 51.
28 Ibid., no, 32.
29 Armand du Châtellier, 1863, p. 202.
30 Armand du Châtellier, 1835–1837, pp. 120–1.
31 Jean Choleau, 1907, and Georges Le Bail, 1913.
32 AD Quimper, Maître Le Deliou, 4 E 205/439 to 482, IAD 21 April 1866.
33 Armand du Châtellier, 1863, pp. 199–200.
34 Ibid.
35 Ibid., p. 202.
36 See the work done on the basis of these documents by Jean Jacquart, 1979; Daniel Roche, 1982; Suzanne Tardieu, 1964.
37 Armand du Châtellier, 1849, pp. 3–4.
38 AD Quimper, Maître Emile Arnoult, 4 E 205/377 to 438, 4 June 1855; Maître Le Deliou, 4 E 205/439 to 482, IAD 1 March 1865.
39 This assumption rests on what is suggested by the rules and regulations of the diocese of Quimper, which placed on the list of cases reserved to the bishop

the fact of having a child sleep in its parents' bed. These rules remained in force until 1851.

40 Pierre Hélias, 1975, p. 416.
41 AD Quimper, Maître Emile Arnoult, 4 E 205/377 to 438, IAD 27 July 1848.
42 ANat, F 20 187.
43 AD Quimper, Maître Le Deliou, 4 E 205/439 to 482, IAD 21 April 1866.
44 AD Quimper, Maître Emile Arnoult, 4 E 205/377 to 438, IAD 17 May 1847.
45 AD Quimper, Maître Flamant, 4 E 212/130. IAD 3 May 1864.
46 AD Quimper, Maître Ronac'h. 4 E 213/168, IAD 29 November 1859.
47 According to the *Enquête agricole* (Agricultural Survey) of 1868, this livestock corresponded to a farm of about twenty hectares.
48 AD Quimper, Maître Kernilis, 4 E 206/71 to 93, IAD 7 October 1856.
49 Jacques Cambry, 1835–1938, p. 154.
50 de Ritalongi, 1894, pp. 61–2.
51 J.-F, Broumische, 1977 (1830), p. 293.
52 Thierry Fillaut, 1983, p. 49.
53 Franklin Mendels, 1978, p. 791.

8 Tradition and modernity: the years 1880–1980

1 Elicio Colin, 1947; Pierre Keraval, 1954. An exception is the flail, which was replaced by the horse-driven mill.
2 Elicio Colin, 1947, p. 65.
3 Ibid. p. 78.
4 This text was drawn up by two pupils from the *lycée* in Pont-l'Abbé, who interviewed Hervé C., recorded his words, typed out the recording, and gave him the typescript to correct. Hervé C., who went into an old people's home in 1983, gave me a copy, from which I have quoted several passages.
5 The term used [*andain*] usually refers to the strip of hay laid out in the field to dry before being brought in. Hervé C. is referring to the small piles of straw made up after the grain had been threshed out.
6 Camille Vallaux, 1905, p. 245.
7 Anne Lebel, 1981, p. 8.
8 AE Quimper.
9 Jakez Cornou and Pierre-Roland Giot, 1977, pp. 373–7.
10 E. Le Doaré, 1896.
11 Anne Lebel, 1981.
12 Paul du Châtellier and Emile Ducrest de Villeneuve, 1893.
13 AE Quimper, pastoral visit of 1909.
14 AD Quimper, Série V, Dépôt 1: *Registre de déliberations du Conseil de la Fabrique* (parochial church council minute book).
15 The numbers of dead for each commune of the two Bigouden cantons in the First World War were as follows: Pont-l'Abbé, 225; Tréméoc, 47; Ile Tudy, 37; Combrit, 137; Loctudy, 103; Plobannalec-Lesconil, 113; Treffiagat, 76; Le Guilvinec, 119; Saint-Guénolé-Penmarc'h, 181; Plomeur, 90; Saint-Jean-Trolimon, 38; Tréguennec, 25; Plogastel-Saint-Germain, 108; Plonéour-Lanvern, 204; Plovan, 73; Tréogat, 31; Peumerit, 89; Pouldreuzic, 107; Plozévet, 201; Guiler, 48; Landudec, 78; Gourlizon, 33; Ploneis, 58. The total was 2,221 (*Dialogue*, 34, 16 October 1975).
16 Youenn Drezen, 1943.

17 Martine Segalen, 1979, pp. 186–9.
18 *L'agriculture en Bretagne*, 1976.
19 Françoise Bonnaby *et al.*, 1973.
20 *Le Télégramme de l'Ouest*, 21 October 1982.

9 The importance of being kin

1 Michel Bozon, 1978, pp. 87–8.
2 AD Quimper, 30 C 15/1–8.
3 AD Quimper, Maître Emile Arnoult, 4 E 205/377 to 438, IAD 4 June 1855.
4 AD Quimper, Maître Emile Arnoult, 4 E 205/377 to 438. IAD 4 July 1860.
5 AD Quimper, Maître F. Kernilis, 4 E 206/71 to 93, IAD 18 February 1837.
6 AD Quimper, Maître F. Kernilis, 4 E 206/71 to 93, *vente à Kerstrat* (sale at Kerstrat) 22 January 1841.
7 AD Quimper, Maître F. Kernilis, 4 E 206/71 to 93, *vente à Keryoret* (sale at Keryoret) 25 November 1851.
8 William J. Smyth, 1982, pp. 37–8.
9 Gérard Delille, 1981–1982. pp. 458–61.
10 Patrick Le Guirriec, 1983.
11 Maurice Godelier, 1973, pp. 8–9, 21.
12 AD Quimper, 3 G 12–78 to 81, Tribunal d'Officialité. 14 April 1784.
13 AD Quimper, Maître F. Kernilis, 4 E 206/71 to 93, CM 25 October 1855.
14 *Le pouvoir au village*, 1976.
15 André Burguière, 1975.
16 Claude Karnoouh, 1980.
17 Louis Roussel, 1976.
18 Claude Macherel, 1983, p. 157.

Bibliography

(*Books and articles dealing specifically with Brittany.)

*Abgrall, J.-M., 1981. 'Voie romaine conduisant de Quimper à l'oppidum de Tronoën', in *Bulletin de la Société archéologique de Finistère*, pp. 223–7

*'L'agriculture en Bretagne. Dynamisme ou domination', in *Ar Falz* (nouvelle série) 13, 14, 15, April-June 1976, 128 pp.

Agulhon, Maurice, Gabriel Désert, Robert Specklin, 1976. *Histoire de la France rurale*, Paris, Seuil, vol. III: *1789–1914,* 568 pp.

*Audouyn de Pompery, Madame, 1884. *Un coin de Bretagne pendant la Révolution. Correspondance,* Paris, Alphonse Lemerre, 2 vols., pp. 319–31

Augustins, Georges, 1977. 'Reproduction sociale et changement social: l'exemple des Baronnies', in *Revue française de Sociologie* XVIII, pp. 465–84

Augustins, Georges, 1979. 'Division égalitaire des patrimoines et institution de l'héritier', in *Archives européennes de Sociologie* XX, pp. 127–41

*Augustins, Georges, 1981. 'Mobilité résidentielle et matrimoniale dans une commune du Morbihan au XIXe siècle', in *Ethnologie française* XI, 4, pp. 319–28

Augustins, Georges, 1981. 'Maison et société dans les Baronnies au XIXe siècle', in *Les Baronnies des Pyrénées*, Paris, Ecole des Hautes Etudes en Sciences Sociales, pp. 21–2

Augustins, Georges, 1982. 'Esquisse d'une comparaison des systèmes de perpétuation des groupes domestiques dans les sociétés paysannes européennes', in *Archives européennes de Sociologie* XXIII, pp. 39–69

Berkner, Lutz, 1976. 'Inheritance, land tenure, and peasant family structure: a German regional comparison', in Jack Goody, Joan Thirsk, and E.P. Thompson (eds.), *Family and Inheritance*, Cambridge, Cambridge University Press, pp. 71–95

*Bernard, Jacques, 1968. *Navires et gens de mer à Bordeaux, 1400–1550*, Paris, Société d'Editions et de Vente des Publications de l'Education Nationale, Ecole des Hautes Etudes en Sciences Sociales, 3 vols.

*Berrou, Pierre-Jean, 1983–1984. 'Le Guilvinec: Comment cette commune se sépara de Plomeur en 1880', in *Cap Caval*, Goanv (winter) pp. 11–21

*Blayo, Yves, Louis Henry, 1967. 'Données démographiques sur la Bretagne et l'Anjou de 1740 à 1829', in *Annales de démographie historique*

*Bonnaby, Françoise, Daniel Chatelin, Philippe Jarreau (with the collaboration of Gilles Poullaouec), 1973. *Pour une mise en valeur et une protection du sud de la baie d'Audierne*, Laboratoire d'Ecologie 1–6, Groupe France, Université de Paris VIII, October (roneoed document)

*Bouchaud-Le Tendre, Hélène, 1967. 'Un rapport de 1829 sur l'état du Finistère. Mémoire du capitaine Conrier sur la reconnaissance de la route de Châteaulin à Quimper', in *Bulletin de la Société archéologique du Finistère* XCIII, pp. 144–214

*Bouët, Alexandre, Olivier Perrin, 1981. *Galerie bretonne ou vie des Bretons d'Amorique 'Breiz Izel'* (with a Preface and notes by Frédéric Le Guyader), Brest, 488 pp.

Bourdieu, Pierre, 1962. 'Célibat et condition paysanne', in *Etudes rurales* V–VI, pp. 83–134

Bourdieu, Pierre, 1972. 'Les stratégies matrimoniales', in *Annales Economies, Sociétés, Civilisations* XXVII, 4–5, July–October, pp. 1105–27

Bourdieu, Pierre, 1972. *Esquisse d'une théorie de la pratique, précédé de trois études d'ethnologie kabyle,* Geneva, Droz

Bozon, Michel, 1978. 'Quelques emplois du concept de la sociabilité', in *Cahiers universitaires de la Recherche urbaine* 4, pp. 87–8

*Broumische, J.–F., 1977 *Voyage dans le Finistère en 1829, 1830 et 1831*, Quimper, Morvran, 2 vols., pp. 181–349

*Burguière, André, 1975 *Bretons de Plozévet*, Paris, Flammarion, 384 pp.

Burguière, André, 1979. 'Endogamie et communauté villageoise: pratique matrimoniale à Romainville au XVIIIe siècle', in *Annales de démographie historique*, pp. 313–36

*Caillon, Marcellin, Guy Riou, 1980. *A la découverte du pays bigouden,* Pont-l'Abbé, 298 pp.

*Cambry, Jacques, 1835–1838. *Voyage dans le Finistère ou état de ce département en 1794 et 1795*, Paris, Librairie du Cercle Social, Year VII, 2 vols.; revised and expanded by E. Souvestre, Brest, Come & Bonetbeau, 2 vols.

*Cambry, Jacques, chevalier de Freminville, 1836. *Voyage dans le Finistère,* Brest, J.–B. Lefournier, XIII, 480 pp.

*Charpy, Jacques, 1972. 'Dénombrements de la population des communes du Finistère (1790–1968)', in *Bulletin de la Société archéologique du Finistère* XCIX, 2, pp. 849–87

Chayanov, A.V., 1866. *The Theory of Peasant Economy* (1925), Irwin, Illinois, 318 pp.

*Choleau, Jean, 1907. *Condition des serviteurs ruraux bretons domestiques à gages et journaliers agricoles,* Vannes, Lafolye, 204 pp.

*Clout, Hugh, 1979. 'Land use change in Finistère during the eighteenth and nineteenth centuries', in *Etudes rurales,* January-March 73, pp. 69–96

*Cognec, Abbé Eugène, 1904. *Plonéour-Lanvern*, Brest, Imprimerie de la Presse libérale du Finistère, 196 pp.

*Colin, Elicio, 1947. 'Evolution de l'économie rurale au pays du Porzay de 1815 à 1930', in *Bulletin de la Société archéologique de Finistère* LXXIII, pp. 60–80

*Collet, Daniel, 1982. 'La population rurale du Finistère au XIXe siècle', in *Mémoires de la Société d'histoire et d'archéologie de Bretagne* LIX, pp. 83–118

Collomp, Alain, 1981. 'Conflits familiaux et groupes de résidence en Haute-Provence', in *Annales Economies, Sociétés, Civilisations*, May–June, pp. 408–25

Collomp, Alain, 1983. *La maison du père*, Paris, Presses universitaires de France, 342 pp.
*Cornou, Jakez, Pierre-Roland Giot, 1977. *Origine et histoire des Bigoudens*, Le Guilvinec, F. Le Signor, 394 pp.
*Cornou, Jakez, 1978. *Ar vro vigouden gwechall*, Le Guilvinec, Imprimerie du Marin
*Cotty, Jacques, Gérald Hamon, 1974. *La population de Penmarc'h au XVIIIe siècle, étude géographique*, master's thesis, University of Rennes II
Couroucli, Maria M., 1987. 'Parentesco y ley: relación entre el sistema rural y el estado', in John G. Peristiany (ed.) *Dote y matrimonio en los paises mediterráneos*, Madrid, Siglo XXI de España Editores, pp. 1–17
Cresswell, Robert, 1969. *Une communauté rurale de l'Irlande*, Paris, Travaux et mémoires de l'Institut d'Ethnologie, LXXIV, 572 pp.
*Creston, René-Yves, 1953–1961. *Les costumes des populations bretonnes*, Rennes, Travaux du Laboratoire d'anthropologie générale de la Faculté des Sciences de Rennes, 4 vols.
*Creston, René-Yves, 1974. *Le costume breton*, Paris, Tchou, 444 pp.
*Croix, Alain, 1981. *La Bretagne aux XVIe et XVIIe siècles. La vie, la mort, la foi*, Paris, Maloine, 2 vols., 1,572 pp.
de Brandt, Alexandre, 1901. *Droits et coutumes des populations rurales de la France en matière successorale*, Paris
*de la Bourdonnais, Mahé, 1892. *Voyage en Basse-Bretagne chez les Bigoudens de Pont-l'Abbé*, Paris, Henri Jouve, 368 pp.
Delille, Gérard, 1981–1982. *Famille et propriété dans le royaume de Naples, XVe – XIXe siècle*, PhD thesis, University of Paris I, Faculté des Lettres et des Sciences humaines, 2 vols., 954 pp.
*Delroeux, Jacques, 1979. *Etude d'anthropologie sociale de trois sociétés occidentales: Goulien, Plogoff et Lescoff de 1800 à 1970. Recherche du principe de réciprocité*, PhD thesis, University of Paris V, typescript, 660 pp. plus appendices
*Divanach, Marcel, 1978. *Le marsouin hableur – Ar moroc'h fougaser,* Brest, Editions du Vieux Meunier breton, 128 pp.
Douglass, W.A., 1975. *Echalar and Murelaga. Opportunity and Rural Exodus in Two Spanish Basque Villages*, Guilford and London, Billing & Sons Ltd., 222 pp.
*Drezen, Youenn, 1977. *Itron Varia Garmez* (1943); translated into French as *Notre-Dame Bigouden*, Paris, Denoël, 228 pp.
*du Châtellier, Armand, 1835–1837. *Recherches statistiques sur le département du Finistère*, Nantes, Imprimerie de Mellinet, 146 pp.
*du Châtellier, Armand, 1849. *De la condition du fermier et de l'ouvrier agricole en Bretagne*, Paris, Imprimerie Bouchard-Huzard, 12 pp.
*du Châtellier, Armand, 1858. *La baronnie du Pont (ancien évêché de Cornouaille)*, Paris (Dentu) and Nantes (Guiraud), 76 pp.
*du Châtellier, Armand, 1863. *L'agriculture et les classes agricoles de la Bretagne*, Paris, Guillaumin, 230 pp.
*du Châtellier, Paul, 1878. *Exploration du cimetière gaulois de Kerviltré en Saint-Jean-Trolimon*, Saint-Brieuc, F. Guyon, 16 pp.
*du Châtellier, Paul, Emile Ducrest de Villeneuve, 1893. *Paysages et monuments de la Bretagne*, Parts 14, 15, and 16: *Pont-l'Abbé, Lambourg-Fouesnant et Plogastel-Saint-Germain*, Paris, May & Motteroz

*du Châtellier, Paul, 1896. 'Une habitation gauloise à Tronoën en Saint-Jean-Trolimon (Finistère)', *Bulletin archéologique*, 6 pp.

*du Châtellier, Paul, 1897. 'Découverte d'un graphium à Tronoën' in *Bulletin de la Société archéologique du Finistère* XXXIX

*Dupuy, A., 1890. 'L'Agriculture et les classes agricoles en Bretagne au XVIIIe siècle', in *Annales de Bretagne* VI, pp. 3–28

*'En pays bigouden', 1982. *Les Cahiers de l'Iroise* 3, July-September, 180 pp.

**Enquête agricole: enquête départementale, 3e circonscription, Morbihan, Finistère, Côtes-du-Nord, Ille-et-Vilaine*, 1868. Ministère de l'Agriculture, du Commerce et des Travaux Publics, Paris, Imprimerie Impériale

**Enquête sur la structure des exploitations agricoles en 1975, région Bretagne, résultats provisoires*, 1976. Ministère de l'Agriculture, Service régional de Statistique agricole, Rennes, 10 pp. (roneoed document).

*Ernault, E., 1896–1897. 'Dictons et proverbes bretons', in *Mélusine,* VIII, 7, p. 167

*Falc'hun, François, Bernard Tanguy, 1979. *Les noms de lieux celtiques, Nouvelle méthode de recherche en toponymie celtique*, Bourg-Blanc, Editions armoricaines, 62 pp.

*Favé, Abbé, 1893. 'Le mobilier et le vêtement dans la classe rurale aux environs de Quimper au XVIIe siècle', in *Bulletin de la Société archéologique du Finistère* XX, pp. 329–38

*Favé, Abbé Antoine, 1995. 'Notes sur l'aspect extérieur d'une ferme cornouaillaise avant 1789', in *Bulletin de la Société archéologique du Finistère* XXII, pp. 33–6

Favret-Saada, Jeanne, 1977. *Les mots, la mort, les sorts. La Sorcellerie dans le Bocage,* Paris, Gallimard, 332 pp.

Fillaut, Thierry, 1983. *L'alcoolisme dans l'ouest de la France pendant la seconde moitié du XIXe siècle*, Paris, La Documentation Française, 248 pp.

Flandrin, Jean-Louis, 1975. *Les amours paysannes*, Paris, Gallimard-Julliard, collection 'Archives', 256 pp.

Flandrin, Jean-Louis, 1976. *Familles, parenté, maison, sexualité dans l'ancienne société*, Paris, Hachette, 288 pp.

*Flatrès, Pierre, 1944. 'Le pays nord-bigouden', in *Annales de Bretagne* 51. pp. 158–205

Fleury, Michel, 1965. Louis Henry, *Nouveau manuel de dépouillement et d'exploitation de l'état civil ancien.* Paris, Institut national d'Etudes démographiques, 182 pp.

Fortes, Meyer, 1971. Introduction to Jack Goody (ed.), *The Developmental Cycle in Domestic Groups*, Cambridge, Cambridge University Press, pp. 8–9

Fox, Robin, 1972. *Anthropologie de la parenté. Une analyse de la consanguinité et de l'alliance*, Paris, Gallimard, 268 pp.

Fox, Robin, 1978. *The Tory Islanders. A People of the Celtic Fringe,* Cambridge, Cambridge University Press, 210 pp.

Godelier, Maurice, 1973. 'Modes de production, rapports de parenté et structures démographiques', in *La Pensée*, December, no. 172, pp. 7–31

Godelier, Maurice, 1975. *Horizons, trajets marxistes en anthropologie*, Paris, Maspero, 396 pp.

Goody, Jack, 1971. *The Developmental Cycle in Domestic Groups*, Cambridge, Cambridge University Press

Goody, Jack, 1972. *Domestic groups*, Addison Wesley Module in Anthropology, module 28–4

Goody, Jack, S. J. Tambiah, 1973. *Bridewealth and Dowry*, Cambridge, Cambridge University Press, 170 pp.

Goody, Jack, Joan Thirsk, E.P. Thompson (eds.), 1976. *Family and Inheritance*, Cambridge, Cambridge University Press, 422 pp.

Goody, Jack, 1976. 'Inheritance, property, and women: some comparative considerations', in Goody, Thirsk and Thompson, pp. 10–36

Goody, Jack, 1976. *Production and Reproduction: a comparative study of the domestic domain*, Cambridge, Cambridge University Press

Goody, Jack, 1983. *The Development of the Family and Marriage in Europe*, Cambridge, Cambridge University Press, 308 pp.

*Gouletquer, Pierre-Louis, 1968. 'Le souterrain de Castellou Peron', in *Annales de Bretagne* LXXV, pp. 85—100

*Gourvil, Francis, 1970. *Noms de famille bretons d'origine toponymique*, Quimper, Editions de la Société d'Archéologie du Finistère, 330 pp.

*Gourvil, Francis 1966. *Noms de famille de Basse-Bretagne*, Société Française d'Onomastique, Editions d'Artrey

*Guilcher, André, 1948. *Le relief de la Bretagne méridionale*, La Roche-sur-Yon, Pottier, 682 pp.

*Guilcher, Jean-Michel, 1963. *La tradition populaire de danse en Basse-Bretagne*, Paris, Mouton, 616 pp.

*Guilcher, Jean-Michel, 1967. 'Un jeu des mariages en Basse-Bretagne', in *Arts et Traditions populaires,* 1, pp. 81–5

*Guilcher, Jean-Michel, 1971. 'Aspects et problèmes de la danse populaire traditionelle', in *Ethnologie française*, 1, 2, pp. 7–48

Habakkuk, H. J., 1974. *Population Growth and Economic Development since 1750*, Leicester, Leicester University Press, p. 8

Hajnal, John, 1965. 'European marriage patterns in perspective', in D.V. Glass and D.E.C. Eversley (eds.), *Population in History*, Chicago, Aldine, pp. 101–47

Hajnal, John, 1983. 'Two kinds of pre-industrial household formation system', in Richard Wall, Jean Robin, Peter Laslett (eds.), *Family Forms in Historic Europe*, Cambridge, Cambridge University Press, pp. 65–104

Hélias, Pierre Jakez, 1975. *Le cheval d'orgueil. Mémoires d'un Breton du pays bigouden*, Paris, Plon, 576 pp.

Henry, Louis, 1982. 'Comment mesurer la fécondité des couples mobiles', in *Population*, 1, pp. 9–27

Héritier, Françoise, 1981. *L'exercice de la parenté*, Paris, Ecole des Hautes Etudes en Sciences sociales, Gallimard/Le Seuil, 200 pp.

*Izard, Michel, 1963 *Parenté et mariage à Plozévet, Finistère*, Laboratoire d'Anthropologie Sociale, 160 pp. (roneoed document)

*Izard, Michel, 1965. 'La terminologie de parenté bretonne', in *L'Homme*, July-December, 3–4, pp. 88–100

Jacquard, Albert, 1970. *Structure génétique des populations*, Paris, Masson

Jacquard, Albert, (ed.), 1976. *Etude des isolats, espoirs et limites*, Ecole des Hautes Etudes en Sciences sociales, Institut national d'Etudes démographique, 334 pp.

Jacquart, Jean, 1979. 'L'utilisation des inventaires après décès villageois. Grille de dépouillement et apports', in *Les actes notariés, source d'histoire sociale, XVIe–XVIIe siècle*, Strasbourg, Istra, pp. 187–96

*Jakobi, Lucienne, Albert Jacquard, 1971. 'Consanguinité proche, consanguinité éloignée. Essai de mesure dans un village breton', in *Génétique et population*, Hommage à J. Sutter, Cahier no. 60, Paris, Institut national d'Etudes démographiques, pp. 263–8

Jolas, Tina, Yvonne Verdier, Françoise Zonabend, 1970. 'Parler famille', in *L'Homme* X, 3, pp. 5–26

Jolas, Tina, Françoise Zonabend, 1970. 'Cousinage, voisinage', in *Echanges et communications, Mélanges offerts à Claude Lévi-Strauss*, Paris, Mouton, pp. 169–80

Karnoouh, Claude, 1980. 'Le pouvoir et la parenté', in Hugues Lamarche, Susan Carol Rogers, Claude Karnoouh, *Paysans, femmes, citoyens*, Le Paradou, Actes Sud, pp. 143–209

*Keraval, Pierre, 1954. 'Fermes du pays de Quimper à la fin du XVIIe et au début du XVIIIe siècle', in *Bulletin de la Société archéologique du Finistère* LXXX, 54, pp. 63–81

*Kernen, Pierre, 1967. *Etude démographique classique de Plozévet (1841–1962)*, typed report, 48 pp. plus tables

*Lailler, Dan,. 1947. *Les coiffes bigoudens de Pont-l'Abbé. Evolution de 1880–1946*, manuscript 1 and 2, pp. 8–13, Archives du Musée national des Arts et Traditions populaires

Lamaison, Pierre, 1979. 'Les stratégies matrimoniales dans un système complexe de parenté: Ribennes en Gévaudan (1650–1830)', in *Annales Economies, Sociétés, Civilisations*, 34, 4, pp. 721–43

Lamaison, Pierre, Elizabeth Claverie, 1982. *L'impossible mariage, Violence et parenté en Gévaudan, XVIIe, XVIIIe, XIXe siècle*, Paris, Hachette, 362 pp.

Laslett, Peter, Richard Wall (eds.), 1972. *Household and Family in Past Time*, Cambridge, Cambridge University Press, 624 pp.

*Le Bail, Georges, 1913. *L'émigration rurale et les migrations temporaires dans le Finistère*, Paris, Giard & Brière, 104 pp.

*Lebel, Anne, 1981. *Les luttes sociales dans la conserverie et le milieu maritime en 1926 et 1927 sur le littoral bigouden*, Faculté des Lettres de Brest, Centre de Recherches Bretonnes et Celtiques, master's thesis, 192 pp. (typescript)

Lebrun, François, 1915. *La vie conjugale sous l'Ancien Régime*, Paris, Armand Colin, 180 pp.

*Le Doaré, E., 1896. *Plonéour-Lanvern*, manuscript

*Le Floc'h, Chanoine J.–L., 1982. 'Les circonscriptions territoriales dans le canton de Pont-l'Abbé, en pays bigouden', in *Cahiers de l'Iroise* 3, July–September, pp. 125–9

*Le Floc'h, Vincent, 1966. 'Le régime foncier et son application pratique dans le cadre de la paroisse de Plonivel au XVIIIe siècle', in *Bulletin de la Société archéologique du Finistère* XCII, pp. 117–205

*Le Gallo, Yves, 1980. *Prêtres et prélats du diocèse de Quimper de la fin du XVIIIe siècle à 1830*, Faculté des Lettres de Brest, Centre de Recherches Bretonnes et Celtiques, 2 vols. (typescript)

*Le Grand, Alain, 1980. 'Paysans cornouaillais (XVIIe–XXe siècle). Robert Huella en Guengat: quatre siècles au moins de continuité familiale', in *Bulletin de la Société archéologique du Finistère* CVIII, pp. 207–24

*Le Guirriec, Patrick, 1983. 'Le bourg et les espaces ruraux dans une commune du Sud Bigouden', in *Ethnologie française*, 2, pp. 163–70
**Le mariage en Bretagne*, 1980. Catalogue de l'exposition organisée par Buhez 1980–1984, Rennes, Buhez, 190 pp.
*Lemoine, Jean, 1897–1898. 'La révolte dite du papier timbré ou des bonnets rouges en Bretagne en 1675', in *Annales de Bretagne*, XII, and XIII
*Lemoine, J., H. Bourde de La Rogerie, 1902. *Inventaire sommaire des Archives départementales antérieures à 1790. t. III. série B. Inventaire des fonds des Amirautés de Morlaix et de Quimper*, Quimper, A. Jouen
*Le Prohon, Roger, 1972. 'La démographie léonarde de 1600 à 1715', in *Bulletin de la Société archéologique du Finistère* XCIX, 2, pp. 705–30
Levi, Giovanni, 1976. 'Terra e strutture familiali in una comunita piemontese del 700', in *Quaderni Storici* XI, no. 3, September–December
Lévi-Strauss, Claude, 1958. *Anthropologie structurale*, Paris, Plon, 454 pp.
Lévi-Strauss, Claude, 1968. *Les structures élémentaires de la parenté*, Paris, Mouton, 592 pp.; the quotations are taken from *Elementary Structures of Kinship*, Boston, Beacon Press, 1969
Lévi-Strauss, Claude, 1983. 'Histoire et ethnologie', in *Annales Economies, Sociétés, Civilisations*, 6, pp. 1217–31
Macherel, Claude, 1983. 'Don et réciprocité en Europe', in *Archives européennes de Sociologie* XXIV, pp. 151–66
*Mazeas, Goulven, 1940. *Petite histoire bretonne de la pomme de terre*, Brest, Imprimerie Commerciale et Administrative, 218 pp.
Mendels, Franklin F., 1978. 'La composition du ménage paysan au XIXe siècle: une analyse économique du mode de production domestique', in *Annales Economies, Sociétés, Civilisations*, July-August, 4, pp. 780–802
*Meyer, Jean, 1966. *La noblesse bretonne au XVIIIe siècle*, Paris, Société d'Editions et de Vente des Publications de l'Education nationale, 1,292 pp.
Mira, Joan F., 1971. 'Mariage et famille dans une communauté rurale du pays de Valence (Espagne)', in *Etudes rurales*, 42, pp. 105–19.
*Morin, Edgar, 1967. *Commune en France, la métamorphose de Plodémet*, Paris, A. Fayard, 288 pp.
Netting, Robert McC., 1981. *Balancing on an Alp. Ecological change and continuity in a Swiss mountain community*, Cambridge, Cambridge University Press, 278 pp.
*Ogès, Louis, 1949. *L'agriculture dans le Finistère au milieu du XIXe siècle*, Quimper, Le Goaziou, 174 pp.
Ott, Sandra, 1981. *The Circle of Mountains: a Basque shepherding community*, Oxford, Clarendon Press, 220 pp.
Ott, Sandra, 1980. 'Blessed bread country' in *Archives européennes de Sociologie* XXI, pp. 40–58
Ott, Sandra, 1987. 'Matrimonio y secundas nupcias en une comunidad vasca de Montaña', in John G. Peristiany (ed.), *Dote y matrimonio en los paises mediterráneos*, Madrid, Siglo XXI de España Editores, pp. 193–223
Pellegrin, Nicole, 1979–1982. *Les Bachelleries, Organisation et fêtes de la jeunesse dans le Centre-Ouest, XVe–XVIIIe siècle*, Poitiers, Mémoires de la Société des Antiquaires de l'Ouest, fourth series, vol. 16, 400 pp.
*Pelras, Charles 1965. *Goulien, commune rurale du cap Sizun (Finistère). Etude d'ethnologie globale*, Paris, 3 vols., 470 pp., PhD thesis (duplicated)

*Peyron, Chanoine, 1901–1910. 'Cartulaire de l'Eglise de Quimper', in *Bulletin de la Commission diocésaine d'architecture et d'archéologie du diocèse de Léon*, Quimper, (I–X)

Philippe, Pierre, Jacques Gomila, 1972. 'Inbreeding effects in a French Canadian isolate', in *Z. Morph. Anthrop.*, June, 64, 1, pp. 54–9

Pingaud, Marie-Claude, 1978. *Paysans en Bourgogne. Les gens de Minot*, Paris, Flammarion, 300 pp.

*Planiol, Marcel, 1896. *La très ancienne coutume de Bretagne*, Rennes, Plihon & Hervé, 566 pp.

Polanyi, Karl, 1983. *La grande transformation. Aux origines politiques et économiques de notre temps*, Paris, Gallimard, 420 pp.

*Postel-Vinay, Sylvie, 1977. *Minihy Treguier. Une aristocratie paysanne*, Ecole pratique des Hautes Etudes, sixth section, PhD thesis, (typescript)

Postel-Vinay, Sylvie, 1981. 'La famille bretonne', in A. Vince, C. Gracineau-Alasseur, S. Postel-Vinay, *Briérons, Naguère*, Saint-Nazaire, Jean Le Fur, pp. 113–29

'Le pouvoir au village', 1976. in *Etudes rurales*, July-December 63–4

Rieu-Gout, Anne-Marie, Marie-Louise Sauzéon-Broueilh, 1981. 'Parenté et alliance dans la vallée de Barèges', in *Ethnologie française* XI 4, pp. 343–58

*de Ritalongi, Gabriel P., 1894. *Les Bigoudens*, Nantes, Librairie Libaros, 548 pp.

Roche, Daniel, 1982. 'Le costume et la ville. Le vêtement populaire parisien d'après les inventaires du XVIIe siècle', in *Ethnologie française* XII, 2, pp. 157–64

Roussel, Louis, 1976. *La famille après le mariage des enfants,* Travaux et documents, Cahier no. 78, Presses Universitaires de France, 258 pp.

Sahlins, Marshall, 1976. *Age de pierre, âge d'abondance*, Paris, Gallimard, 410 pp.

Salitot-Dion, Michelle, 1977. 'Evolution économique, cycle familial et transmission patrimoniale à Nussey', in *Etudes rurales*, October–December, 68, pp. 23–53

Salitot-Dion, Michelle, 1978. 'Régime matrimonial et organisation familiale en Franche-Comté'. in *Ethnologie française* 4, October–December, pp. 321–8

Sauvé, L. -F., 1878. *Lavarou kos a Vreiz Izel*, Paris, Champion

*Savina, J., 1920. 'Essai d'histoire économique d'une paroisse rurale: Plogastel-Saint-Germain au XVIIIe siècle', in *Bulletin de la Société archéologique du Finistère*

*Savina, J., D. Bernard, 1927. *Les cahiers de doléances des sénéchaussées de Quimper et Concarneau*, Paris, E. Leroux, 2 vols.

*Sebillot, Paul-Yves, 1968. *Le folklore de la Bretagne*, Paris, Maisonneuve & Larose, 408 pp.

*Sée, Henri, 1906. *Les classes rurales en Bretagne du XVIe à la Révolution*, Paris, Giard & Brière, 544 pp.

Segalen, Martine, 1972. *Nuptialité et alliance. le choix du conjoint dans une commune de l'Eure*, Paris, G.-P. Maisonneuve & Larose, 142 pp.

*Segalen, Martine, 1976. 'Evoluzione dei nuclei familiari di Saint-Jean-Trolimon, Sud-Finistère, a partire dal 1836', in *Quaderni Storici*, September–December, 33, pp. 1122–82

*Segalen, Martine, 1977. 'Household structure, the family life cycle over five generations in a French village', in *Journal of Family History*, 2, 3, pp. 223–36
*Segalen, Martine, 1978. 'Cycle de la vie familiale et transmission des biens; analyse d'un cas', in *Ethnologie française*, 4, pp. 271–8
*Segalen, Martine, 1978. 'L'espace matrimonial dans le pays bigouden Sud au XIXe siècle', in *Gwechall*, 1, pp. 109–22
Segalen, Martine, 1979. *Mari et femme dans la société paysanne*, Paris, Flammarion, 212 pp.
*Segalen, Martine, 1980. 'Le nom caché. Le dénomination dans le pays bigouden Sud', in *L'Homme*, October–December, XX (4), pp. 63–76
Segalen, Martine, 1981. *Sociologie de la famille*, Paris, A. Colin, collection 'U', 282 pp.
*Segalen, Martine, 1983–1984. 'Saint-Jean-Trolimon, un territoire communal divisé en deux parties', in *Cap Caval*, Goanv (winter) pp. 3–5
*Segalen, Martine, Philippe Richard, 1986. 'Marrying kinsmen in Pays Bigouden Sud', in *Journal of Family History*, 11, 2, pp. 109–30
*Selle, Anna, 1936. *Thumette Bigoudène*, Paris, Figuière, 192 pp.
*Signor, Alain, 1969. *La Révolution à Pont-l'Abbé (1789–1794)*, Paris, 422 pp.
'Situation démographique de la France', 1978. In *Population*, March–April
Smith, Daniel, 1977. 'A homoestatic demographic regime: patterns in West European family reconstitution studies', in Ronald Lee (ed.), *Population Patterns in the Past*, New York, Academic Press, pp. 19–52
Smith, Richard M., 1983. 'Hypothèses sur la nuptialité en Angelterre aux XIIIe–XIVe siècle', in *Annales Economies, Sociétés, Civilisations*, January–February, 1, pp. 107–36
Smyth, William J., 1982. 'Nephews, dowries, sons and mothers: an analysis of the geography of farms and marital transactions in a South Tipperary parish', in *Conférence franco-irlandaise sur les communautés rurales*, Paris, 24–26 March
*Souvestre, Emile, 1838. *Le Finistère en 1836*, Brest, Come & Bonetbeau, 252 pp.
Sutter, Jean, Léon Tabah, 1948. 'Fréquence et répartition des mariages consanguins', in *Population*, 3, pp. 481–98
Sutter, Jean, Léon Tabah, 1955. 'Evolution des isolats de deux départements français: Loir-et-Cher et Finistère', Paris, Institut national d'Etudes démographiques, *Population*, 4, pp. 645–73
*Tanguy, Bernard, 1981. 'Les paroisses primitives en plou et leurs saints éponymes', in *Bulletin de la Société archéologique du Finistère* CIX, pp. 121–55
*Tanneau, Yves, 1958. 'Pont-l'Abbé', in *Bulletin de la Société archéologique du Finistère* LXXXIV, pp. 68–159
Tardieu, Suzanne, 1964. *La vie domestique dans le Mâconnais rural préindustriel*, Paris, Institut d'Ethnologie, 546 pp.
*Trepos, Pierre, 1962. *Enquêtes sur le vocabulaire breton de la ferme*, Rennes, Imprimeries Réunies, 156 pp.
*Vallaux, Camille, 1905. *La Basse-Bretagne, Etude de géographie humaine*, Paris, E. Cornely, 312 pp.
*Vallaux, Camille, 1907. *Penmarc'h aux XVIe et XVIIe siècles*, Paris, Cornely & Co., 44 pp.

Van de Walle, Etienne, 1978. 'Alone in Europe: the French fertility decline until 1850', in Charles Tilly (ed.), *Historical Studies of Changing Fertility*, Princeton University Press

Van de Walle, Etienne, 1979. 'France', in W.R. Lee (ed.), *European Demography and Economic Growth*, London, Croom Helm

Van Gennep, Arnold, 1909. *Les rites de passage*, Paris, E. Nourry

Van Gennep, Arnold, 1943–1958. *Manuel de folklore français contemporain*, Paris, A. Picard, 7 vols.

Verdon, Michel, 1973. *Anthropologie de la colonisation au Québec*, Presses de l'Université de Montréal, 284 pp.

*Villermé et Benoiston de Châteauneuf, 1982. *Voyage en Bretagne en 1840 et 1841*, new edition with an Introduction by Fanch Elegoët, Tud ha Bro, sociétés bretonnes, Université de Rennes I, 160 pp.

Vu Tien Khang, Jacqueline, André Sevin, 1977. *Choix du conjoint et patrimoine génétique. Etude de quatre villages du pays de Sault de 1740 à nos jours*, Paris, Centre National de la Recherche scientifique, 160 pp.

Wachter, K.W., E. Hammel, P. Laslett (eds.), 1978. *Statistical Studies of Historical Social Structure*, New York, Academic Press, 230 pp.

Weber, Eugen, 1976. *Peasants into Frenchmen. The Modernisation of Rural France 1870–1914*, Stanford, Stanford University Press, 616 pp.

Wrigley, E.A., 1966. 'Family reconstitution', in E.A. Wrigley (ed.), *An Introduction to English Historical Demography*, New York, Basic Books

Yanagisako, Sylvia Junko, 1978. 'Family and household: the analysis of domestic groups', in *Annual Review of Anthropology*, 8, pp. 161–205

Yver, Jean, 1966. *Egalité entre héritiers et exclusion des enfants dotés*, Paris, Sirey

Zonabend, Françoise, 1980. *La mémoire longue*, Paris, Presses Universitaires de France, 314 pp.

Zonabend, Françoise, 1981. 'Le très proche et le pas trop loin', in *Ethnologie française* XI, 4, pp. 311–18

Index

Cambridge Studies in Social and Cultural Anthropology

Editors: JACK GOODY, STEPHEN GUDEMAN, MICHAEL HERZFELD, JONATHAN PARRY

1 The Political Organisation of Unyamwezi
R.G. ABRAHAMS
2 Buddhism and the Spirit Cults in North-East Thailand*
S.J. TAMBIAH
3 Kalahari Village Politics: An African Democracy
ADAM KUPER
4 The Rope of Moka: Big-Men and Ceremonial Exchange in Mount Hagen, New Guinea*
ANDREW STRATHERN
5 The Majangir: Ecology and Society of a Southwest Ethiopian People
JACK STAUDER
6 Buddhist Monk, Buddhist Layman: A Study of Urban Monastic Organisation in Central Thailand
JANE BUNNAG
7 Contexts of Kinship: An Essay in the Family Sociology of the Gonja of Northern Ghana
ESTHER N. GOODY
8 Marriage among a Matrilineal Elite: A Family Study of Ghanaian Civil Servants
CHRISTINE OPPONG
9 Elite Politics in Rural India: Political Stratification and Political Alliances in Western Maharashtra
ANTHONY T. CARTER
10 Women and Property in Morocco: Their Changing Relation to the Process of Social Stratification in the Middle Atlas
VANESSA MAHER
11 Rethinking Symbolism*
DAN SPERBER. Translated by Alice L. Morton
12 Resources and Population: A Study of the Gurungs of Nepal
ALAN MACFARLANE
13 Mediterranean Family Structures
EDITED BY J.G. PERISTIANY
14 Spirits of Protest: Spirit-Mediums and the Articulation of Consensus among the Zezuru of Southern Rhodesia (Zimbabwe)
PETER FRY

15 World Conqueror and World Renouncer: A Study of Buddhism and Polity in Thailand against a Historical Background*
S.J. TAMBIAH
16 Outline of a Theory of Practice*
PIERRE BOURDIEU. Translated by Richard Nice
17 Production and Reproduction: A Comparative Study of the Domestic Domain*
JACK GOODY
18 Perspectives in Marxist Anthropology*
MAURICE GODELIER. Translated by Robert Brain
19 The Fat Shechem, or the Politics of Sex: Essays in the Anthropology of the Mediterranean
JULIAN PITT-RIVERS
20 People of the Zongo: The Transformation of Ethnic Identities in Ghana
ENID SCHILDKROUT
21 Casting out Anger: Religion among the Taita of Kenya
GRACE HARRIS
22 Rituals of the Kandyan State
H.L. SENEVIRATNE
23 Australian Kin Classification
HAROLD W. SHEFFLER
24 The Palm and the Pleiades: Initiation and Cosmology in Northwest Amazonia*
STEPHEN HUGH-JONES
25 Nomads of Southern Siberia: The Pastoral Economies of Tuva
S.I. VAINSHTEIN. Translated by Michael Colenso
26 From the Milk River: Spatial and Temporal Processes in Northwest Amazonia*
CHRISTINE HUGH-JONES
27 Day of Shining Red: An Essay on Understanding Ritual*
GILBERT LEWIS
28 Hunters, Pastoralists and Ranchers: Reindeer Economies and their Transformations*
TIM INGOLD
29 The Wood-Carvers of Hong Kong: Craft Production in the World Capitalist Periphery
EUGENE COOPER
30 Minangkabau Social Formations: Indonesian Peasants and the World Economy
JOEL S. KAHN
31 Patrons and Partisans: A Study of Two Southern Italian Communes
CAROLINE WHITE
32 Muslim Society*
ERNEST GELLNER
33 Why Marry Her? Society and Symbolic Structures
LUC DE HEUSCH. Translated by Janet Lloyd
34 Chinese Ritual and Politics
EMILY MARTIN AHERN

35 Parenthood and Social Reproduction: Fostering and Occupational Roles in West Africa
ESTHER N. GOODY
36 Dravidian Kinship
THOMAS R. TRAUTMANN
37 The Anthropological Circle: Symbol Function, History*
MARC AUGE. Translated by Martin Thom
38 Rural Society in Southeast Asia KATHLEEN GOUGH
39 The Fish-People: Linguistic Exogamy and Tukanoan Identity in Northwest Amazonia
JEAN E. JACKSON
40 Karl Marx Collective: Economy, Society and Religion in a Siberian Collective Farm*
CAROLINE HUMPHREY
41 Ecology and Exchange in the Andes Edited by
DAVID LEHMANN
42 Traders without Trade: Responses to Trade in two Dyula Communities
ROBERT LAUNAY
43 The Political Economy of West African Agriculture*
KEITH HART
44 Nomads and the Outside World
A.M. KHAZANOV. Translated by Julia Crookenden
45 Actions, Norms and Representations: Foundations of Anthropological Inquiry*
LADISLAV HOLY AND MILAN STUCKLIK
46 Structural Models in Anthropology*
PER HAG AND FRANK HARARY
47 Servants of the Goddess: The Priests of a South Indian Temple
C.J. FULLER
48 Oedious and Job in West African Religion*
MEYER PORTES
49 The Buddhist Saints of the Forest and the Cult of Amulets: A Study in Charisma, Hagiography, Sectarianism, and Millennial Buddhism*
S.J. TAMBIAH
50 Kinship and Marriage: An Anthropological Perspective (available in paperback/in the USA only)
ROBIN FOX
51 Individual and Society in Guiana: A Comparative Study of Amerindian Social Organizations*
PETER RIVIERE
52 People and the State: An Anthropology of Planned Development*
A.F. ROBERTSON
53 Inequality among Brothers; Class and Kinship in South China
RUBIE S. WATSON
54 On Anthropological Knowledge*
DAN SPERBER
55 Tales of the Yanomami: Daily Life in the Venezuelan Forest*
JACQUES LIZOT. Translated by Ernest Simon

56 The Making of Great Men: Male Domination and Power among the New Guinea Baruya*
MAAURICE GODELIER. Translated by Rupert Swyer
57 Age Class systems: Social Institutions and Politics Based on Age*
BERNARDO BERNARDI. Translated by David I. Kertzer
58 Strategies and Norms in a Changing Matrilineal Society: Descent, Succession and Inheritance among the Toka of Zambia
LADISLAV HOLY
59 Native Lords of Quito in the Age of the Incas: The Political Economy of North-Andean Chiefdoms
FRANK SALOMON
60 Culture and Class in Anthropology and History: A Newfoundland Illustration*
GERALD SIDER
61 From Blessing to Violence: History and Ideology in the Circumcision Ritual of the Merina of Madagascar*
MAURICE BLOCH
62 The Huli Response to Illness
STEPHEN FRANKEL
63 Social Inequality in a Northern Portuguese Hamlet: Land, Late Marriage, and Bastardy, 1870–1978
BRIAN JUAN O'NEILL
64 Cosmologies in the Making: A Generative Approach to Cultural Variation in Inner New Guinea*
FREDRIK BARTH
65 Kinship and Class in the West Indies: A Genealogical Study of Jamaica and Guyana*
RAYMOND T. SMITH
66 The Making of the Basque Nation
MARIANNE HEIBERG
67 Out of Time: History and Evolution in Anthropological Discourse NICHOLAS THOMAS
68 Tradition as Truth and Communication
PASCAL BOYER
69 The Abandoned Narcotic: Kava and Cultural Instability in Melanesia
RON BRUNTON
70 The Anthropology of Numbers
THOMAS CRUMP
71 Stealing People's Names: History and Politics in a Sepik River Cosmology
SIMON J. HARRISON
72 The Bedouin of Cyrenaica: Studies in Personal and Corporate Power
EMRYS L. PETERS edited by Jack Goody and Emanuel Marx
73 Bartered Brides: Politics, gender and marriage in an Afghan Tribal Society
NANCY TAPPER

* available in paperback